THE MAN WHO SAVED
SEA TURTLES

THE MAN WHO SAVED SEA TURTLES

*Archie Carr and the Origins of
Conservation Biology*

FREDERICK ROWE DAVIS

OXFORD

UNIVERSITY PRESS

2007

OXFORD
UNIVERSITY PRESS

Oxford University Press, Inc., publishes works that further
Oxford University's objective of excellence
in research, scholarship, and education.

Oxford New York
Auckland Cape Town Dar es Salaam Hong Kong Karachi
Kuala Lumpur Madrid Melbourne Mexico City Nairobi
New Delhi Shanghai Taipei Toronto

With offices in
Argentina Austria Brazil Chile Czech Republic France Greece
Guatemala Hungary Italy Japan Poland Portugal Singapore
South Korea Switzerland Thailand Turkey Ukraine Vietnam

Published by Oxford University Press, Inc.
198 Madison Avenue, New York, New York 10016

www.oup.com

Oxford is a registered trademark of Oxford University Press

Library of Congress Cataloging-in-Publication Data
Davis, Frederick Rowe, 1965–
The man who saved sea turtles: Archie Carr and the origins of conservation
biology/Frederick Rowe Davis.
    p.   cm.
Includes bibliographical references.
ISBN 978-0-19-531077-1
1. Carr, Archie Fairly, 1909–1987   2. Herpetologists—United States—Biography.
3. Sea turtles.   I. Title.
QL31.C34D38 2007
590.92—dc22 [B]   2006029638

9 8 7 6 5 4 3 2 1

Printed in the United States of America
on acid-free recycled paper

For my son,
Spenser Lake Davis

FOREWORD

David Ehrenfeld

Archie Carr, one of the greatest biologists of the twentieth century, played a leading part in finding a new and critical role for natural history and systematics in a post-1950s world that had come to be dominated by the glamorous science of molecular biology. Coincident with the rise of molecular biology, a growing popular awareness of species extinction has been accepted as a grim fact of modern life. This awareness was brought home by scientists like Rachel Carson, who explained the extinction crisis in a way that the public could understand and remember. What Carr did, using endangered sea turtles as his model, was to show that successful conservation, the next step after becoming aware of the extinction problem, was utterly dependent on knowing the life histories and the evolved relationships and species classification of the creatures you were trying to save.

The story of Carr's life is both fascinating and uplifting—from his childhood passion for reptiles and other animals, to his dramatic talent and linguistic abilities (translating for African-American speakers of Gullah and Geechee on the docks of Savannah), to his emergence as the world's greatest authority on turtles and a pioneering conservationist. Archie Carr was that rare scientist who could communicate with the public as easily as with his fellow biologists. A truly charismatic figure, even the Central American turtle poachers whom he was effectively and openly putting out of business liked and respected him, as indeed he enjoyed their company. Part of his ability to reach out to people was based on his remarkable ear for language and dialect, he once delivered a series of lectures on evolution to a Costa Rican radio audience in flawless Spanish.

Carr's scientific and popular writings were as compelling, elegant, and wryly humorous as his speech. Here is an extract from his *Handbook of Turtles*, which won the Daniel Giraud Elliot medal of the U.S. National Academy of Sciences in 1952:

> The first of the innovations made by the stem reptiles was in a way the most extraordinary and ambitious of all.... By a cryptic series of changes, few of which are illustrated in the fossil record, there evolved a curious and improbable creature [with] a bent and twisted body encased in a bony box the like of which had never been seen....The new animal was a turtle. Having once performed the spectacular feat of getting its girdles inside its ribs, it lapsed into a state of complacent conservatism that has been the chief mark of the breed ever since.

Another example of Carr's engaging style comes from his popular work *The Windward Road* (1955), about Caribbean nature and culture: "It was dawn and the wind was down. Curls of mist rose and drifted, thinned and dissolved into the calm confusion of sea and air.... The cook rested his belly and elbows on the rail and looked sleepily after the catboat sliding away in the fog, leaving behind it the only streak of difference between sky and water."

In this book you will come to know, as I did during the exciting years when I was his student and friend, a unique and inspiring human being. Frederick Davis has done a fine job of capturing the intertwined personal, public, and scientific lives of the complex, extraordinary person who, if anyone deserves the title, really was the man who saved sea turtles.

PREFACE

It is so dark that I cannot see Sebastian Troëng, who is the Caribbean Conservation Corporation's scientific director and our team leader, or any of the other people in the group. I can't even see my hand in front of my face. Nor can I seem to shake off the sleep I've recently left for my midnight to 4 a.m. shift. Dressed entirely in dark clothes, we speak in hushed whispers, barely audible above the crashing of the surf. Each of us has come to Tortuguero with one question in mind, Will we find turtles? Sebastian has pointed out the turtle tracks in the sand, and he demonstrates how to tell the direction of the tracks by feel. Somehow he discerns a second set returning to the water. We start to move down the beach again, theoretically at the surf line, but my soaked socks and shoes indicate we're below it. Just when I ease into a comfortable pace, Sebastian freezes, oblivious to the minor train wreck that results from the rest of us bumping into one another. "Can you see the turtle?" he asks. At first, I can barely differentiate the shell of the turtle from the driftwood, but gradually I manage to see it. It's a green sea turtle (*Chelonia mydas*). We stand motionless until the turtle passes by.

Just a little farther along the beach, Sebastian locates the first nest of the night. We set to work by attaching the tags (a sequential set, one to each flipper), measuring the carapace (straight and curved), and check for abnormalities (especially fibropapillomas). Throughout the process, we stay behind the turtle where our red-tinted light won't disturb her as she completes the nesting process. I try to write the measurements according to the guidelines, while the turtle inadvertently pummels me with flipperfuls of black sand as she covers the nest. Then it starts to rain. Thunder and lightning follow. No one wants to hold the 6-foot-long aluminum calipers, now respectfully called the "lightning rod." At that moment, it dawns on me that if Archie Carr were with us, he would be having a great time.

At Tortuguero, one of Carr's favorite places, I feel a greater sense of connection with him than I have over the course of the ten years I have spent studying his life and work. I try to imagine what it must have been like to visit Tortuguero fifty years ago. It occurs to me that without the attention that Carr's research, writing, and activism brought to Tortuguero, there might be many fewer turtles nesting on the beaches (instead, 2005 was a near-record year).

In his now classic book, *The Windward Road*, Carr wrote, "Adventure is just a state of mind." For me, writing this book has been an adventure of sorts. Given the breadth of Carr's interests, the range of his scholarly and literary activities, and the duration of his career (fifty years), exploring his life has been a monumental task at once challenging and exhilarating. Over the course of this journey, I have had the opportunity to meet many people whose lives were touched by Carr. In addition, I have learned about the biology and conservation of sea turtles, both from Carr's writings and from the many scientists who continue to fight to save these fascinating animals. Along the way, I have incurred many debts, so it is a pleasure to acknowledge the people and institutions that made this study possible.

The Department of History at Florida State University has proved to be an excellent base from which to complete my study of Archie Carr. As chair of the department, Neil Jumonville has given me the benefit of his considerable knowledge and experience, and many other members of the department have offered advice: Michael Creswell, Max Friedman, Bob Gellately, Jim Jones, Edward Gray, Joe Richardson, Suzy Sinke, and Heike Schmidt. I am very grateful for research assistants provided by the department: Seth Weitz, Lindsey Harrington, Patrick Hightower, Jen Bishop, and Victoria Penziner. Robin Sellers and the staff of the Oral History Program were very helpful in transcribing interview tapes. India Van Brunt read the full manuscript twice, once for enjoyment and once with a proofreader's critical eye. Colleagues in the Program in the History and Philosophy of Science at Florida State University have also been supportive, especially Michael Ruse and Matt Day.

The Man Who Saved Sea Turtles has been significantly improved by discussions with Archie Carr's family, friends, and colleagues. Karen Bjorndal, Alan Bolten, Marjorie H. Carr, Mimi, Chuck, Tom, and David Carr, David Auth, Joshua C. Dickinson, Jr., Joshua C. Dickinson, III, David Ehrenfeld, Brian McNab, Anne Meylan, Jeanne Mortimer, Charles Myers, Larry Ogren, Peter C. H. Pritchard, Jack Putz, Jonathon Rieskind, Hugh Popenoe, Ernst Mayr, and E. O. Wilson all permitted me to interview them about Carr and related topics.

At the University of Florida, I am grateful to Betty Smocovitis and Brian McNab for detailed comments on several drafts of the manuscript. Jack Putz, Fred Gregory, Bob Hatch, Mark Thurner, and Laura Snook all provided feedback on earlier versions.

The Caribbean Conservation Corporation (CCC) has provided significant assistance during the project, especially in facilitating a research trip to Tortuguero. I am particularly grateful to David Godfrey, Dan Evans, Sebastian Troëng, and Andrea de Haro, as well as a wonderful group of CCC research assistants: Victor

Huertas Martín, Stacey Kennealy, Yazmín Monro y García, Ricardo Morales, Jonathan Quan, Marcela Rodrígues Sánchez, Alejandro Paz Torres, and Mauricio Contreras Vasquez.

My sincere gratitude goes to the following individuals for their assistance in procuring images and publishing rights for this biography: Mimi Carr, David Godfrey, Dan Evans, Larry Ogren, Simón Malo, Joyce Dewsbury, Florence Turcotte, and Larry Crowder. Chuck McCann, head of the Digital Library at Strozier Library, Florida State University, was helpful in assisting me to repair damaged photographs and produce the best images possible. Peter Krafft, Director of Cartography at the Florida Resources and Environmental Analysis Center, prepared the maps of the Caribbean and Ascension Island. The Department of Special and Area Collections at the George A. Smathers Library at the University of Florida greatly facilitated my study of Archie Carr's papers. I appreciate the efforts of Frank Orser, Joyce Dewsbury, and Florence Turcotte. The office of the Florida Defenders of the Environment provided me with complete access to files concerning their early history, before most of the files were transferred to the University of Florida. The staff of the Claude W. Pepper Library at Florida State University provided assistance with Marjorie Carr's letters in opposition to the Rodman Dam and the Cross-Florida Barge Canal. At the Harvard University Archives, Megan Sniffin-Marinoff (Harvard University Archivist) and Tim Driscoll fielded queries and assisted with permissions. The late Louisa Barbour Parker kindly granted permission to publish excerpts from her father's correspondence with Archie Carr held in the Harvard University Archives. I appreciate permission to publish revised portions of papers published by *Endeavour*, *The Florida Historical Quarterly*, and the University Press of Florida.

The research for *The Man Who Saved Sea Turtles* was supported by several organizations: the National Science Foundation (SES-0526422), the Center for Research and Creativity at Florida State University, and the Department of History at Florida State University.

Audiences for papers presented at many conferences, including the American Society for Environmental History, the History of Science Society, the Latin American Studies Association, the International Society for the History, Philosophy, and Social Studies of Biology, the Joint Atlantic Symposium for the History of Biology, the William H. and Lucyle T. Werkmeister Conference at Florida State University, and the Organization of American Historians, provided useful insights.

At Oxford University Press, I am grateful for Peter Prescott's sage advice and for the constructive criticism of four distinguished scholars, including Mark Barrow and Paul Farber. In the process of writing the foreword, David Ehrenfeld offered countless valuable comments on the entire book. The readers' suggestions have improved this book in innumerable ways. Any errors remaining in the text are my sole responsibility.

I feel such profound gratitude to my family that words seem limited to express it. Sincere thanks go to my sister and brothers (Amy, Chip, and Tim).

My son Spenser is interested in turtles, and his questions always clarify my ideas and bring joy to my heart. My parents, Dan and Judy Davis, have provided unlimited support and encouragement throughout this study and my life. To anyone I have left off this list, my apologies.

CONTENTS

CHRONOLOGY: THE LIFE OF ARCHIE CARR

Years	Professional/ Personal	Selected Publications	Selected Awards
1909 (June 16)	Born, Mobile, Alabama Moved to Ft. Worth, Texas Moved to Savannah, Georgia		
1932	B.S., University of Florida (UF)		
1934	M.S., UF	"A Key to Frog Calls"	
1937	Married Marjorie Harris		
1937	Ph.D., Biology, UF		First Ph.D. granted in biology at UF
1937–41	Instructor, UF		
1937–43	Fellow, Museum of Comparative Zoology, Harvard University	"Notes on Sea Turtles," 1942	
1939	Expedition to Mexico		
1940	Expedition to Mexico	*A Contribution to the Herpetology of Florida*	
1941–44	Assistant Professor, UF		
1944–49	Associate Professor, UF		
1945–49	Professor, Escuela Agricola Panamericana		
1949–59	Professor, UF		
1952	Biologist, Shire Valley, Nyasaland	*Handbook of Turtles*	Daniel Girard Elliott Medal, National Academy of Sciences
1953		*High Jungles and Low*	
1956		*The Windward Road*	O'Henry Memorial Award
1956–57	Technical Advisor, University of Costa Rica		
1957			John Burroughs Medal (American Museum of Natural History)
1959–87	Graduate Research Professor, UF		

1959–87	Technical Director, Caribbean Conservation Corporation (CCC)		
1963		*The Reptiles*	First Annual Honors Medal, Florida Academy of Sciences
1964		*Ulendo*	
1964		*Africa*	
1966–84	Chairman, Marine Turtle Specialist Group, SSC, IUCN		
1966–79	Organization for Tropical Studies, Board of Directors		
1967		*So Excellent a Fishe*	
1968–87		Consulting Editor, *Biological Conservation*	
1971–87	Honorary Consultant, World Wildlife Fund		Annual Outstanding Alumni Award, UF, 1971
1972–87		Contributing Editor, *Audubon*	
1973		*The Everglades*	World Wildlife Fund Gold Medal
1975			Edward W. Browning Award, Smithsonian Institution
1978–87	West Atlantic Turtle Survey, CCC, for National Marine Fisheries Service		
1978	Research vessel *Alpha Helix*, Green Turtle Expedition in Costa Rica and Nicaragua		Order of the Golden Ark (Netherlands), New York Zoological Society Gold Medal
1979	World Conference on Sea Turtle Conservation, Washington, D.C.		
1983			Archie F. Carr, Jr. Postdoctoral Fellowship, Department of Zoology, UF
1984			Fairfield Osborn Lecturer Hal Borland Award (National Audubon Society)
1987 (May 21)	Dies, Micanopy, Florida		Eminent Ecologist, Ecological Society of America
1994		*A Naturalist in Florida*	

THE MAN WHO SAVED
SEA TURTLES

CHAPTER 1

Introduction

About a hundred years ago, when Archie Carr was born, a large female green turtle (*Chelonia mydas*) hauled herself onto a beach in Costa Rica. The locals (Afro-Caribes who had emigrated from Bluefields, Nicaragua), called the beach Turtle Bogue or by the Spanish name Tortuguero. The turtle knew nothing of the name of the beach. She knew only that she was drawn to lay her eggs on it. No other beach in the world inspired the same nesting urgency as the very beach that she had left behind many years before, after emerging from the nest and avoiding the wide array of predators that had congregated to consume the hatchling sea turtles.

Scientists at the time knew little about green turtles. Most of their information was limited to sightings of hatchlings leaving their natal beaches or adult females returning to nest. Between these observations, the life history of the green turtle represented a biological enigma with far more questions than answers. Better informed were the Caymanian captains, who fished for the species. For generations, Caymanians had known that the turtles migrated from Tortuguero to the Mosquito Cays, where they fed on the abundant beds of turtle grass. The status of green turtles was clear to the Caymanians because their ancestors had seen the disappearance of sea turtle colonies all across the Caribbean: Bermuda, the Bahamas, the Cayman Islands, and Cuba. But they took heart in the spectacular abundance of the colony that moved between Turtle Bogue and the Mosquito Cays. They sailed confidently from Grand Cayman to Tortuguero, where they could harpoon as many turtles as they wanted before traveling to Key West to sell their bounty by the pound. From Key West, the turtles would be shipped (as soup) to New York, London, and other markets around the world. Given all that they knew about the life cycle of sea turtles, it may seem strange that the Caymanians were so confident in the continued abundance of their prey, but as the turtles had always arrived to nest, so too did the Caymanians to harpoon them and haul them onto catboats. Stranger still, the Caymanian's collective knowledge of sea turtles remained unknown to the scientific community until it was almost too late (most scientists found little of value in the anecdotal observations of nonscientists). All over the world, sea turtles of all species were threatened either directly through hunting or indirectly through the transformation of their nesting beaches and foraging grounds.

The general demise of regional and global populations of sea turtles might have continued unabated until it reached a tragic conclusion in extinction, first of local populations and then throughout a species' range. It was fortunate, therefore, that in the 1950s a zoologist named Archie Carr undertook a detailed study of sea turtles. Unlike most of his scientific colleagues, he sought out local insights regarding the behavior of sea turtles. In the stories of Caymanians and other Caribbeans, the "man who saved sea turtles" found testable hypotheses that ultimately led to an understanding of sea turtle ecology and conservation. Through it all, the turtle continued to nest, each time narrowly avoiding the harpoons, and her offspring continued to emerge for a frantic scramble to the sea.

Throughout his life, Archie Carr found turtles and nature fascinating. As a ten-year-old child in Fort Worth, Texas, Carr kept a significant menagerie in his backyard. Later in his life, he recalled specific encounters with turtles (especially sea turtles) and other wildlife. His recollections of duck hunting were similar to Aldo Leopold's. Even as a young man, Carr had already developed an intense interest in the natural world and in turtles specifically.

Anne Meylan, one of Carr's graduate students, has a vivid memory that captures the essence of Archie Carr. One early morning, during the late 1970s, she watched in horror as poachers spotted a turtle's float (used for tracking) and dragged it into a *cayuca* (canoe). The distraught young biologist ran to *casa verde* (where the Carrs slept). Archie and his wife Marjorie emerged from their cabin in their nightshirts. Marjorie began screaming for the poachers to desist their activities. When Meylan turned her attention back to the turtles, she heard a distant "pop." Back at the station, Archie had fired a small pistol into the air in the hopes of frightening off the turtle hunters. Carr's action indicated his single-minded devotion to the conservation of sea turtles.[1] Meylan, now a leading sea turtle conservationist in her own right, has never forgotten that morning at Tortuguero. The significance of these two anecdotes is that from the time he was a child until he was a septuagenarian, Carr had an affinity for turtles. Near the end of his life, when someone asked Carr why he was so interested in turtles, he offered a simple response: "I just liked the look on their faces. There is an old, wise, sort of durable, aboriginal look about turtles that fascinates people."[2]

When Carr died on May 21, 1987, at his home in Micanopy, Florida, naturalists and biologists reacted with an outpouring of intense feeling and emotion. Brian McNab, Carr's colleague in the Department of Zoology at the University of Florida, eulogized Carr as follows: "As sad as this occasion is, we all should rejoice for having had the opportunity to know Archie Carr and, if we are wise, for having

Figure 1. Archie Carr and hatchling sea turtles, 1961. Courtesy of the Department of Special and Area Studies Collections, George A. Smathers Libraries, University of Florida.

had our lives thereby enriched. He set a standard against which each of us should judge the purity of our scholarship, the joy in our life and knowledge, the active commitment in our protection of the planet earth, and the compassion in our humanity."[3] David Ehrenfeld, one of Carr's former graduate students, described Carr eloquently in an obituary: "At the time of his death, he was the world's leading authority on sea turtles, a tropical field ecologist of exceptional skill and experience, a brilliant writer for audiences of both scientific and popular literature, a distinguished taxonomist and evolutionary biologist and an internationally acclaimed advocate of conservation."[4] McNab and Ehrenfeld captured the many facets of Carr as an individual. The details of Carr's life also provide a window to the development of biology in America over the course of the twentieth century.

Archie Carr and Biology in the Twentieth Century

Early in his career as a herpetologist, Carr extensively revised taxonomic classifications of turtles and described several species new to science. He also published many papers on the natural history of frogs, snakes, fish, and turtles. Through these activities, Carr developed a network of many of the American herpetologists who contributed to his education and professional development. Carr's magnum opus in herpetology was the *Handbook of Turtles* (1952), and it contributed significantly to knowledge of the biology and natural history of turtles. In recognition of this book, the National Academy of Sciences awarded Carr the Daniel Giraud Elliott Medal.

Carr published more than one hundred scientific papers on taxonomy and ecology. Many of these were devoted to the ecology and migrations of sea turtles. In addition to technical monographs, Carr was a prolific nature writer. His subjects included Honduras, the Caribbean, Africa (two books), reptiles, sea turtles, the Everglades, and Florida (published posthumously). A careful study of Carr's publications reveals that, unlike many scientists, he was particularly attentive to local knowledge in developing his research agendas and in his writing, technical and popular alike. In his earliest papers on the reptiles and amphibians of Florida, Carr incorporated local stories. As he began to study sea turtles, interviews with Caymanian turtle-hunting captains provided valuable clues about turtle ecology that Carr developed as testable hypotheses. While exploring the beaches of the Caribbean for evidence of sea turtle nesting, Carr encountered a cultural diversity as great as the natural diversity. Caribbean culture was the focus of two of his books. Similarly, while in Africa Carr relished local stories and myths regarding nature.

As an ecologist, Carr's study of the ecology and migrations of sea turtles formed the basis for further investigations. Carr's conservation ethic developed out of his work in natural history. He drew upon his sea turtle life histories and scientific studies to target principal areas for conservation efforts. Carr received numerous awards for his conservation efforts with the Caribbean Conservation Corporation, including the World Wildlife Fund Gold Medal (1973), the New York Zoological Society Gold Medal for biological conservation (1978), and an official post in the Order of the Golden Ark (Netherlands) for biological research and conservation (1978). The Ecological Society of America awarded Carr its highest honor, Eminent Ecologist, just weeks before his death in 1987. A National Wildlife Refuge in Florida bears Carr's name as a tribute to his pioneering work.

Throughout his career, Carr tried to reach a larger audience than the community of scientists and conservationists. To that end, he wrote ten nature books for the public (plus one published posthumously). In addition to the prize given to the *Handbook of Turtles*, his publications received several major awards for writing: the O'Henry Memorial Award for a short story (1956), the John Burroughs Medal of the American Museum of Natural History for nature writing (1957), and the Hal Borland Award of the National Audubon Society for making a lasting contribution to the understanding, appreciation, and protection of nature through his writing and publications (1984).

During his fifty-year affiliation with the University of Florida, Carr taught thousands of students about natural history, evolution, and ecology. As an advisor and mentor, he supervised the doctoral dissertations of eighteen students, many of whom continued to work in biology and conservation. In honoring one of its most illustrious graduates and professors, the University of Florida promoted Carr to graduate research professor (its highest academic rank) in 1959, granted him the Annual Outstanding Alumni Award in 1971, established the Archie Carr Jr. Postdoctoral Fellowship in the Department of Zoology in 1983, and named one of the zoology buildings in his honor.[5] Many of Carr's former students and colleagues cherish memories of the charismatic professor.

In studying the life and work of Archie Carr, I explored the institutional history of the Department of Biology at the University of Florida and the Museum of Comparative Zoology at Harvard, the disciplinary history of herpetology, popular travel narratives, and the social history of popular conservation movements. Carr's work and interactions with colleagues have served as a guide to general trends in the history of science. At the same time, the examination of developments in natural history has produced a rich image of Carr as naturalist, herpetologist, ecologist, and conservationist. In following Carr through these various contexts, I strove not to lose sight of his personality.

Archie Carr as a Person

There is no question that Carr's achievements in natural history, ecology, and conservation were tremendous. Yet, even a thorough account of such successes may fail to capture who Carr was as a person. Like most people, Carr was rife with contradictions. Although he could be shy and was nonconfrontational by nature, he had a dramatic flair, and students flocked to his lectures on evolution and ecology. Carr was renowned as friendly and charismatic; he could delight informal and formal groups alike with his stories and his renditions of the "Jabberwocky," and he even enjoyed practical jokes. At the same time, however, he could be withdrawn and distant (particularly as writing deadlines approached). At such times, he would avoid interactions with colleagues, students, and especially well-wishers at all costs. His students and colleagues marveled at his uncanny ability to seemingly disappear into thin air in order to avoid conflict or even interaction. Rumors of secret writing locations abounded (a second office, his car, a quiet spot in the woods near his house).[6] Carr's gift with the written word shines through virtually everything he wrote, but he could be shockingly profane. Admirers have suggested that Carr used profanity for its shock value. For many years after Carr had become known as one of the world's leading sea turtle conservationists, he continued to eat sea turtle meat and eggs, including turtle soup. As early as the 1940s, Carr fretted about the unbridled growth of population around the world and its inevitable consequences for wildlife conservation. Yet he and his wife Marjorie raised five children of their own. Many Afro-Caribes fondly recall their friendship with Carr, and they respect his efforts for the people of Tortuguero. But, even in print, Carr made statements that would today be considered racist. In his early private correspondence, Carr occasionally indicated that he held the racist views prevalent in the South before the civil rights movement. To the extent possible, I have tried to place Carr's many contradictions in context. By including these elements of his personality, I create a richer portrait of Carr as a person, warts and all, rather than a simple collection of achievements in science and conservation. One important element of Carr's development as a scientist and as a person was his marriage.

A Natural Collaboration

Carr shared one of his accomplishments with his wife Marjorie Harris Carr: the two naturalists were married for fifty years (1937–1987), and their marriage shaped many aspects of their careers. Before they came together, their lives followed parallel paths. In the early years of marriage, Archie and Marjorie Carr's interests continued to develop in tandem as Marjorie completed her master's degree studying the breeding habits of freshwater fish. Moreover, during the course of seven summers at the Museum of Comparative Zoology at Harvard, Marjorie worked in the Ornithology Collection (birds), while Archie studied the taxonomy of turtles with Thomas Barbour. After several trying years during World War II, when both were distracted from their mutual interest in nature, the Carrs spent five years exploring the tropical forests of Honduras together on horseback while their cook prepared meals for their growing family and a *niñera*, or nurse, looked after the children (four in all by the time they returned to Gainesville; another child arrived a few months later).

Though Marjorie's interest in ecology and conservation continued to develop, she was also primarily responsible for the children's upbringing. Each of the Carr children remembers their mother's efforts to "keep them out of daddy's hair," while he was working on an endless stream of writing projects. Marjorie continued her activities as part of the Gainesville Garden Club and the Alachua Audubon Club (which she cofounded). Through the latter group, she spearheaded the fight to save the Ocklawaha River from the ravages of the Cross-Florida Barge Canal (one of the first successful environmental campaigns to stop a project planned by the Army Corps of Engineers; see chapter 10). Among Florida environmental activists, Marjorie is better known than her husband (in fact, her efforts have received greater attention in the popular and scholarly literature). In addition, once the last of their children had started college, Archie and Marjorie resumed their collaboration. Marjorie's efforts produced a suite of important papers regarding the nesting ecology of sea turtles at Tortuguero (see chapter 9). Of their five children, their four sons became directly involved with conservation, while their daughter contributed to several projects regarding her parents' impact on the Florida environment. Archie and Marjorie Carr shared a remarkable relationship for its lasting mutual support of related interests, beginning with their passion for natural history.[7]

The Naturalist Tradition

Close examination of Carr's life and collaborative efforts reveals that natural history and naturalists contributed to the development of biology in the twentieth century, as the field of natural history morphed into new disciplines such as ecology and conservation. To fully appreciate Carr's significance in this larger story, it is useful to review the rich secondary literature in the history of biology. In *Life Science in the Twentieth Century*, Garland Allen mapped many of the trends in biology in the twentieth century. Allen argued that nineteenth-century biology was descriptive and speculative, while twentieth-century biology became experimental, rigorous, and integrative. That is, natural history gave way to methodologically rigorous disciplines such as genetics and molecular biology during the twentieth century.[8] Accepting Allen's call for new studies of twentieth-century biology, a new generation of historians of biology clarified what became known as the "Allen thesis."[9] One effect of the Allen thesis has been a lasting impression among historians of science that with the rediscovery of Mendelian genetics at the beginning of the twentieth century, scientists abandoned the practices of natural history for rigorous, quantitative, yet reductive study of genes. Those historians who studied genetics generally accepted the geneticist's view that their science had dominated biology. After Watson and Crick's discovery of the structure of DNA in 1953, molecular biologists claimed the cutting edge of biology. Many universities supported molecular biology over natural history. According to this view, the naturalists had become antiquated and dusty, like the specimens they studied. In effect, the scientific community and later the science studies community marginalized naturalists such as Carr.

Other historians of the life sciences refined the "Allen thesis." For example, Keith Benson argued that as American biologists became more interested in European methods of experimental biology, academic biology at universities such as Johns Hopkins shifted from natural history to experimental laboratory research. According to Benson, during the twentieth century, natural history was restricted to museums and was no longer taught in academic biology programs.[10] Until recently, historians of science continued to limit American natural history either in time or space, to the nineteenth century or to museums, respectively. To a certain degree, Carr's early career reflected the move of naturalists to natural history museums (see chapters 2 and 3). Modern-day naturalists view natural history in a very different light, predictably in opposition to the view that it is an antiquated science.

Some scientists who have written about the history and philosophy of science have presented natural history in a very positive light. Ernst Mayr, for example, placed natural history at the center of biological thought: "Natural history is one of the most fertile and original branches of biology. Is it not true that Darwin's *Origin of Species* was essentially based on natural-history research and that the sciences of ethology and ecology developed out of natural history? Biology would be an exceedingly narrow discipline if restricted to experimental laboratory researches, deprived of the contact with the continuing, invigorating input from natural history."[11] Similarly, E. O. Wilson acknowledged the centrality of natural history to his own development as an evolutionary biologist in his autobiography, *Naturalist*.[12] Wilson's *Biophilia* is a rhapsody on themes of natural history.[13] No doubt Carr would have allied himself with Mayr, Wilson, and other scientists who acknowledge the importance of natural history to their disciplines.

Recently, Paul Farber demonstrated remarkable continuity between the naturalists of the eighteenth and nineteenth centuries and the practice of twentieth-century biologists, ecologists, and conservationists. The "naturalist tradition," according to Farber, refers to the continuation of practices of natural history and views of nature that developed in the eighteenth century. Farber argued that "In the discipline of natural history, researchers systematically study natural objects (animals, plant, minerals)—naming, describing, classifying, and uncovering their overall order. They do this because such work is an essential first step before other, more complex analyses can be undertaken. We cannot start discussing a wetland, or the interactions within it, until we know something about what is there."[14] Beyond questions of identification, description, and classification, naturalists seek to answer more fundamental questions of biology: "How do all the pieces fit together? What interactions can we discover? What changes? What responsibilities does our knowledge confer upon us?"[15] With these questions, Farber suggests how an interest in natural history led to the study of ecology and conservation.

Other historians have accepted natural history on its own terms to produce rich histories of biology.[16] Philip Pauly drew connections between natural history, biology, and such quintessentially American ideological trends as the Progressive Era in his sweeping cultural history of American biology.[17] Mark Barrow extended the role of naturalists to conservation activism, concluding:

> Before the Second World War countless American naturalists were active in efforts to establish state and national parks, wilderness areas, and wildlife preserves; to gain protective legislation for plants, animals, and landscapes; and to mobilize the public in support of their conservation initiatives.... Swept up in the reform ethos of the progressive era, anxious to bolster the standing of their nascent professions, informed by an emotional attachment to nature and deep sense of civic duty, naturalists deployed their expertise on behalf of more measured, efficient, and sustainable uses of the natural world.[18]

Aldo Leopold was one of the central figures in the development of conservation during the decade before World War II. Leopold's writings influenced a generation of conservationists, including Carr, who exemplified the trend Barrow described.

In examining Carr's life and work, I further elucidate the significance of the naturalist tradition in twentieth-century conservation and biology. Carr's career reveals and reflects many of the elements of the naturalist tradition: a lifelong interest in nature (especially reptiles and amphibians), taxonomic studies of animals, attempts to explain biological phenomena, such as migration and orientation, the search for funding for such studies, his conviction that scientists must contribute to the conservation of species, reflections on the role of naturalists in universities; efforts to educate the public through popular writings; and participation in regional and international conservation campaigns. Expanding still further on the works of Farber and Barrow, my study of Carr reveals that natural history enjoyed a previously unappreciated success during the twentieth century as it transformed into ecology and conservation. Carr himself moved seamlessly from museum taxonomy to ecological studies to advocacy for conservation. Experiment played an important role in Carr's scientific method. His tagging program to determine whether green turtles migrated could be considered a large experiment, and determining the role of vision and olfaction in finding natal nesting beaches necessitated experimentation. In this way, Carr integrated many of the concepts and practices of the relatively new discipline of conservation biology.

A Note on Chronology

For the first two decades of his career, more or less, it is possible to chronicle Carr's life and work in an orderly fashion. His projects, commitments, and contacts were few enough to facilitate a fairly linear chronological account of his activities. By the mid-1950s, however, the demands on Carr's time increased exponentially. Most of the time, he was preparing one or more technical papers regarding his research for grants from the National Science Foundation and the Office of Naval Research, supervising several graduate students, researching, writing, or revising one or more popular books, directing international conservation efforts in his capacity as technical director for the Caribbean Conservation Corporation and head of the Marine Turtle Specialist Group of the International Union for Conservation of Nature. At this point strict chronology would become repetitive and confusing. For this reason, in chapters 5–10, my approach is topical. Many of the chapters overlap in time, but distinct topics define them. In addition, Carr's career spanned fifty years (more if one includes his graduate studies). He worked on a range of projects simultaneously, some which lasted twenty years or more. For example, Carr spent more than two decades searching for nesting Kemp's ridley sea turtles. In this case and others, I follow the research trail across many years (and many pages) along with the rest of Carr's research program.

Science and Narrative

Carr's use of narrative to understand nature set him apart from most other scientists. As I noted earlier, he incorporated local stories and myths into his technical and popular writings. When he began to study sea turtles, he interviewed turtle captains. The stories they told became Carr's working hypotheses. Everywhere he went in search of nature, Carr found culture, and the stories of people eking out a living in distant and often desolate places became a significant part of the narratives he related in his popular books. Unlike most scientists, Carr readily acknowledged nonexpert contributions to his research. Moreover, natural history and ecology, as mastered by Carr, represent narrative exercises. One of Carr's central goals through much of his career was to complete life histories, which is to say the story of life from birth to death, for each of the turtle species of the world. Thus, Carr's use of narrative provides the theoretical tool with which to understand his life and work. One of the few narratives Carr left untold was his own life story. Here is my attempt to fill that void. The story of the man who saved sea turtles begins with a portrait of the naturalist as a young man.

CHAPTER 2

Parallel Paths in Nature

I have waked, I have come, my beloved! I might not abide:
I have come ere the dawn, O beloved, my live-oaks, to hide
In your gospelling glooms,   to be
As a lover in heaven, the marsh my marsh and the sea my sea.
—Sidney Lanier, *Hymns of the Marshes*

W hen Archie Carr was a young boy, he spent entire days of family vacations in McIntosh County, Georgia, trying to catch sheepshead from a dilapidated dock by a small saltwater creek. In the hot sun, he waited for bites only half awake. Decades later, Carr recalled his impression that every time he actually dozed off a huge loggerhead turtle swam up the creek and sounded right by the dock. When the turtle exhaled just feet from his face, Carr was startled awake, just in time to see the turtle slowly swimming up the channel (presumably in search of oysters, he would later note). Other children might recall seeing a giant turtle, but the young naturalist's observations (and recollections) were much clearer. Carr remembered seeing the turtle over six consecutive years. In addition to its monstrous size (more than 500 pounds according to fishermen who had caught it), there was a patch of barnacles over one eye that Carr used to identify the turtle. Once Carr managed to "catch" the turtle with his fishing line, though he suspected that the turtle surfaced only because the hook irritated its mouth. It snapped the line with an impatient swipe of its foreflipper.[1] Thirty years later, he drew upon these memories for his account of loggerhead turtles in his book, *Handbook of Turtles*.

Carr's childhood turtle encounter illustrates several aspects of his personality. That Carr could recall specific details so many years later suggests that the event was important to him at the time. Like E. O. Wilson, Ernst Mayr, Thomas Eisner, and earlier generations of naturalists, Carr's interest in the natural world and in turtles in particular emerged at an early age. When the turtle surfaced, Carr experienced fear, fascination, and humor simultaneously.[2] Throughout his life, through many similar interactions with the natural world, Carr found humor and fascination even in frightening situations. Equally important was the context in which Carr developed his interest in the natural world. As a child, his family moved from Mobile, Alabama, to San Antonio, Texas, to Savannah, Georgia, and finally to Umatilla, Florida. Each of these southern towns offered access to the natural world. Carr spent his childhood (and the rest of his life) in the South. How did place shape the genesis of naturalist? Finally, the loggerhead anecdote suggests that naturalists are born, not made. In other scientific disciplines such as chemistry and physics, scientists may develop passion for a subject they chose intellectually; in other words, the heart follows the mind. For most naturalists, the process is reversed. The mind follows the heart. Carr studied turtles because they fascinated him even as a child. Similarly, Carr's wife Marjorie loved wildlife and wildlands from an early age. As we will see, the two naturalists followed parallel paths before a fateful intersection.

A Portrait of the Naturalist as a Young Man

Raised in backwoods Mississippi, Carr's father, Archibald Fairly Carr, Sr. (1868–1958), learned to hunt and fish at a young age. Along with his brothers and sisters, he provided food for the table on a regular basis by using his skills as a woodsman. When school was in session, the Carr children lived in a rented house near the school. During the week, the children lived on their own, and Archibald's eldest sister cooked for the family. The children returned home to the country on the weekends.

Archibald Sr.'s father, Lafayette Carr (1814–1896), was a doctor. Dr. Carr and his wife Jane Fairly Carr (1837–1929) had immigrated to Snow Hill, North Carolina, from Scotland. It is possible that Dr. Carr served as a doctor for the Confederate Army during the Civil War. From Snow Hill, the Carrs moved to Mississippi, where they raised a large family. After high school, Archibald left Mississippi to attend college. Eventually, he found his way to Southwestern Presbyterian College in Clarksville, Tennessee, where he studied for his doctor of divinity degree.

It was at Southwestern that Archibald met and married Archie's mother, Louise Gordon Deaderick (1884–1968), who was related to the southern poet Sidney Lanier. Louise was studying to be a classical pianist. Her studies had taken her to Chicago, where she spent more than a year under the tutelage of one of Franz Liszt's pupils. Tragically, her aspirations to perform ended abruptly when she suffered from temporary paralysis in one arm and was forced to stop playing. Later in life, Louise returned to music as a piano teacher. Her father, Thomas Oakley Deaderick (1852–1928), was a professor of classics at Southwestern Presbyterian. Once Archibald earned his degree, the young couple moved to Mobile, Alabama, where Archibald became minister of the Government Street Presbyterian Church.

Archie Fairly Carr, Jr. was born on June 16, 1909. In Mobile, the Carrs had a big yard and dogs. "Parson," as Archibald Sr. became known, would ride his bicycle to the edge of town to hunt wild turkeys. More often than not, Parson returned with a turkey hanging from his handlebars. In his spare time, Archie's father would

Figure 2. Archie Carr with two turtles (ca. 1919). Courtesy of Mimi Carr.

take his son fishing in a rowboat, and some of Archie's earliest memories were of these trips.[3] From the time he first began to talk, young Archie demanded that his father read to him from Rudyard Kipling's *The Jungle Book* before he went to sleep, and this became a nightly ritual that father and son shared for many years.

After a few years in Mobile, the Carr family moved to Fort Worth, Texas, where Archie's brother, Thomas Deaderick Carr, was born in 1917. Archie's interest in reptiles blossomed in Fort Worth. In the backyard, he kept a small collection of animals, including snakes, turtles, lizards, and an armadillo. Carr's brother recalled that when he was about three years old, he once retaliated against Archie's teasing by releasing all the animals from the cages. Rather than face Archie's distress, his mother recaptured all the animals and returned them to their cages (despite her considerable fear and dislike of the creatures).[4] In Fort Worth, Carr's father continued to hunt and fish. Often two or three families would travel by wagon to camp beside a river and hunt quail, dove, turkey, and deer. Thus was Carr introduced to the landscapes and wildlife of the South.

In 1920, when Archie was eleven years old, the Carr family moved to Savannah, Georgia. Archibald Sr. became the pastor at his third large church, First

Presbyterian, in downtown Savannah. As a minister, Parson tended to be quietly persuasive and avoided fire-and-brimstone sermons.[5] Although Archie and Tom Carr attended their father's church on a weekly basis, their religious upbringing fell mostly to their mother. The Carr family values appear to have been staunchly Presbyterian: sobriety, stability, and a life of discipline, both individual and social. Moreover, Presbyterian moralism was central to Parson's belief system, and he and his wife shared this outlook with their children. Though Carr appears to have eschewed organized religion after he left home, his staunch Presbyterian values always supported his worldview, particularly as he began to argue for the conservation of sea turtles.[6]

Archibald Sr. never lost his passion for hunting and fishing. In fact, finding good locations for hunting and fishing was second only to locating a church among his interests. Once the family was established in Savannah, Parson eventually found a cottage in Darien, Georgia. Darien is near Brunswick on the coast of Georgia. Sidney Lanier memorialized this region in his poem, *The Marshes of Glynn*:

> How still the plains of the waters be!
> The tide is in his ecstasy;
> The tide is at his highest height;
>   And it is night.
>
> And now from the Vast of the Lord will the waters of sleep
> Roll in on the souls of men,
> But who will reveal to our waking ken
> The forms that swim and the shapes that creep
>   Under the waters of sleep?
> And I would I could know what swimmeth below when the tide comes in
> On the length and the breadth of the marvelous marshes of Glynn.[7]

It is not difficult to imagine Carr's father finding in the marshes around Darien a similar satisfaction, linking his spirituality and his love of the outdoors. Each summer, he would take Archie and Tom fishing for a month or so. Many of the brothers' fondest memories of childhood were of times spent fishing or hunting with their father.

During hunting season (November to March), Parson would take his sons on trips to hunt turkey, quail, duck, and deer. Archie's brother Tom remembered hours spent in a boat freezing while waiting for ducks to fly within range.[8] The boys' mother was an excellent chef, and she skillfully prepared the ducks they brought home. Years later, Carr recalled duck hunting with his father:

> My earliest memory of live wild ducks is of mallards sheering wisely away from our cunningly grouped patch of Montgomery Ward decoys, set out at dawn on a Texas lake by my father and my seven-year-old self when a norther [sic] was on and, in the style of Texas winters, the temperature had suddenly dropped to 19°F. We were crouched by a little charcoal burner, and I remember the thrill and the letdown as one flock after another veered into a climbing turn away from some apparent flaw

in the design or spread of our make-believe mallards or from the shine of a shotgun shell overlooked on the floor of the boat. At times my father did not come home empty-handed, and my mother would take over, burdening me with lusts I would one day have to kick: She cooked duck like an angel.[9]

Archie compared the Savannah of his childhood memory to the Nicaraguan port of Bluefields: "The town looks as the Yamacraw district of Savannah did when I was a boy. Most of the houses are of unpainted wood, with shingles of split laurel and wide verandas, all with a silvery patina imparted by time and sun and blown sea-salt."[10] In high school, Carr developed a reputation not as a student but as a dramatist. He appeared in the leading role in a number of plays at Savannah High School including *Seventeen*, by Booth Tarkington. Carr also took up sailing with his high-school friends in the coastal waters, which gave him another means of accessing the natural world. In one harrowing trip, Carr and a few of his friends decided to sail from Tybee at the mouth of the Savannah River to Brunswick, Georgia, a distance of about 80 miles. When the weather turned foul, the young

Figure 3. Archie Carr with his pointer (ca. 1928). Courtesy of Mimi Carr.

men fought to stay afloat in the Atlantic Ocean in an open boat (probably about 22 feet long) with no motor, only sail and oars. They rationed their limited food for the time they were sailing (they caught a shark and presumably ate it raw). During the worst of the storm, they counted numerous waterspouts between their small craft and the shore.[11]

When Carr lived in Savannah, one coming-of-age ritual for young men was working as a stevedore in the Savannah seaport. Loading and unloading ships as they came into port was an extremely demanding job. Most of the stevedores were strong African-American men, some of whom were the sons of former slaves. The challenge for high-school boys was to test their mettle by obtaining temporary employment on the Savannah docks. Most found the work too taxing, and their careers as stevedores lasted days, not weeks. Carr stayed at the job most of the summer, despite his rather slight build. Working alongside men only a generation removed from slavery, Carr learned phrases and some songs of the Gullah language still spoken along the Atlantic coast between Savannah and Charleston. To synchronize their efforts to lift the heaviest loads, the stevedores would sing Gullah chants, which he learned. It was the language rather than the intense physical work that stuck with Carr, and hearing Mosquito Indians conversing inspired fond memories. Carr later recalled:

> I grew up in Savannah, where two of the more distinctive Negro dialects of the southeast, 'Geechee and Gullah, meet. The first of these is English patois and the latter is an Anglo-African hodgepodge, and although so dissimilar that Gullahs and 'Geechees often cannot understand each other at all (on my honor I several times interpreted for them on the Savannah docks), from a distance they sound much the same. The speech of the Black Caribs of Honduras and the linguistically very different talk of the old and middle-aged Mosquito people are the same quiet music and their English similarly modulated.[12]

Carr's appreciation for the rhythms of language continued throughout his life.

On a fishing trip during the summer of 1927, Carr contracted osteomyelitis in his right arm and became very ill. Over the next two years, he required seven operations in which doctors attempted to remove all of the remaining infection without the benefit of antibiotics. The infection was localized in the ulna, and surgeons removed more of the bone with each operation. Eventually, they took out all but a sliver of the ulna on which new bone could grow. During the final surgery, the doctors discovered that the infection had spread to Carr's elbow joint. The only option remaining was to immobilize the joint permanently. The doctors gave Carr the choice of immobilizing the arm in a straight position or bent at the elbow. Carr chose a bent elbow, hoping that he would still be able to shoulder a rifle and continue to hunt. In fact, he became a better shot in the aftermath of the ordeal. Carr's brother recalled that in later years Archie's students would unconsciously adopt their mentor's bent-arm stance.

With the exception of this extended bout of illness, childhood and adolescence for Carr were idyllic times filled with time in the out-of-doors near small southern cities where wildlife was still abundant. Certainly, Carr inherited his love of nature and the outdoors from his father, the avid hunter and fisherman. Nevertheless, when Carr began college, he pursued his interest in language by studying English. Before he contracted osteomyelitis, Archie had completed his freshman year at Davidson College in North Carolina, where he joined a fraternity (Pi Kappa Phi), but the ongoing infection and numerous operations kept him out of college for two years. After the surgeries, Carr stayed with his aunt in Weaverville, North Carolina, and attended classes at Weaver College. Once fully recovered, Carr moved to a dormitory at Weaver. One of his roommates was Cuban, and Carr quickly learned Spanish. In a relatively short time the two friends could converse entirely in Spanish.[13]

In January 1930, Carr's father moved the family to Umatilla, Florida. Having gone into semi-retirement, Parson chose Umatilla for its still undisturbed wilderness and the 1500 lakes of Lake County. Archie applied to the University of Florida but could not be admitted mid-year. During this time he attended Rollins College for a quarter. There, he studied English, and one of his professors had a profound influence on him by encouraging him to write and by praising his compositions. But at the end of the quarter, Carr was admitted to the relatively new University of Florida, where he completed his B.S. in 1932.[14]

Figure 4. Archie Carr with his parents, Archibald (Parson) and Mimi Carr (ca. 1930). Courtesy of Mimi Carr.

The University of Florida

In the fall of 1930, Carr enrolled at the University of Florida (UF; founded in 1906) in Gainesville, Florida. At the age of 21, Carr most likely did not stop to consider that he might remain at UF for the rest of his life, more than fifty-six years. Initially, his major was English, but he soon changed to zoology. Still, it would be difficult to exaggerate the significance of this training in language to the development of a gifted writer-naturalist, who would become renowned for his skill with the written word. In graduate school, Carr would develop complementary skills in biology and natural history. Years later, Carr recalled his switch to zoology: "As I look back on my junior year in college, when I changed my major from English to zoology, I realize that one of the factors that influenced my decision was the hyacinth fauna—the diverse assortment of self-effacing little animals that most people didn't even know existed."[15] In ecology classes, students rolled large mats of water hyacinths to determine what diverse species lurked within. Throughout his long career, Carr supervised field trips in which students rolled water hyacinths.

Carr continued the pursuit of natural history by studying for a master of science degree in biology (also at the University of Florida). Biology was a new subject at the young university. The program was directed by James Speed Rogers (1891–1955), who had been a professor at Grinnell College in Iowa (since studying entomology at the University of Michigan). Having been appointed head of the Department of Biology at the University of Florida, Rogers set out to build

Figure 5. Archie Carr on collecting trip (ca. 1932). Courtesy of Mimi Carr.

the department. One of Rogers's first hires was Theodore H. Hubbell (1897–1989). Rogers offered Hubbell the position while he was working on his doctorate under William Morton Wheeler at Harvard. Rogers knew Hubbell from the University of Michigan, where the two had conducted fieldwork together. In addition to Hubbell, Rogers hired a mammalogist and a botanist. All were acquaintances from Rogers's graduate study in Michigan. For this reason, the University of Florida program became known as "the University of Michigan, Florida Branch."[16] Charles Francis Byers (b. 1902) joined the faculty later. Rogers, Hubbell, and Byers were all entomologists, and each of them contributed to their subfields: Rogers, crane flies (Tipulidae); Hubbell, crickets (Orthoptera); and Byers, dragonflies (Odonata). Both Rogers and Hubbell eventually left the University of Florida to return to the University of Michigan, where both would eventually hold the directorship of the Museum of Zoology, but they helped establish the Department of Biology at the University of Florida.

The clearest indication of the form and structure of biology as taught by Rogers, Hubbell, and Byers can be gleaned from the textbook Rogers and Hubbell wrote with Byers's assistance in 1937 and published in 1940: *Man and the Biological World*, which was used for the general biology course for nonmajors in University College (the general college at UF).[17] This volume provides a window on Carr's graduate training in biology and, more generally, the teaching of biology in the broader context of the southern United States.[18] In the introduction, the authors described the scope and content of biology under the following headings: "The Organism as an Isolated Individual," (i.e., the simplest possible viewpoint) "The Organism as a link in a Sequence of Generations," (i.e., a temporary unit in a sequence of individuals) "The Organism as a Product of Evolution," (i.e., its relationship to other organisms) and "The Organism as a Unit in a Social-Economic Complex" (i.e., a member of society). Even in its nascent form, biology at the University of Florida emphasized the study of organisms.

But biology for Rogers, Hubbell, and Byers was more than description; it was also a rigorous science, and they developed this belief in their textbook under the heading "Biology as a Science: The Scientific Method and its Limitations." The notion of the hypothesis (defined as a generalization or relationship that is suggested but not proven by the facts already known) rested at the center of this discussion. Further observations and experiments "decide the fate of the hypothesis," and whether it would prove to be untenable, partially tenable, difficult to test, or highly probable, in which case it would become one of the accepted principles. The authors thereby clarified the significance of hypothesis testing to biology. After noting the importance of the "objective viewpoint" (i.e., the ideal, and largely the practice, of making all observations and all proposals and tests of hypotheses without any personal bias), the Florida biologists summarized their understanding of science: "the testimony of checked and repeated observation and experiment is the final authority on which truth or falsity must rest."[19] Such a statement suggests a strong commitment to experimental biology. Other university biology programs emphasized theoretical approaches, while UF remained committed to hypothesis testing and experimentation. It is clear that Carr knew this book and its approach because the authors cited him in the acknowledgments of subsequent editions. Furthermore, Rogers and Hubbell were Carr's main professors in his graduate study of biology. Thus the theories and approaches presented by Rogers and Hubbell undoubtedly influenced Carr's education and his approach to science.

Man and the Biological World provides more than an explanation of scientific method, however. Beginning with nearly a hundred pages on the human body, the authors lead the reader through the complexities of contemporary biology by examining such topics as the varied patterns of animal life, the structure and functioning of plants, the major patterns of plant life, reproduction, the continuity of racial qualities: inheritance, the development of the concept of evolution, the evidence for the fact of evolution, theories concerning the mechanism of evolution, the energy cycle in the organic world, the physical environments in which organisms live, and the biotic environment. In other words, Rogers, Hubbell, and Byers had explained human anatomy and physiology of animals and plants, inheritance (genetics), evolution, and ecology as those disciplines stood in 1941. This brief synopsis of *Man and the Biological World* indicates the range of biological topics considered at the university when Archie Carr was a student. Indeed, Carr would have learned a similar suite of subjects in most biology programs in the United States at the time. At UF, however, the professors emphasized hypothesis testing over theoretical approaches, field natural history over laboratory biology, and above all regional subjects, such as life histories of Florida species.

Carr's own notes from courses provide another view of the Department of Biology at the University of Florida. He studied animal ecology (Biology 402) with Rogers in the spring of 1931. The University of Florida Catalogue for 1927–1928 describes Biology 402 as follows: "Studies on the local fauna as an introduction to the methods of animal ecology."[20] One of the first assignments of the course was an analysis of the relationships among ecology, physiology, and zoogeography. After defining the three sciences and describing how each was pursued, Carr suggested two approaches to ecology: "Since ecology deals with the response of the organism to its environment, we have two ways of approach. We can select an organism and study it in relation to all the factors of its environment or we can select certain environments and study them as composed of interrelated physical and organic factors. Theoretically either road should lead us to the same destination, an understanding of the existing relationship between each organism and its environment."[21] Both approaches would lead to the completion of life histories (that is, a delineation of the animal's range and distribution, habitat, diet, reproduction, and behavior).

In the early twentieth century, life histories of particular animals or groups of animals provided the basis of natural history. With a well-developed life history, the naturalist could assess taxonomic relationships between species.[22] Hubbell recalled that he and the other members of the biology faculty were "in love with Florida and passionately fond of field work."[23] Their passion for the natural history of Florida transferred to expectations of their students: "We were fired by the belief that we had a unique opportunity and mission—to make known the fauna of Florida by our own efforts and those of our students. Each student was encouraged to work on a different group, or, as an alternative, to study the biota and ecology of some characteristic Floridian environment."[24] Even as an undergraduate, Carr caught the fervor for Florida fauna. When he was taking a course in herpetology and ichthyology with Professor Leonard Giovannoli, he and his friends enjoyed hunting for new specimens to expand the list of Florida species or revise their known ranges in Florida. Perhaps because neither women nor alcohol were permitted on campus, collecting reptiles, amphibians, and fish for the museum was an exciting part of Carr's undergraduate life. Among other finds, he caught the first two specimens of brick-red water snakes (*Natrix erythrogaster*) in a fearless over-water leap.[25]

Hubbell recalled that he and Rogers had a higher objective in mind as they structured the new department:

> We hoped that its [the biology department's] faculty and students would become a cooperative, self renewing body of field oriented biologists with interests focused on Florida. And do you know, after we got going this plan worked pretty well for quite a while. It produced a crop of students each of whom soon knew far more about his chosen group than did his major professor, a series of theses on various Florida taxa and environments, and eventually a number of distinguished biologists.[26]

A review of the titles of the first graduate theses and dissertations produced in the UF Department of Biology suggests the success of the approach laid out by Hubbell. H. K. Wallace, Carr's contemporary and an arachnologist, wrote a dissertation entitled "The habitat distribution of the spiders of the family Lycosidae in the Gainesville region" (1938). Another contemporary of Carr's, Horton H. Hobbs, Jr., wrote his master's thesis on the crayfishes of Gainesville (1936) and extended his study to cover the crayfishes of Florida for his Ph.D. dissertation (1940). Lewis Berner and Frank Young developed life histories of Florida mayflies and water beetles, respectively, for their master's theses and doctoral dissertations. Along with Carr, these individuals were among the first recipients of doctorates in biology at the University of Florida.[27] When Carr was a student at UF, the Department of Biology thus provided grounding in experimental biology, organismal biology, and natural history. Carr continued to develop his knowledge in both areas as he began to contact acknowledged experts in his chosen field—of herpetology.

In earning his master of science degree in biology from UF in 1934, Carr received a strong introduction to biology and particularly to limnology (the study of freshwater ecosystems, especially lakes). He wrote a master's thesis on plankton and the carbon dioxide–oxygen cycle in a Florida lake.[28] Carr's study of freshwater ecosystems provided a solid basis for his later studies of turtles and their natural history. But given that his main professors (Rogers, Hubbell, and Byers) were all entomologists, Carr studied herpetology with Leonard Giavannoli, his friends, and through correspondence with noted herpetologists throughout the United States.

One of the people who greatly influenced Carr in his study of herpetology was fellow graduate student Oather C. Van Hyning. At the time, the Department of Biology was housed in Science Hall (now Keene Hall), which was near the center of the UF campus. In addition to the biology department, Science Hall housed the Florida Museum of Natural History, which was directed by Oather's father, Thompson Van Hyning. Oather was the consummate naturalist (other students believed he knew more about the natural history of Florida than almost anyone else, including his father). Oather Van Hyning was building a large collection of the snakes, turtles, and fish from all over the state of Florida and had published several papers in *Copeia*, the journal of the American Society of Ichthyologists and Herpetologists (see below).[29] Van Hyning showed Carr that the study of herpetology in Florida offered a rich subject for scientific research. Unfortunately, Van Hyning was forced to leave the university prematurely when a campus policeman caught him drinking lab alcohol (in violation of university policies) mixed with Coca-Cola in the basement of Science Hall. He was expelled and never completed his degree.[30] Nevertheless, the publications Van Hyning wrote at UF contributed to the knowledge of regional herpetology.

Carr's early publications were similar to Van Hyning's in that their insights were of regional significance. Carr's first publication and his early correspondence provide insight not only to his development as a herpetologist but also to herpetology and its practitioners in the early twentieth century. In 1934, Carr published "A Key to the Breeding-Songs of the Florida Frogs" in the second issue of a relatively new journal called *The Florida Naturalist*.[31] The simplicity of his directions and the twenty-four steps that followed belied the considerable effort behind producing such a key. In the first place, Carr did not have recording material at his disposal to compare frogs' breeding calls. Naturalists also knew that most frogs (even males in the throes of breeding season) stop calling upon the approach of large mammals like humans. To make matters even more difficult, frogs have a tendency to give their calls from the safety of extremely wet areas. In creating his key, Carr had to isolate a given call, track it to a specific frog without disturbing the caller, and identify it to species. Carr was able to determine some of the species by capturing a few individuals and holding them in terraria until they resumed calling. Though labor intensive, this strategy proved effective. Carr warned of particular difficulties with tree frogs (genus *Hyla*): "Be certain that you have limited your attention to the call of one frog and are not listening to the sounds made by several. It is sometimes difficult to isolate an individual call, especially in large choruses of *Hylas*."[32] Despite Carr's disclaimer about the provisional nature of the key, it was a valuable contribution to the herpetology of Florida, which, like the University of Florida, was still in its early stages of development. Moreover, the key to frog calls served as Carr's introduction to prominent herpetologists in America.

Forming a Network of Herpetologists

Carr sent his "Key to the Breeding-Songs of the Florida Frogs" to several of the established herpetologists in the United States. He also began a correspondence with other herpetologists on issues related to his study of amphibians and reptiles. Throughout history, naturalists have developed correspondence networks.[33] Carr's correspondence around the time of his first publication reflects the state of herpetology at the beginning of his career. By reviewing these letters, one develops a sense not only of who was studying herpetology but also the universities and museums that served as the centers of study. I analyze Carr's correspondence here to give readers a better understanding of the history of herpetology and its practice in the United States, of the growth of museum collections and research, and of the role of universities in the definition and practice of natural history and ecology.[34] Equally important, the individuals to whom Carr wrote came to serve as a network for the young naturalist and herpetologist.

About the time his first publication appeared in 1934, Carr initiated correspondence with Thomas Barbour (Museum of Comparative Zoology, Harvard University), Leonhard Stejneger (U.S. Museum), Helen Gaige (Museum of Zoology, University of Michigan), Edward H. Taylor (University of Kansas), Albert Hazen Wright (Cornell University), and M. Graham Netting (Carnegie Museum, Pittsburgh). One of the most enthusiastic responses to the "frog key" came from Thomas Barbour (1884–1946), who wrote: "Thank you ever so much for the copy of 'The Florida Naturalist' containing your article, 'A Key to the Breeding-Songs of the Florida Frogs.' This is very interesting and I am glad to have it."[35] Barbour also asked Carr for any information regarding subscription to the journal as well as back issues.

Thomas Barbour served as director of the Museum of Comparative Zoology at Harvard University from 1927 until his death in 1946.[36] Like Carr, he became interested in reptiles and the natural history of Florida at an early age, while recovering from typhoid fever in Florida in 1898. All of his degrees came from Harvard University (A.B. 1906, A.M. 1908, and Ph.D. 1911), where he remained for his entire career. Barbour's father was a wealthy businessman who left him a large inheritance that funded his pursuits as a naturalist and herpetologist, which included contributing to the development of Barro Colorado Island in Panama as a research station (now part of the Smithsonian Tropical Research Institute). Barbour also used his own funds to support promising herpetologists like Carr and to support the herpetological journal *Copeia*. In the final years of his life, Barbour wrote several lengthy travel narratives on his experiences as a naturalist that indicate considerable commitment to the works of Alfred Russell Wallace, Thomas Belt, and W.H. Hudson, as well as the tropical botanist David Fairchild (*The World Was My Garden*) and the tropical ornithologist Frank M. Chapman, who helped found the research station on Barro Colorado Island (*My Tropical Air Castle, Life in an Air Castle*).[37] Given his interests, his position at the Museum of Comparative Zoology, and his affiliation with Harvard University, Thomas Barbour was a good choice to direct Carr's doctoral studies. As I illustrate below, few people had greater influence on Carr than Thomas Barbour.

Carr corresponded with Leonhard Stejneger (1851–1943) at the U.S. National Museum about the speciation of *Pseudemys*, a turtle genus in Florida: "I am making an ecological study of the reptiles and amphibians of Florida, and have met with several very puzzling problems. . . . My conclusion has been that

North Florida is the region of intergradation between the more northerly *P. concinna* and the peninsular *P. floridana*. Do you believe this to be the case?"[38] Stejneger responded immediately with considerable enthusiasm, noting that Carr might solve some of the problems that had stymied him for lack of complete series of specimens: "What is the nature of the intergrading of these forms? Is it due to hybridization? Do their ranges overlap? Are specimens carried about? Do they wander to any extent? I am of the opinion that only resident naturalists having ample living material of all ages will be able to decide.... Please keep up the good work."[39] Stejneger thus encouraged and engaged Carr's early efforts. He also noted the advantage of Carr's location in Florida, which would work to Carr's advantage repeatedly throughout his career as a scientist.

Leonhard Stejneger was born in Bergen, Norway, in 1851. He studied law, assisted with his father's business, and published twenty-two papers on ornithology before moving to the United States.[40] Hired by Spencer Fullerton Baird as an assistant in ornithology at the U.S. Museum (now the National Museum of Natural History of the Smithsonian Institution in Washington, D.C.) in 1911, Stejneger became head curator of biology at the museum, a position in which he was able to complete numerous publications including *A Check List of North American Amphibians and Reptiles* with Thomas Barbour in 1917 (and four subsequent editions).[41] The *Check List* significantly stabilized the nomenclature of herpetology in the United States, provided the basis for life history studies, and indicated many problems for further examination (hence, the correspondence with Carr). Though he did not hold a university position, Stejneger influenced numerous herpetologists, including Thomas Barbour, Frank N. Blanchard, Emmett R. Dunn, and Alexander G. Ruthven. Like Barbour, Stejneger managed one of the largest collections of reptiles and amphibians in America and would prove to be an invaluable contact to Carr.[42]

As Curator of Herpetology at the Museum of Zoology at the University of Michigan, Helen Thompson Gaige (1890–1976) managed a sizable collection of herpetological specimens. One of her early letters to Carr warmly supported his activities and his correspondence with Stejneger and Barbour: "It was pleasant to hear from you and to know that you are enjoying work at the laboratory.[43] I am glad that you wrote to Dr. Stejneger about your theories of the intergrading of *Pseudemys*. If you prove to be right, a number of our difficulties in the identification of these species will be solved. And you will be a public benefactor if they are. The concurrence of Dr. Stejneger on any theory is almost as good as proving it, for he is very cautious about expressing his opinion."[44] Like Barbour and Stejneger, Gaige indicated considerable interest in the young herpetologist. She suggested that Carr might complete his doctorate in herpetology at Harvard, where Thomas Barbour rated his chances of receiving admission and funding as good.[45] Both Michigan and Harvard offered Carr practicing herpetologists and large specimen collections. When Carr remained in Florida to complete his Ph.D., Gaige continued to support his education as a herpetologist.

Although she never completed a doctorate, Helen Thompson Gaige became one of the most influential figures in American herpetology during the twentieth century.[46] After finishing her A.B. and M.A. at the University of Michigan in 1909 and 1910, respectively, Gaige joined the Museum of Zoology as scientific assistant in the division of herpetology. At that time, Alexander G. Ruthven was curator, but in 1918 he became director of the museum and relinquished many of his responsibilities (including the training of graduate students) to Gaige. She became assistant curator of reptiles and amphibians in 1918 and curator in 1923. Her publications covered a range of areas including the Midwest, Texas, Washington, Florida, Colorado, and the neotropics. Besides training graduate students (nominally under the supervision of Ruthven), Gaige greatly influenced herpetology as editor of *Copeia*, a position she held for twenty years. After retiring from the University of Michigan in 1945, Gaige relocated to Gainesville, Florida, where she died in 1976.

Some of the herpetologists recognized in Carr the opportunity to obtain specimens (the stuff of material practice in natural history) from northern Florida, which was generally under-represented in collections. M. Graham Netting, curator of herpetology at the Carnegie Museum in Pittsburgh, Pennsylvania, expressed his gratitude for the "frog key" and asked to subscribe to the new journal. Later, Netting requested that Carr collect specimens for the Carnegie Museum. Apparently, Carr saw this as an opportunity and almost immediately sent several examples. Further correspondence between Carr and Netting is filled with references to species sent and received and to collecting trips in Florida.

Other prominent herpetologists such as Albert Hazen Wright of Cornell University fondly recalled collecting trips with Carr. Wright (1879–1970) and his wife, Anna Allen Wright (1882–1964), were married in 1910 and collaborated with each other on natural history projects until Anna's death in 1964.[47] Together, they wrote and illustrated several books in the series, *Handbooks of American Natural History*, published by Comstock Press of Cornell University. *The Handbook of Frogs and Toads of North America* was published in 1933 and the *Handbook of Snakes* (3 volumes) followed in 1952, the same year that Carr published his *Handbook of Turtles* in the same series.[48]

In reading Carr's correspondence, one is struck by the number of people with whom he developed a casual familiarity after only a few letters—that is, his aptitude for the social practice of natural history. The few people with whom he did not develop such familiarity stand out in contrast. Carr's correspondence with Edward H. Taylor (1889–1978) of the Department of Zoology at the University of Kansas is one example.[49] Taylor was helpful in assisting Carr with a particularly difficult turtle identification problem, but the professional correspondence remained formal.

Carr's early efforts to develop contacts in herpetology would form the basis of his professional network. Many of these individuals continued to participate in Carr's development as a herpetologist. But one of the central questions regarding Carr's education centered on how his professors defined the study of biology. What do biologists do? To understand how Carr developed as a herpetologist, we need to explore a similar question, What do herpetologists do? At least in terms of reptiles and amphibians, Florida represented a biological *terra incognita* and was an ideal location for biological exploration. Such fields offered rich opportunities to eager students like Carr. The problems that Carr discussed with his network of herpetologists had as much in common with the interests of the great naturalists of the nineteenth century as they did with the concerns of his mentors at the University of Florida or even the early ecologists of the twentieth century. A brief history of American herpetology underscores this point.

American Herpetology

Herpetology has a rich history in the United States.[50] The study of reptiles and amphibians developed along the same lines as ornithology and mammalogy. The great explorer-naturalists of the eighteenth and nineteenth centuries first recorded reptiles and amphibians along with many other animals and plants. As biological exploration continued, museum naturalists conducted taxonomic studies of the many specimens and wrote extensive catalogs describing North America's flora and fauna, including reptiles and amphibians. With the expansion of collections, naturalists concentrated on particular groups, and the formal study of herpetology emerged from the more generalized study of natural history. Late in the nineteenth century, herpetologists founded a society and started a technical journal to share their findings. By the early twentieth century, herpetologists who held university appointments began to train students. This new generation of biologists began to examine the ecology and evolution of amphibians and reptiles.

Reptiles and amphibians were among the many organisms noted by the great explorer-naturalists of the eighteenth and nineteenth centuries. Mark Catesby (1683–1749), John James Audubon (1785–1851), and William Bartram (1739–1823) recorded encounters with reptiles in their efforts to explore the natural history of North America. Bartram made a drawing of several alligators that looked like mythical sea monsters, thrashing about in the water, fish trapped in their jaws, and "smoke" flowing from their nostrils. Largely descriptive, the work of these naturalist-explorers provided glimpses of uncharted territory. Later naturalists such as Thomas Say (1787–1834), John Eaton Le Conte (1784–1860), and Richard Harlan (1796–1843) began the painstaking process of collecting, preserving, and describing species of frogs and amphibians as well as birds, mammals, invertebrates, and plants. Each of these naturalists maintained an affiliation with the Philadelphia Academy of Natural Sciences, where they deposited their specimens, reported their findings, and named new species after family members and friends. Museums had emerged as the center for the study of reptiles and amphibians along with other fauna. Drawing on the foundational taxonomic work of these collector-naturalists, later scientists were able to produce comprehensive works.

John Edwards Holbrook (1796–1871) synthesized contemporary knowledge of reptiles and amphibians in his five-volume work, *North American Herpetology* (1842). With Holbrook's opus as a core of knowledge, the study of American herpetology became more focused and sophisticated. After Holbrook, two prominent museum naturalists defined the practice of herpetology in America: Louis Agassiz at the Museum of Comparative Zoology at Harvard University and Spencer Fullerton Baird at the Smithsonian Museum in Washington, D.C. When he arrived in the United States from Switzerland in 1846, Jean Louis Rodolphe Agassiz (1807–1873) already had a considerable reputation in Europe as a naturalist, anatomist, and embryologist.[51] Agassiz taught zoology at Harvard and gave popular lectures in natural history throughout the United States. But it was his efforts to develop the Museum of Comparative Zoology at Harvard that contributed to the science of herpetology. Agassiz completed four volumes of his *Contributions to the Natural History of the United States of America*, including two volumes on turtles.[52] Although Agassiz never accepted Darwin's theory of natural selection, his taxonomic work on turtles provided a research basis for herpetologists such as Thomas Barbour (see chapter 3) into the twentieth century.

Like Agassiz, Spencer Fullerton Baird's (1823–1887) close affiliation with the Smithsonian Institution, from shortly after its establishment in 1846 and until his death in 1887, contributed greatly to the development of modern herpetology.[53] Through various surveying explorations (including the Lewis and Clark expeditions), Baird created a museum that represented the fauna of all of North America, whereas other collections exhibited a considerable bias for eastern specimens. With his massive network of collectors and correspondents, Baird described many species of reptiles and amphibians, especially from the western states.[54] Although his contributions to the study of herpetology in America consist of species lists, Baird managed to create an unparalleled research collection. He left the more detailed analysis of taxonomic relations to collaborators and assistants such as Edward Drinker Cope (1840–1897), Charles Girard (1822–1895), and Leonhard Stejneger.

On December 27, 1913, the journal *Copeia* was launched, at the personal expense of John Treadwell Nichols (the title honored the herpetological contributions of E. D. Cope). Several short contributions upheld the journal's raison d'être, which was as follows: "Published by the contributors to advance the science of coldblooded vertebrates."[55] In 1916, a publication committee formed, and the members founded the American Society of Ichthyologists and Herpetologists (ASIH). Another society, The Herpetologist's League, formed two decades later.

As the journal and society offered members publishing opportunities and new status, the formal study of herpetology continued to evolve at several universities and museums. Under the direction of Alexander G. Ruthven, the University of Michigan became the leading center of graduate training in herpetology. In his research, Ruthven produced a taxonomic revision of the genus *Thamnophis* (the garter snakes). Graduate students at the University of Michigan and elsewhere tended to follow Ruthven's example by focusing their research on a well-defined group of amphibians or reptiles. As curator of herpetology, Gaige also fostered this approach (see above). Michigan graduates influenced the development of similar programs at the Chicago Academy of Sciences and the University of Florida.

Even a brief history of herpetology reveals the material practices of American herpetologists: developing collections, describing species, and constructing taxonomies; in short, natural history. Major museums served as the centers of such work. At two of the largest American museums (the Museum of Comparative Zoology and the Smithsonian Institution), scientists concentrated on taxonomic relations and species description.[56] Opportunities for graduate study in herpetology, the journal *Copeia*, and the ASIH all paved the way for a new generation of specialists. As herpetology lagged behind ornithology and mammalogy in its evolution into a distinct discipline, there were still strong opportunities for young scientists, even students, to make significant contributions to knowledge of the natural history and taxonomy of reptiles and amphibians, particularly in the South. As Carr was becoming acquainted with the herpetologists of America, his alter ego was studying zoology and botany about a hundred miles to the northwest.

Marjorie Harris's Parallel Path

Born in Massachusetts in 1914, Marjorie Harris's love of the natural history of Florida began early. She recalled that her father had always thought: "Why should anybody stay up in New England where you can only be out-of-doors for a few months of the year, when there is Florida where you can be out-of-doors all year round?"[57] A schoolteacher in Massachusetts, Harris's father dreamed about establishing a small farm. Along with several associates, he purchased a piece of property south of Bonita Springs in Lee County, Florida, in 1918. The young Marjorie began to suffer colds during the winters at about this time, so beginning in 1918 her mother took her to Florida each winter, after which the family moved to Bonita Springs permanently. To get to school, Harris rode her horse 3 miles each way. She remembered that her parents could answer her questions regarding plants and animals and that this instilled in her an appreciation of the natural environment. Harris's father died in 1929 when she was just fifteen years old. To support the family, her mother taught school, first in a one-room school house on Sanibel Island, then in Bonita Springs, and finally in Ft. Myers Beach.

Even at a young age, Harris became aware of environmental destruction by observing that populations of herons and egrets had been decimated by the plume trade. Her home was on the banks of the Imperial River, and at least a few times each week, she would go canoeing with her parents. She remembered that there was little in the way of wildlife: "You didn't see a living thing of any size. And why? Because it was the custom to come down, hire a boat, use a gun and stand up in the front of the boat and shoot anything that made a moving target. Alligator, red bird, heron, what have you. Any thing that moved was the sport. . . . That outrage, it wasn't just the fellows hunting—it was the stupidity of killing everything."[58] The wanton slaughter of animals which was a popular form of tourism on the rivers of Florida during the early twentieth century.[59] This kind of hunting stood in sharp distinction to the measured sportsmanship in which Archie Carr participated as a young man.

Harris's interest in nature continued to develop in college as a student at the Florida State College for Women (FSCW), now Florida State University.

In choosing FSCW, Harris had initially focused on zoology, but she embraced botany as well:

> My aim was to become a zoologist. I wanted to work with whole, live animals, preferably birds, in their natural surroundings. Fortunately, I came under the direction of an outstanding teacher, Dr. Edza Mae Deviney, who guided and advised me during my undergraduate years. The zoology department offered only a few field courses, such as ornithology and marine invertebrates, but the botany department under the direction of Dr. Herman Kurz provided a stimulating, even exciting, experience.[60]

Through Kurz's courses, Harris learned about forest types and their evolution as well as the importance of fire to these ecosystems. She found her studies "enormously satisfying." "What a pleasure it was to go into the woods and fields and, by recognizing a set of characteristic key plants, be able to put a name to a particular association of plants. It was thrilling to look at a landscape and think perhaps you knew its past history and its future. The ability to 'read' a landscape provides the kind of pleasure that comes from a knowledge of Bach or Shakespeare or Van Gogh."[61]

To pay for tuition, room, and board, Harris worked during the summers. The New Deal South offered new opportunities for people who were knowledgeable about the flora and fauna of Florida. In 1934 and 1935, Harris designed and taught a three-month summer field course for children throughout Lee County with support from the National Youth Administration (a New Deal Program). At the end of each summer she received enough money to cover her college expenses for the following year. Harris recalled that the summers had a profound influence on her:

> I learned to have great respect for all sorts of people and not to have contempt for people, whatever their background. You find naturalists or people who are interested in the environment or sensitive to the environment in all walks of life. I think I learned that from way back there. You see, this was not a program that appealed to everybody in the community, but in every place—these little, tiny places—there would be nine or ten who were faithful and absolutely devoted to this program. They had stars in their eyes.[62]

After Harris earned her bachelor of science degree in 1936, she planned to continue her studies by accepting a fellowship at the University of North Carolina–Chapel Hill, but her fellowship was delayed. In the meantime, she was hired as a wildlife technician at a fish hatchery in Welaka as part of the Resettlement Administration, also under the New Deal (the first woman in the United States to be employed in this capacity). When a covey of sick quail arrived at the hatchery, Harris took them in a box to the University of Florida in the hopes she might get some advice. In the Department of Biology, Marjorie met Archie Carr. By all accounts, it was love at first sight. Over the next few months,

Figure 6. Archie and Marjorie Carr (ca. 1937). Courtesy of
Thomas D. Carr.

Archie courted Marjorie with the assistance of his brother's car. The two young
naturalists married in January 1937. Marjorie Harris Carr went on to complete a
master's degree in zoology at UF, completing her thesis on the breeding biology
of bass. She later published her findings in the *Proceedings of the New England
Zoological Club.*[63]

Conclusion

In studying the early lives and education of Archie Carr and Marjorie Harris, it
is remarkable to see how much the two naturalists had in common. Both
learned to love nature from a young age, both were educated at Florida state
universities, both studied biology at the graduate level, and both spent much of
their youth studying the flora and fauna of the southeast. The months after their
marriage were particularly demanding for Carr, who was completing work on his
doctoral dissertation.[64] Like his fellow graduate students, Carr had developed into

a regional authority on reptiles and amphibians. When he was awarded his doctorate in May, it was the first Ph.D. to be granted in biology by the University of Florida.

Most naturalists discover the passion for nature and the out-of-doors as children. Early in life, Carr developed an appreciation of nature through his own initiative and through hunting and fishing trips with his father. Carr's interest in natural history and herpetology evolved at the University of Florida, where a dedicated group of professors taught the scientific method and emphasized the relationship between humans and nature while encouraging organismal studies of Florida taxa (such as turtles, spiders, and mayflies). Beyond Carr's passion for field work in Florida and his courses with a dedicated group of professors, he benefited considerably from correspondence with several established herpetologists in America. Such interactions gave him a clear sense of the predominant concerns of American herpetologists and valuable insights about his potential to contribute to the field. In combination, Carr's childhood hunting trips, formal university study, and extensive correspondence with American herpetologists provided him with a strong background to begin his career as an instructor at the University of Florida.

CHAPTER 3

Dear Dr. Barbour

Intellectual creativity rarely develops in a vacuum. In a study of seven geniuses in a broad range of disciplines, Howard Gardner showed that genius requires a nurturing environment of emotional and intellectual support.[1] Marjorie Carr shared Archie Carr's interests in natural history, and it is clear that the young couple supported each other emotionally. Nevertheless, Gardner argued that for genius to develop, a person needs cognitive or intellectual support. As a young lecturer in zoology, Carr needed to publish his findings while formulating his larger research agenda. Thomas Barbour provided exactly the kind of intellectual support Carr needed to develop as a scholar and advance professionally. Beyond critical support, Barbour had both the means and the connections to assist the Carrs in myriad of ways, both minor and major, which further facilitated Archie's ability to focus on his research and publications. After a few years of Barbour's light-handed guidance, Carr felt confident in challenging Leonhard Stejneger, who was widely acknowledged as one of the leading herpetologists in America, regarding the taxonomy of several genera of turtles. Despite his conviction in his assessment of the issue at hand, Carr diffidently directed his arguments to Barbour, who served as mediator in the debate. Carr's debate with Stejneger provides insights into Carr's scholarly development and personality.

Summers at the Museum of Comparative Zoology

As Archie Carr was completing his dissertation on the herpetology of Florida in January 1937, he wrote to Thomas Barbour about the possibility of visiting the collection at the Museum of Comparative Zoology (MCZ) in Cambridge, Massachusetts.[2] Barbour responded enthusiastically to Carr's letter, extending an invitation for him to visit the collection. In April, he sent another invitation and offered to fund Carr's trip. As a result of this correspondence, Archie and Marjorie spent their first of seven summers with Thomas Barbour at the MCZ in Cambridge. Barbour possessed a powerful personality, and he welcomed the Carrs into the museum he directed as well as his home and even his family (he would later recount that he presumed to feel *in loco parentis* to the Carrs).[3] Barbour, who had an independent income from his family's holdings, devoted a small portion of both museum and personal resources to facilitating the work of promising herpetologists. From the summer of 1937 on, Carr was a primary beneficiary of Barbour's intellectual and financial patronage. Although the actual dollar value of Barbour's financial support was fairly minimal, his support made it possible for

Figure 7. Archie Carr and Arthur Loveridge, Cambridge, Massachusetts (ca. 1941). Courtesy of Mimi Carr.

the Carrs to spend their summers between 1937 and 1943 at Harvard. With Thomas Barbour as a friend and patron, Carr matured as a herpetologist, naturalist, and writer.[4]

Thomas Barbour quoted a letter he received from Henry James that summarized his view of biology and natural history in discussion about the small number of Harvard College graduates who pursued biology (as opposed to medicine) at Harvard:

> What used to be called Natural History ought to be one of the best cultural studies. A man who cannot use his eyes and ears as he goes about in his physical environment and cannot learn about the universe except by digging himself into the stacks of the Widener Library and putting on a pair of spectacles is only half a man. But for some strange reason two generations of scientists have chosen to treat amateur naturalists as triflers, the systematist as a pedant; and the school teachers have failed pretty completely to do much with natural science. Crazy and deplorable!![5]

James made several important points in this statement. First, Harvard students were largely interested in biology for future work in medicine. Second, he noted the value of the natural history study for its own sake. Finally, and most significantly, the study and practice of natural history had lost prestige in the eyes of other scientists (even though it was still pursued actively at the MCZ).[6] If Barbour wanted to support naturalists like himself, he would have to find them outside of Cambridge. The study of organismal and experimental biology was central to the biology curriculum at the University of Florida. Given Barbour's views, it is not surprising that Carr's turtle studies at the MCZ were largely taxonomic in nature and of considerable interest to Barbour.[7]

Carr Revises *Pseudemys*

In the course of work on his doctoral dissertation, Carr observed that two species of turtles in the genus *Pseudemys* intergraded in north Florida (recall the correspondence between Carr and Stejneger in chapter 2). *Pseudemys* is the generic name for the species of aquatic turtles commonly called "cooters." It was this genus that provided Carr's first major intellectual challenge, and he published several papers on its taxonomy. Most taxonomists function as "lumpers" or "splitters." Lumpers tend to lump separate races under a single species, whereas splitters split distinct races into species. In his revision of the *Pseudemys* genus, Carr acted as both a lumper and a splitter. In 1935, Carr lumped *Pseudemys floridana* and *P. concinna* into the single species *P. floridana* with two subspecies. His second paper on the topic (1937) split *P. floridana suwanniensis* (a new subspecies) from *P. mobiliensis*.[8] In his next paper (1937) Carr combined *P. scripta* and *P. troostii*, but he declined to suggest a nomenclatural change.[9]

It is clear that Carr's research and publishing benefited from the extensive collections at the MCZ. Near the end of 1937, he published another analysis of

P. scripta, drawing on a specimen collected by Helen T. Gaige in the Rio Grande in Brewster County, Texas, and the collections of the MCZ. In establishing that this individual was distinct from other races found in the United States, Carr displayed his growing knowledge of the biogeography of *Pseudemys*.[10] After examining more specimens, Carr suggested that geographical races of *Pseudemys* created a gradient from one species to another. More than a mere taxonomic claim, this observation amounted to a statement of evolutionary systematics.

As Carr became more immersed in subspecies of *Pseudemys*, he began to consider the possibility of revising the classification of all of the species and subspecies in the genus. Leonard Stejneger considered this genus to be his pet project, so Carr wrote to Barbour to avoid conflict:

> You may recall that some time ago I asked your advice regarding the propriety of my publishing remarks on extra-Floridian *Pseudemys*. The opinion which you expressed and with which I quite agreed, was that it might be the more decent course to lay off the genus, in view of Dr. Stejneger's fondness for it.
>
> I have recently had a letter from Clifford Pope requesting my ideas on the forms of *Pseudemys* to be used in his popular book on North American turtles. I have certain fairly definite ideas which are at variance with the current classification of the group, but I have intended to withhold anything of a revisional nature, for the sake of decorum.[11]

In handwritten notes on Carr's letter, Barbour considered his comments: "L. S. has been working on these forms for years. When he will publish—if ever— no one knows. On thinking it over, I see no reason why Carr should not list his ideas on paper + S. can accept or reject when the time comes. Carr's views are certainly more plausible than any other. They may or may not represent the final conclusions on all N.A. herps, but they are highly interesting!"[12] Barbour's comments regarding Stejneger's priority would prove prophetic.

On May 23, 1938, Barbour invited the Carrs to Cambridge for another summer: "I am very keen to have you work on the Antillean *Pseudemys* this summer and would be delighted to offer you $200 so that you and your wife could come here and be reasonably comfortable while working on the critters.... I should be glad to see the manuscript of *P. nelsoni* when it arrives. Have no doubt it can be placed with promptitude."[13] "Promptitude" seems rather conservative given that Carr's paper appeared on June 9, 1938, little more than two weeks after Barbour sent his letter. The paper described (or, more accurately, "split") yet another turtle species, *Pseudemys nelsoni*. Carr named it for the preparator-in-chief at the MCZ, George Nelson, in recognition of his extensive contributions to museum collections of Florida reptiles and amphibians.[14]

The Carrs did spend the summer of 1938 with Barbour at the MCZ, and Archie's work on the Antillean turtles resulted in several collaborative publications. Barbour and Carr described a new freshwater turtle (*P. malonei*) from the Bahamas. With the assistance and guidance of the commissioner for the island of Inagua, the honorable J. V. Malone, two MCZ staff members collected a series of eleven specimens (including the "type," the specimen from which they wrote the description, the "male holotype," a male of the same species, and nine "paratypes," additional specimens collected at the same location).[15] In addition to his work on Antillean turtles, Carr completed a description of a new subspecies of *P. floridana*, which included a review of the *floridana* subspecies. The resulting paper, which appeared in the most prominent journal for herpetologists, *Copeia*, gave Carr the opportunity to display his gift for infusing his scientific papers with natural history and history. Beginning with Le Conte's descriptions of the species of North American tortoises, Carr wondered why the great naturalist failed to assign a race of *P. floridana* to the Atlantic coastal plain. Rather than relying exclusively on museum specimens, Le Conte had extensive field experience with turtles. Nevertheless, he neglected to describe a subspecies for the very place where his plantation stood: Riceboro, Georgia. Carr delved into the history of the area but found little to explain Le Conte's oversight. He explained: "The Riceboro Plantation was established in 1810 by Louis Le Conte, brother of John, who was a New Yorker. Louis and his sons were all ardent naturalists, and although I can find no record of John's having visited his brother prior to the publication of his paper, it seems unreasonable to suppose that correspondence regarding the local fauna had not been exchanged."[16]

Describing a New Salamander

On May 19, 1939, an exciting find virtually fell into Carr's lap when H. K. Wallace, his colleague and the biology department's resident arachnologist, presented him with a strange salamander. Carr could barely contain his excitement when he wrote to Barbour:"I have something very delightful. I have a blind, white salamander from the 200-foot well in Albany, Georgia. It arrived, casually, yesterday in a bottle of water sent parcel post by the sanitary engineer there. It is alive, apparently pregnant, and apparently neotenic. There is no vestige of eyes."[17] By "neotenic," Carr suggested that this species reached sexual maturity with juvenile characteristics. Both blindness and neoteny would become significant themes for analysis within evolutionary biology. Carr requested from the U.S. National Museum (Smithsonian) specimens of what he guessed were the salamander species most closely related to the one in hand, but he believed the specimen represented an undescribed species and genus.[18]

To write a complete description including its designation into a family, Carr knew that he would have to conduct a complete dissection. With only a single specimen, he was hesitant to mutilate it so he asked for Barbour's advice. Barbour replied with his usual speed and decisiveness: "I would get out a preliminary description of your new salamander and then hold up the final decision as to family allocation until we can talk the situation over up here. I suggest however, that it might be a very good idea to have an X-ray made if that can conveniently be done in Gainesville."[19] Fortunately, Carr's brother Thomas had just completed his master's degree in physics on X-ray diffraction and was able to gain access to

X-ray equipment to photograph the salamander, and this resulted in an acceptable image of its skeleton. Unfortunately, in the process of taking the X-ray, the salamander fell off the table and died.

By naming the new salamander, *Haideotriton wallacei*, Carr honored his friend H. K. Wallace, who had given him the specimen. Carr also allowed himself a measure of poetic license in constructing the animal's generic name; *Haideotriton* meant "salamander from Hades" (i.e., the underworld). The name was especially fitting given the salamander's home in underground caves. Published July 8, 1939, Carr's description meticulously distinguished the new species from other salamander genera and explained its evolutionary significance: "The extreme simplicity of the skull and atlas of the present form would seem to indicate a more degenerate animal, and thus, a more confirmedly subterranean (and possibly a more ancient) form than *Typhlomolge*. The more complete degeneracy of the eyes may further support this view. The astonishingly elongate and spatulate head is, as far as I know, unique among amphibians."[20]

Revising the Antillean Terrapins

A rchie and Marjorie Carr spent most of June and July 1939 in Cambridge with Barbour. As planned, Carr devoted his time to the analysis of Antillean turtles, particularly *Pseudemys*. During the course of the summer, Barbour became very attached to the Carrs, and his letters reflected the fondness he felt for the young naturalists. At the end of the summer, he wrote a brief note to Archie: "It was a great pleasure to have you here, as always, and I'm glad you enjoyed your stay. My love to you both."[21] The Carrs also began to express their affection for Barbour. When they returned to Gainesville, the young couple moved into a new house, the first they had purchased. Archie shared his anxiety over home ownership with Barbour: "Although we are tremendously pleased with the new location the new and terrific responsibility involved in owning a home still makes me vaguely uneasy. For example, I haven't the remotest idea how, where or when taxes are paid. And a hound which we are trying to keep in the house pending the building of a dog-yard keeps getting loose and trailing rabbits and a neighbor's cocker spaniel around the subdivision, and although this hound has an inimitable voice I can't help wondering if the neighbors really appreciate it."[22] Throughout August, Carr and Barbour exchanged letters regarding their joint paper on the Antillean turtles. One letter indicates that their intellectual relationship had reached a new level during the summer as well. Barbour wrote: "I have fussed quite a little with the manuscript and been a little bold about making changes hither and yon, all of which will come out in the wash when you see proof and, if you don't like what I have done, you can borrow your hound's voice for a spell and use it.... As a matter of fact I don't think you'll give a damn about what I have done."[23]

Carr was aware of his debts to Barbour, both intellectual and financial, and he actively sought ways to repay Barbour, such as providing specimens of Florida species. Barbour capitalized on this offer and asked Carr to obtain specimens of the round-tailed muskrat (*Neofiber alleni nigricans*). Unfortunately, this species was not as common in Florida as Barbour thought, and Carr could not produce specimens. Nevertheless, when Barbour requested specimens of a newly described subspecies of the spring peeper (*Hyla crucifer bartramiana*), Carr quickly sent one to the MCZ.

Two other papers written by Carr in 1939 signaled a departure from his publishing pattern, which consisted primarily of taxonomic notes and descriptions of new species and races of reptiles and amphibians. He wrote a brief paper on escape behavior in the Florida marsh rabbit (*Sylvilagus palustris paludicola*) based on responses of the species to the two hounds that Carr kept on the grounds of the University of Florida Biology Station at Lake Newnan (also known as Newnan's Lake) in Gainesville. As a nongame species, the marsh rabbit had not received the scrutiny of other rabbits such as the cottontail (*Sylvilagus floridanus*).[24] Carr also examined the breeding behavior of the warmouth bass (*Chaenobryttus gulosus*) and found that the males built the nests and served as parents. In addition, he surveyed the wide range of predators that fed on the eggs of the warmouth.[25] Though minor contributions, these papers suggested Carr's growing confidence as a biologist and his willingness to explore questions outside the realm of herpetology.

As work on these papers and Antillean terrapins progressed, Carr continued to revise his dissertation for publication as a separate monograph. He revealed his growing frustration with the work in a letter to Barbour dated October 21, 1939: "I have reached a point, however, where I can't see much good in the paper. I wrote it so long ago and have piddled with it so much in trying to keep it up to date that reading it always brings on a mild nausea."[26] Barbour assisted Carr by providing him with copies of recent books and publications. Given the constraints of purchasing at state university libraries during the depression, Carr often found it difficult to access the most recent scholarly publications.

Proofs of the Antillean terrapins paper arrived at the MCZ in February, and Barbour's secretary forwarded them to Carr. When it was published in April 1940, the paper described one new species and three new subspecies. In the introduction, Barbour noted that he had been collecting specimens and having them illustrated for many years but that his administrative duties and lack of expertise in *Pseudemys* had precluded him from preparing an analysis of the specimens. As junior author, Carr provided exactly the expertise Barbour lacked. More than an overview of the various species of *Pseudemys* occurring in the West Indies, Barbour and Carr evaluated the characters used in earlier taxonomic studies of the complex of turtles. They found shell proportions to be useful taxonomic characters, while noting the shortcomings of previous studies: "The many unsuccessful attempts to distinguish forms of *Pseudemys* on the basis of shell proportions have been due in large measure to a failure to compare comparable stages and sexes."[27] Because they failed to recognize sexual dimorphism (morphological differences between males and females) and melanism (the darkening of certain forms or stages), earlier researchers had misclassified various forms as separate species. Having clarified the role of both epiphenomena, Barbour and Carr laid out a key to the West Indian terrapins and described each species and subspecies in detail. Following the descriptions were nine plates of *Pseudemys*, eight in color showing several of the species discussed in the paper. Like the collecting trip to Mexico, studying and writing the review of West Indian turtles had provided Carr with valuable exposure to the diversity of life outside the United States.

Collecting in Mexico

Just as Carr had begun to look beyond reptiles and amphibians for projects, Antillean turtles had given him exposure to the diversity of turtles outside the United States. In November 1939, he planned an expedition to Mexico to collect specimens for the Florida Museum of Natural History and the MCZ. Barbour offered a small contribution ($75) and copious advice regarding collecting localities and limitations. In addition, he sent Carr a letter on Harvard stationery that he hoped would facilitate Carr's collecting activities in Mexico. He wrote facetiously: "I send you herewith a dago dazzler which may help."[28] "Dazzlers" are any official looking document on impressive stationery, in this case, The Museum of Comparative Zoology, Harvard University. In effect, a dazzler legitimizes a scholar's efforts by association. The term is still in use today. "Dago" is more problematic as a racist epithet for Mexicans. In light of the ongoing correspondence between Carr and Barbour, the term "dago" takes on a more specific connotation as a reference to bureaucrats or border guards who could impede the

Figure 8. Archie Carr and Mexican border guards inspecting his Harvard "dazzler," 1939. Courtesy of Mimi Carr.

collection and export of scientific specimens. In another letter, Barbour provided an extensive list of species that interested him including bats, shrews, and small, inconspicuous snakes as well as freshwater fish. He was particularly interested in axolotls (genus name: *Ambystoma*) and other salamanders (*Urodeles*): "I don't have to tell you to watch for *Urodeles*, both in Bromelias, under logs and stones and aquatic. I don't know whether you will be any where near Lake Patzcuraro but there's one there which occurs in deep water which is very interesting and there is a regular fishery for it as it is eaten by the Indians (so I am told). It is probably called 'Axalote,' the 'x' being pronounced rather like a hard 'ch.'"[29]

Carr's first trip to Mexico was very successful, as he reported (on MCZ stationery) to Barbour in January 1940. By contacting local residents, Carr and his group were able to collect large numbers of reptiles and more than one hundred mammals. Particularly helpful was Dyfrig Forbes, who provided guides, a patio on which the group camped for three nights, and even an old car for collecting trips. Forbes also prepared a list of the species of reptiles and amphibians in the region. To a significant degree, Carr credited the success of the trip to Forbes's list: "His annotated list enabled us to get several species of *Oedipus* which otherwise we probably would have missed entirely. He gave us a number of specimens from his own collection and directed us to a miserable little water-hole at a ruined hacienda in the desert in Puebla where we seined over a hundred axolotls and some stink-jims [musk turtles]."[30] Carr regaled Barbour with the beauty of Mexico: "The country around Orizaba, Cordoba and Fortin far surpasses in beauty anything I ever saw. I have rarely been so upset at having to leave a place."[31] There was a violent hurricane while Carr's group visited Metlac. The storm (locals said it was the worst to strike the region in fifty years) blew down trees, destroyed coffee and banana groves, blocked roads, and even caused several deaths in nearby Orizaba. Fortunately, Carr, his students, and their tents were protected from the wind by a large spur of rock. Carr recalled that they spent the worst of the storm enjoying the hospitality of an engineer in the power plant of the Moctezuma Beer Company, a regional brewery.[32]

Regardless of the scenery, hospitality, and adventure, the real measure of the success of the trip was the quantity and quality of the specimens, with which Carr was quite satisfied:

> Altogether we got a pretty good lot of reptiles and amphibians, considering the briefness of our sojourn. Lizards, *Oedipus*, *Hyla*s and *Eleutherodactylus* make up the greater part of the collection. We found a few snakes, but didn't do much with turtles. We brought back something over a hundred mammal skins and skulls, mostly mice, bats, and possums. We also got a slim intelligent-looking marsupial with beautiful fur, big eyes and a half-bare tail. It looks something like a panda.[33]

This last specimen was most likely a woolly opossum (*Caluromys derbianus*). After Carr shipped all the specimens to the MCZ, Barbour wrote an enthusiastic response: "The collections have come and everyone is much pleased with them. In fact they add some new locality records to some of our departments."[34]

Figure 9. Archie and Marjorie Carr at Camp in Mexico, 1939. Courtesy of Mimi Carr.

Digging for Fossils at Thomas Farm

Just as Barbour opened his house to the Carrs each summer while they were in Cambridge, they began to return the favor by hosting Barbour in their home during some portion of each winter. In 1938, Barbour heard of a remarkable site of paleontological significance located near Gainesville: Thomas Farm. Thomas Farm was first discovered in 1931 by Clarence Simpson of the Florida Geological Survey. When Simpson discovered fossils from an extinct three-toed horse at the site, he sent them to George Gaylord Simpson at the American Museum of Natural History in New York City. George Simpson further identified the fossils as Miocene land animals, which were rarely found in the eastern United States. Barbour purchased the surrounding 40-acre tract of land after examining some of the Florida Geological Survey fossils in 1938.[35] There is no record of correspondence between the Carrs and Barbour between January 25 and May 4, 1940, presumably because Barbour spent most of that time in Gainesville digging fossils at the Thomas Farm with his colleague Ted White. After his return to Cambridge, Barbour began reviewing the fossils along with his staff. He wrote enthusiastically to Carr about their findings: "The Thomas Farm material is turning out very interesting indeed and a good many specimens will be prepared by the time you get here. One pair of jaws which we took to be camel evidently represent some other genus of Artiodactyl which none of us here recognize at all."[36] Thomas Farm continued to produce excellent fossils, and Barbour eventually donated the site to the University of Florida.

A Contribution to the Herpetology of Florida

Even as he began to explore faunal regions beyond Florida, Carr continued his study of Floridian herpetology. In 1940, he published his dissertation, "A Contribution to the Herpetology of Florida." It opened with a literary flourish for which Carr would later become known:

> Since the days of the earliest explorations the herpetological fauna of Florida has evoked spirited comment. Hardly a mosquito-bitten Spaniard writing home for supplies or a French sea-captain recording in his log the adventures of a shoe-party but mentions "vipers" or "crocodiles," or the shocking noise the frogs make, or the Indian who tried to feed him snake. Colonizing Florida was such a strenuous matter, however, that zoological observation was considerably tainted with emotion, and only those forms of life which bit people, or which people could eat, elicited any enthusiasm in the early reports. And since the colonizing is mercifully not quite complete, current reports, too, retain enough of the old emotional taint to warm the soul of any decent herpetologist.[37]

Carr's unique ability to blend culture and nature, history and science, past and present reveals remarkable insights into the natural history of Florida. He proceeded to present several unexplained mysteries of herpetology in Florida before reviewing the history of the study in the state. The first published illustration of an alligator in North America, he noted, was Jaques Le Moyne's of "crocodiles," (the American alligator). However, Carr recognized William Bartram's writings (published two centuries after Le Moyne's) as the first reliable observation of reptiles and amphibians. Carr wrote: "Although primarily interested in plants, Bartram was fascinated by the diversity of the southeastern fauna, and commented with pertinence on all that he saw. His discussion of the Florida frogs, lizards, snakes, and turtles includes much accurate information on habits and habitats, and in most cases leaves little doubt as to the identity of the forms described."[38] Carr's brief review of the historical literature provided a concise analysis of the two-dozen previous reports on the herpetology of Florida.

To establish his credentials, Carr summarized his considerable efforts: "The studies on which this paper is based have included six years of field work throughout the state, the collecting of some fourteen thousand specimens with detailed ecological data, and the examination of some twenty-five thousand additional specimens in the collections of various institutions."[39] By any measure, collecting 14,000 specimens and examining 25,000 more represents an impressive (if not Herculean) effort. Carr underscored his interest in ecological data as opposed to strict taxonomic information. Biogeographical information was central to Carr's study as well. He noted that the Florida fauna was derived from two major sources: invasions from the southeastern coastal plain and arrivals from the West Indies (generally apparent only in the southern part of the state). Further, Carr placed endemic forms (those species or subspecies occurring only in Florida) into two groups: "arising through isolation" and "relict forms." Carr's knowledge of ecology can be gleaned from his descriptions of the ecosystems of Florida under the heading "Habitat Distribution." With a remarkable economy of words, Carr characterized twenty-four major ecosystems with notes on vegetation, soil, topography, and drainage. After each entry, he listed the "frequent" and "occasional" species of reptiles and amphibians.

As with most taxonomic works, Carr produced a general key to the orders and suborders of amphibians and reptiles as well as individual keys to the salamanders, frogs and toads, lizards, snakes (prepared by Coleman Goin), and turtles. After the literature review, consideration of biogeography, description of habitats, and taxonomic keys, Carr provided an annotated list of species, in which he described each species' Florida range, habitat, abundance, habits, breeding, and the primary reference on the species.

For some of the accounts, Carr also included a note in which he recounted personal encounters with the species or questioned previous records or relayed historical tales from colonists and Native Americans. For example, regarding the diamond-back rattlesnake (*Crotalus adamanteus*) in Florida, Carr noted that "from the stomach of a five-foot specimen from Citrus County I took an adult marsh-rabbit, and Dr. Thomas Barbour tells me of having taken a king-rail from the stomach of each of two specimens collected at Jupiter."[40] Other notes related to biogeography and evolution. Regarding the subspecies of the Florida box turtle, Carr noted, "There is a definite and fairly regular geographic gradient from *bauri* to *carolina* in North Florida and South Georgia and in the eastern end of the Panhandle."[41] Carr's notes dated back to 1934 (the year he received his master's degree). One early note referenced an invasion of Eastern spadefoots (*Scaphiopus holbrookii holbrookii*):

> From my journal, September 1, 2, and 3, 1935; Pinellas County: "Gray rainy weather and hurricane winds on [September] 2nd. Streets of Tarpon Springs flooded and *Scaphiopus* everywhere. In lower sections of town, where backyards and vacant lots are under water, the toads float about among the chickens and ducks; their snarling croaks amount to a deep, incessant roar which the citizens complain keeps them awake at night and drives them crazy by day. Several species of gulls have come inland and dive among the spadefoots, occasionally carrying one away."[42]

Occasionally, Carr included prescient comments regarding the conservation status of a species, as in the case of the alligator, which had declined precipitously since Bartram suggested that it would have been possible to cross the St. John's River by stepping on their heads if they were harmless.

Out of all the 162 species included in the annotated list, some of the most limited accounts are those Carr wrote for the five species of sea turtles. Carr noted that the green turtle was a staple in Key West, where unripe eggs sold for as high as 90 cents a pound. His direct experience consisted of brief encounters with green, "hawk-bill" (hawksbill), and loggerhead turtles (all off Key West). As of 1940, Carr had had no field experience with the "Bastard" loggerhead (Kemp's ridley sea turtle) or the "trunk-back" (leatherback) sea turtle. Carr's only other personal note cited an encounter with a loggerhead turtle dating from his childhood (see chapter 2). Almost as intriguing as the lack of direct experience and data regarding sea turtles was the single, antiquated reference he cited for sea turtles: G. A. Boulenger, Catalogues of the British Museum (Natural History) from 1889.[43]

Barbour's Generosity

As the summer of 1940 approached, Barbour once again invited the Carrs to spend the summer in Cambridge. On May 15, Carr told Barbour that he had to delay the trip to Cambridge because he could not find a tenant to rent their house during the time up north. As a result of this postponement, Carr accepted a teaching position for the first five weeks of the summer term. Plainly disappointed, Barbour replied: "Needless to say it is too late now for me to suggest that you change your plans again. I am sorry that you did not let us handle the financial details at this end but we will be glad to see you after the 20th in any case and for as long as may be."[44] In the intervening time, Barbour finalized negotiations for the Thomas Farm and bought and refurbished a used station wagon. On September 6, Barbour sold the station wagon to the Carrs for one dollar. After he and Marjorie returned to Gainesville, Carr compared the station wagon to their old car: "The station wagon performed beautifully and the Plymouth miserably; the latter used twenty-four quarts of oil and left a pall of smoke along the entire Atlantic seaboard."[45]

Late in 1940, the Carrs planned another collecting trip to Mexico with Barbour's advice. Barbour wrote that he wanted material from Veracruz or Oaxaca, with a slight preference for the former, and again offered to contribute financially to the expedition. A serious problem developed when Barbour discovered that the Mexican office responsible for issuing collecting permits was no longer functioning. Through his secretary, Barbour advised Carr not to collect birds or mammals without a permit and instead concentrate on invertebrates, reptiles, and amphibians, which were less subject to scrutiny by customs officials.[46] A few days later, on December 2, the Carrs received a gift from Barbour: one of the original prints of John James Audubon's Carolina parakeets (*Conuropsis carolinensis*). Carr immediately wrote to thank his friend, first noting that he was having a bushel of oranges sent to Cambridge, and then turning to the remarkable present:

> I don't understand how I could have begun a letter otherwise than by thanking you for the wonderful Audubon print. It arrived last night in perfect shape, and is the delight of all who see it, including most especially Margie, who is in a perfect frenzy. I'm afraid her determination to provide a proper place to hang it may involve remodeling the house. We were about to hang the same picture in the small print which you sent along with Audubon's America, which would have been a relatively

simple undertaking, but this new project will demand at the very least extensive shifting of furniture. We are both most exceedingly grateful to you.[47]

Carr also mentioned that they would leave for Mexico on December 13. He gently reminded Barbour about his $100 contribution and the letter of support from Harvard: "The money and dago dazzler will be gratefully received whenever it is convenient to send them."[48] Marjorie recorded her thoughts in a letter on December 3 as well: "The print arrived yesterday. We are overwhelmed. It is gorgeous and as you say typically Floridian. We hope to frame it this week and hang it at the end of the living room on the cypress-paneled wall. Needless to say, the entire living room is going to be centered around the picture. I'd like to know more about the history of the print. Dates and all."[49]

The Carolina parakeet had great significance because it was driven to extinction by intense hunting pressure at the beginning of the twentieth century. The ornithologist Frank Chapman may have been the last biologist to see a flock of these stunningly beautiful birds. For the Carrs, the Audubon print brought together their love of Florida natural history and Floridiana. It hung prominently in their home from then on. For his part, Barbour responded to the Carrs' gratitude with characteristic understatement: "Miss Wilder wrote Margie yesterday, or will write her today, about the date, etc., of the Audubon. Certainly glad that you both like it. I have always thought it was one of the finest plates in the Elephant Folio. It was just luck that we had a duplicate copy."[50] The Carr's trip to Mexico was a success. Despite the arrival of a hurricane while they were there, they collected a variety of amphibians and mammals, some of which were new to science.

"Dopeia"

When the Carrs returned from Mexico, the American Society of Ichthyology and Herpetology (ASIH) held its annual meeting in Gainesville. In celebration of the event, Carr created a witty parody of a basic part of a taxonomist's work: the taxonomic key. Having published several such keys, he knew the form well. From the title of the mythical journal of publication "Dopeia," after *Copeia*, to the least significant footnote, the parody is filled with inside jokes and mockeries of the practice of natural history. In the "Forward!" Carr deflated the pretensions of taxonomic keys: "The zoological key of today is perhaps the most unstimulating and oppressive of all literary forms. Stripped of all the more succulent verbiage, its style desiccated and uninspired, and its technical arrangement often so cryptic as to render it wellnigh incomprehensible, the key may not, in any sense, be called good reading."[51]

As the key progresses, it becomes increasingly complicated with long series of reference letters, gratuitous quotations in foreign languages, cross-references, unintelligible contractions, and bad advice, such as "to aid in separating the forms to follow it is almost mandatory that the reader carefully jab either of the side spines of the specimen into the fleshy part of his thumb, recording his sensations in detail."[52] Most ichthyologists would be able to relate to the experience recalled by such advice. The key also called on the user's common sense in several references to the identification of the catfish: "Any damn fool knows a catfish."[53] Above all, the key suggested Carr's growing confidence in his expertise. Carr became renowned for his wit among ichthyologists and herpetologists. Ironically, Carr's discomfort with large gatherings emerged as well. A group photo of the one hundred or so scientists at the ASIH meeting shows Carr standing at the back on the far left ("the best position from which to escape," as his friends later commented wryly).

Another Summer at the MCZ

As in the previous four summers, the Carrs traveled to Cambridge to work at the MCZ from May to July. Carr and Barbour co-wrote a description of a new species of turtle from Grand Cayman. The resulting paper offered an overview of the history of the turtle on the island. The Caymanians referred to the turtle as "Hig-a-tee," which appeared to be a corruption of the Cuban word for turtles: "Jicotea," although Barbour and Carr acknowledged the possibility of an aboriginal term unknown to them. Although an introduced freshwater turtle was reported as early as 1877, Barbour and Carr first received news of the species in 1939, when a collector sent a letter to the MCZ with a drawing and a query regarding the taxonomic status of the turtle. On the basis of the drawing, they did not believe the turtles represented a new form. Nevertheless, Barbour received live specimens at the Havana Zoo in Cuba, which he carried to Gainesville. Barbour also sent four prepared specimens to the MCZ. Having examined the specimens, Barbour and Carr agreed that the turtle was an undescribed species. They named it *Pseudemys granti*, after the collector who sent the specimens. Barbour and Carr mapped the biogeography of *Pseudemys* species throughout the Caribbean, including suspected paths of dispersal. This map showed that most of the *Pseudemys* species in the Caribbean had descended from the Cuban species (*P. decussata decussata*). Their paper revealed Carr's growing interest in biogeography.[54]

After they returned to Gainesville for the second summer session, Carr wrote to Barbour to ask him to watch for a seine for netting animals on field trips. Carr envisioned something 15 to 25 feet long, which at forty to fifty dollars was beyond his budget (or that of the department, for that matter). In Carr's words: "If you could keep your eye out for a factory throw-out it would be most pious, as you say."[55] Barbour immediately called the American Net and Twine Company Office in Gloucester and wrote to Carr. One month after Carr's original request, the seine arrived. In a thank you note, Carr expressed his sense of guilt for accepting the seemingly new net. Barbour immediately conveyed to his friend that he had made a good deal: "I worked a little graft on you. You see we own the factory that makes those nets so I do get them at what might be called a right smart discount. I am glad it was satisfactory. As a matter of fact I bought it for just $14.00. So there you are!"[56]

Graptemys barbouri

On October 18, 1941, Carr wrote to Barbour regarding his recent activities, including his first foray into conservation and a new discovery. The chair of the Department of Biology at UF, J. Speed Rogers, had asked Carr to attend the meetings of the Florida Wildlife Federation: "Back after a good trip and off again to Jax [Jacksonville] to two-day meetings of the Fla. Wildlife Federation, of which I'm to be put up as vice-president to attempt to establish a liaison between the department and state conservation matters. It's a job I hate but Rogers thinks it a good idea. The sportsmen approached us about it."[57] (The word "conservation" here refers to the protection of a resource for future use [in this case hunting] rather than preservation for aesthetic or ethical reasons, a trend that did not emerge until after World War II).[58] Carr continued:

> Among other things I want to tell you hurriedly that I took my fat turtle student on the trip and together we got 16 specimens of *Graptemys pseudogeographica* out of the Chipola R. I had collected there six times before without even suspecting their presence. Marchand's water-goggling did the trick.
>
> These turtles are possibly Baur's *pulchra*, but more probably new. I'd like to keep them a few days to compare carefully with some Louisiana material, but they're yours and I think you should describe them. I'll send them and any dope we can dig up shortly after my return from Jax.[59]

For some time, Carr had suspected that there might be a species of *Graptemys* occurring in the rivers of western Florida. In 1890, G. Baur claimed that two *Graptemys* had been collected in Pensacola, but he did not provide any evidence.[60] Nor could Carr verify it. About the time that Carr received a juvenile *Graptemys* from the Appalachicola River at Chattahoochee, one of his students (Lewis J. Marchand) had adapted a homemade face mask to study the underwater behavior of turtles and had used it to observe *Pseudemys*. Armed with the new equipment, Carr and his student went to Merrit's Bridge on the Chipola River near Marianna, Florida. Fifteen minutes after their arrival, Marchand caught a *Graptemys* turtle. Over the next two hours, they collected fifteen more, for a total of sixteen. Later, they completed an extended collecting trip of four days from Bellamy's Bridge, 10 miles above Marianna to Clarksville in Calhoun County, which produced sixty additional specimens.

The discovery of this turtle surprised Carr for several reasons. First, the Marianna region had long been a popular collecting location for members of the Department of Biology at the University of Florida. Moreover, Carr's old friend Oather Van Hyning collected extensively in the region (he lived only two miles from the Chipola). Despite the popularity of Marianna among herpetologists, this species of *Graptemys* had escaped detection. Carr proposed a possible explanation:

> It is now apparent that the Chipola *Graptemys* escaped notice, in what we had naively come to regard as a well-collected area, because of

47

its preference for relatively deep, swiftly flowing, rock-bottomed and snag-ridden reaches of the river, into which sunning individuals dropped at the slightest disturbance along the bank. Such individuals, not obtainable by the ordinary collector, were readily found and taken by the swimmer equipped with water goggles.[61]

Carr debated whether the turtles were new or were actually one of the species described by Baur. He wrote to colleagues in Alabama regarding any *Graptemys* specimens they might have. None of the Alabama turtles compared to the series Carr had collected from the Chipola River, so he suggested that Barbour and his student describe the new turtle:

> A canvas of sundry prospects has failed to uncover any Alabama *Graptemys* to compare with the Chipola River series. Marchand is simply rabid to be co-author of the new race. Since the turtles belong to you and since any comparing of them with the types (apparently the only specimens in existence from the Montgomery area) would have to be done by you on one of your trips to Washington, I wonder if you would consider a joint paper with Marchand. My only finger in the pie is the hunch that *Graptemys* would live somewhere in the Panhandle. I would willingly withdraw it in your favor, less so in Marchand's, except that he did have the gumption to goggle after them.[62]

In response to Carr's suggestion of a joint paper with Barbour and Marchand as authors, Barbour politely demurred: "Now, with regard to your other letter. You borrow the turtles from Stejneger, compare them yourself and you and Marchand describe the race. There's no reason in the world why you should withdraw in my favor."[63] Carr followed Barbour's advice and sent a report two weeks later. It appeared that the *Graptemys* was in fact a full species that had escaped description. Carr wrote: "Doctor Stejneger sent excellent photographs of the types of *Graptemys pulchra*. The Chipola River specimens aren't it, nor are they *oculifer*, *kohni* or *pseudogeographica*. In some ways they are more like *geographica* but are quite distinct from that too. I think we'll wind up calling it straight *G. barbouri*."[64] As unlikely as it sounds, Carr and his student had found a new species of turtle in Florida just a short drive from Gainesville. *Graptemys barbouri* struck Carr as a particularly interesting turtle: "You'll be astounded at the weird sexual dimorphism and the tremendous heads of the females. . . . It's really a most extraordinary thing to find right under our noses."[65] Sexual dimorphism refers to differences between males and females of the same species.

Several distractions kept Carr from completing the description of the new species. The annual trip to Mexico had to be planned and executed. The Carrs received a minor scare when they heard that all tourist visas to Mexico would be canceled. Collecting permits were, as usual, extremely difficult to obtain. Nevertheless, Archie, Margie, and several students made the trip, which was among the most successful. In her Christmas card to Barbour sent from Mexico, Marjorie described their location (Rancho Santiana, Tamaulipas): "This is the

mildest and most beautiful place we've found yet. Everyone is having good hunting—even Zep [the Carr's dachshund]. He searches and digs for skunks all day."[66]
Archie was more measured in his assessment of the collecting: "We're still having
a wonderful time. Herp collecting not as good as last year, but that was to be
expected, and we are getting some good things. I wonder if you would have someone send back to Gainesville, collect, the turtle box in which I sent the *Graptemys*.
We're getting a number of turtles which are too big for our pickling cans and
which we'll bring home alive. We've made up 100 mammal skins."[67] As usual,
when they returned to Gainesville, Carr sent the collections to Barbour at the
MCZ. Barbour was impressed with the results of the latest trip: "I have refrained
from expressing to you the gratitude of the Museum for the splendid collections
you made in Mexico until we really had a chance to unpack them and see whether
they are splendid or not. They are, so that's that."[68]

Figure 10. Thomas Barbour and Zep (ca. 1942). Courtesy of
Mimi Carr.

Shortly after their return from Mexico, Marjorie successfully defended her thesis for a master's degree in zoology at the University of Florida. She wrote to Barbour to tell him the good news and to request his advice regarding publication. Initially, Barbour was noncommittal, but a colleague thought the thesis was worth publishing, with some minor revisions to reduce printing costs. Barbour encouraged her to complete the revisions and send the paper to his secretary. It appears that Marjorie was not able to revise the paper immediately because Barbour had his secretary, Helene Robinson, resend the text of the original letter with an additional note of encouragement. For myriad reasons (most frequently inertia and fear), many people fail to publish their theses, so Barbour provided Marjorie with access to a journal for publication. As further incentive, Barbour, through his secretary, offered to simplify the Carrs' trip to Cambridge for their annual summer visit: "Also, if you're afraid to try driving up this summer (and I should be) the MCZ is flusher than it has ever been in years owing to deaths and resignations 'for the duration' and Pullman transportation would gladly be provided. You'd better get a compartment when the time comes then you can keep Zep in the basket when the conductor may be coming around."[69]

In May 1942, Carr explained that the final description of the new turtle had been delayed, but he prodded Barbour for his reaction to the honorific name: "What do you think, by now, about the name of the varmint? I can see a lot of logic in calling it *barbouri*."[70] Three days later, Barbour replied: "I would be what White would call a liar, perhaps even a damned liar, if I should tell you that I would not be deeply complimented if you cared to use the name *barbouri* because naturally anything which you did with a kindly idea in mind is naturally much appreciated."[71] When Carr finished the manuscript, he again wrote to Barbour and asked whether he should submit the description to *Copeia* to avoid any hint of impropriety, but Barbour casually dismissed such concerns: "I have a hunch that *Copeia* is pretty well filled up and as you know I don't give the smallest continental damn what other people think so we can publish it in the P. Z. New Eng. C. [*Proceedings of the New England Zoological Club*] where it will come out promptly."[72] At the end of the article, Carr and his co-author paid tribute to Barbour: "It seems altogether proper and pleasant to associate with this new species the name of Doctor Thomas Barbour, who has contributed extensively to our knowledge of Florida reptiles and amphibians."[73]

Delving into the Taxonomy of Sea Turtles

During his time at the MCZ during the summer of 1942, Carr completed his first paper on sea turtles. The title of the article was "Notes on Sea Turtles," and it included, like many of his early publications, a taxonomic key. Carr took issue with previous classifications of loggerhead-type sea turtles into a single genus, *Caretta*, which included three species (*caretta*, *kempi*, and *olivacea*). A more rational approach, according to Carr, would divide the three species between two genera: *Lepidochelys* and *Caretta*.

But this article served as more than a taxonomic review. Carr also raised the point that the common name of the species in the genus *Lepidochelys* should be switched from the bastard turtle to the ridley turtle, as fisherman and turtle hunters along the Gulf Coast commonly referred to it. Like Carr's investigation of water moccasins on Seahorse Key, this article reflected his respect for local, anecdotal evidence and, moreover, his belief that scientists should defer to or at least recognize indigenous knowledge: "Thus, although the term 'bastard turtle' is admittedly a venerable one... I nevertheless propose that it be relegated to synonymy. It is my conviction that in such cases priority should bow to prevalence, that a common name which is not common is a mockery; and henceforth I shall use ridley."[74] Carr thus declared his view that in developing common names for species, scientists should draw on local knowledge.

Throughout his research, Carr repeatedly returned to the anecdotal claims of locals both to gather evidence and to develop hypotheses. This first sea turtle article revealed Carr's commitment to the wisdom of nonscientists and his interest in sea turtles, which would become the focus of his life's work. After the Carrs had left Cambridge for the summer and returned to Gainesville, Barbour made an arresting discovery that merited a telegram: "Young specimen of Kemp's turtle from No-Mans-Land near Martha's Vineyard just found in the Boston Society Natural History. STOP. I am adding this fact to your manuscript. STOP. I assume you have no objection. STOP."[75] Carr responded: "I was tickled to hear of the discovery of the BSNH ridley. The small size is in accord with previous northern specimens, including the European ones, and further supports the interpretation of these northern range extensions as fortuitous."[76]

Although Barbour's relationship with the Carrs was that of a mentor and was somewhat paternal and parochial, he occasionally called upon them to support his new endeavors. Such was the case when Barbour published his first article in a popular magazine. Barbour exulted about the development of this new achievement in a letter to Archie:

> I feel absolutely as if I were a kid with a new toy this morning. The other day I got thinking about all the funny things that have happened in connection with my cleaning up various and sundry museums and I wrote a facetious article—dictated it in about fifteen minutes, certainly not more—just as it happened to flow off the tongue, so to speak. Its text was the finding of Mrs. Chase's gall stones "loaned to the Peabody Museum." I animadverted on why she had not given them and philosophized on pride and possession. The whole thing was pretty nonsensical, and in a moment of rashness I sent it in to Weeks of the *Atlantic Monthly*. To my surprise, he grabbed it, and wouldn't even send it back to me for fear I would "tinker it up and spoil it." So you see, in my old age I have joined the William Beebe class. But if you don't laugh when you see it you will never come to Cambridge again.[77]

Carr responded enthusiastically that he and Margie were anxious to see Barbour's article.

Whenever Barbour inquired about the Carr's dachshund Zep, Archie and Marjorie obliged with anecdotes regarding his exploits. Archie described one such event: "The other day I took Zep and Rhoda for a walk in the woods by the house. Rhoda struck a trail and after a bit treed something in a hollow sweet gum. She couldn't get through the hole but Zep could and did. He quickly reappeared dragging by the scruff of the neck a big lady possum into whose pouch eight rat-sized young were trying to scramble. For twenty-four hours thereafter Zep was too stuck on himself to live with. He walked about with a haughty air that was really revolting."[78] Reveling in his new career as a popular writer, Barbour identified with Zep:

> I feel just like Zep and the possum! I had four articles for the Atlantic.... I wound myself up by getting up at Beverly Farms, 3 a.m. one morning, and composing three of these squibs in my mind; dictated them off here in under an hour. Two got by Weeks' eagle editorial eye without a single verbal change.... Weeks said he never before accepted a dictated manuscript or been offered a dictated manuscript which had not been revised. He wants ten more, and more than that will pay damned good money for them.[79]

Carr must have been impressed by the ease with which Barbour established himself as a writer for a popular magazine like the *Atlantic Monthly*. Surely, he found inspiration for his own efforts to write for the public a few years later.

Debating Leonhard Stejneger

Between his dissertation, scientific papers, and his first book on the herpetology of Florida, Carr had established considerable authority regarding the reptiles and amphibians of the southeastern United States. Barbour, who was preparing the fifth revision of *A Check List of North American Amphibians and Reptiles* (1943), incorporated Carr's taxonomic revisions and sent them to his co-author, Leonhard Stejneger at the Smithsonian. Recall that early in his career, Carr had asked Barbour how to best avoid stepping on Stejneger's toes because Stejneger was one of the most prominent of American herpetologists and had been studying Carr's preferred genus (*Pseudemys*) for many years. For the most part, Carr had confined his papers to the turtles of Florida and the Southeast, with minor departures to study *Pseudemys* in the Caribbean, Texas, and Mexico. Nevertheless, despite Carr's caution, Stejneger disagreed with his taxonomic revisions.

On August 27, 1942, Barbour forwarded a letter he had received from Stejneger to Carr, with a palliative note: "The enclosed letter to me is a most extraordinary and interesting document, and for many reasons I want you to have a copy.... If and when you reach Stejneger's age and can still turn out a letter of this sort—especially if you write it in a language which is not your own—I will be even prouder of you than I am now."[80] From the outset, Stejneger adopted an instructive, if not pedantic, tone: "A checklist of a faunal area (for practical reasons often limited to a political section) is a taxonomic nomenclatorial tool intended to help the student of the fauna. It is not a textbook of genetics, nor should it be an essay on the phylogeny of the forms enumerated. Neither should it be an expression of anybody's personal opinion about the various processes of 'speciation' or the more or less complicated relationships of the forms included."[81] Stejneger was criticizing the efforts of naturalists to construct checklists in a way that reflected evolutionary relationships. His letter continued by tracing the rise of the trinomial (trinominal in his terminology) in reference to subspecies. Stejneger argued that the extent to which such designations reflected evolutionary history was highly subjective, reflecting the whim of the individual taxonomist. He also explained his standards for intergradation, or the gradual geographical transition of one species to another.[82]

With his position clearly established, Stejneger systematically rejected many of Carr's taxonomic revisions. In *Contributions to the Herpetology of Florida*, Carr lumped the Florida box turtles into one species with two subspecies, *Terrapene carolina bauri* and *Terrapene carolina major*, and noted a gradient between the two species. Based on his specimens, Stejneger thought there should be three separate species of box turtle. Carr had lumped *Pseudemys floridana* and *concinna*. Stejneger recommended splitting them again. Carr lumped the snapping turtles (in 1918, Stejneger had described the Florida snapping turtle as a separate species; Carr downgraded it to a subspecies in *Contributions*). Stejneger argued that snapping turtles should be treated as a full species. But it was Carr's revision of the genus *Pseudemys* that created the greatest problems for Stejneger, as revealed in this critique:

> I have the greatest respect for Carr's work: it is intelligent, apparently accurate, well presented and hence exceedingly useful. But—read his argument for the trinominal *Pseudemys scripta hiltoni*: "It's distinguishing features are, for the *most* part, accentuations of trends demonstrable *elsewhere* in the complex would appear to indicate that intergrades may eventually be found in areas from which specimens are at present lacking. Thus, it seems justifiable to designate *hiltoni* as another of the races of *P. scripta*." Nomenclatorial anticipation![83]

Finally, Stejneger questioned Carr's revision of the sea turtles, in which he established a new genus for the ridley sea turtles (*Lepidochelys*), thereby splitting them from the loggerheads (*Caretta*). Stejneger concluded, "I may be old fashioned, but a Check-List is a Check-List or it is something else. Yours in the old way."[84]

Carr dimissed all of Stejneger's claims in a letter to Barbour:

> Thank you also for the copy of the letter from Dr. Stejneger. I agree that there certainly are no signs of senility there. However, as you perhaps anticipate, I butt-headedly refuse to agree with any of the specific statements made therein and with some of the generalizations. Some eight years ago, when several of the ideas which Dr. S. rejects began to jell [*sic*] I had the opportunity of trying them out on A. H. Wright who

paid us a short visit. He came via Washington and arrived with opinions and was quite adamant. But when he left he admitted that if Dr. Stejneger could stay a while in Gainesville and look at dozens of live specimens from single ponds he would go away not only converted but convinced that Alachua County is one of the most critical zoogeographic areas in North America. I should dearly love to know how Dr. S. goes about deciding whether Columbia Springs *Kinosternon* are *subrubrum* or *steindachneri*. Etc.[85]

Barbour replied, "Of course, you're entirely right about Stejneger, but it's a waste of time to argue with him and at his age, of course, you can do nothing but respect his opinions."[86]

The demands of the fall semester and then the holiday season distracted both Carr and Barbour from the debate with Stejneger. The Carrs spent the Christmas holiday with Archie's parents in Umatilla. Archie spent most of the time deer hunting with his father and brother. Marjorie spent most of her time reading *Van Loon's Lives*. The deer hunting was slow, but Marjorie did encounter a large buck in the yard. The dogs, including Zep, chased the deer for more than a mile, and it was finally shot. In a letter to Barbour, Marjorie coyly saved the biggest piece of news for last: "I am becoming fairly domestic these days for I'm due to produce an offspring around the first of June. I feel grand and have all along. I'm still a nurse's aide but I guess I'll stop in a couple of months."[87] A few days later, Archie added his own wry comment on this exciting news: "I think Marge told you about her determination to replenish the earth; she ain't swole' much yet."[88] Barbour was characteristically exhilarated by this news: "Gosh, that's some news! I hope it's A.F.C. III. I wonder if it is reading Van Loon's Lives that did it. Some of them almost had that effect on me, although I swear I think that he batted a pretty high average after all."[89]

Two days later, however, Barbour's thoughts returned again to subspecies, trinomials, and Stejneger's call upon Carr to be more specific regarding the region of intergradation in the snapping turtles. Carr replied definitively:

> Everyone who has lived in Gainesville and has had occasion to examine *Chelydra* here has reached the same conclusion as myself, to wit, that northern Florida (and probably southern Georgia) is a region in which the characters by which *osceola* is contrasted with *serpentina* cannot be relied upon in all (or even a majority of) cases. . . . That phrase, '*seems* to be northern Florida' which Dr. Stejneger mentions meant simply that the intergradation may well extend far up into middle Georgia, whence I had no material. There is absolutely no question in my mind but that the two snappers are subspecies and that some merging of the two occurs about Gainesville. The only question is the extent of the intergradation.[90]

Barbour accepted Carr's argument: "We are forwarding your letter regarding the snappers which I think clinches the situation. Much obliged to you for answering so promptly."[91] In the same letter, Barbour noted that Stejneger had not previously seen Carr's paper on sea turtles and that he apparently liked it.

Stejneger refused to be swayed regarding snapping turtles, however. Stejneger wrote to Barbour:

> Miss Robinson was kind enough to send me a copy of Carr's letter re *Chelydra osceola*. I confess it is not convincing to me. It is not a question of strikingly different form; it is not a question of the numerosity of intermediate specimens (specimens with mixed characters); nor is it a question of the size of the area in which this intermixture occurs not of "geographic" similarity of distribution; the parallelism with *Lampropeltis* is irrelevant. It is the character of the intergradation, not of its quantity. I can well believe that some "merging" of the two occurs about Gainesville. "The only question is the extent and limits of this region of intergradation!" Yes, for the statistically inclined! But I am one of those that do not "swear" by the figures, they can lie as much as the photographs. To me *osceola* has not been proven—or even made probable—to be a case of trinominal intergradation.[92]

Stung by Stejneger's vehemence, but convinced of his clear concept of the problem, Carr responded at length to Barbour. In a six-page letter, Carr answered each of Stejneger's criticisms meticulously, but one specific defense clarifies his commitment to natural history:

> Finally, as to Dr. Stejneger's implication that I am one of the "statistically inclined," words cannot express my amazement. If he, perchance, means that I believe that zoogeographic work, to be respectable, should be based upon as large series of specimens as can be gathered I cheerfully accept the appellation. Otherwise I can only plead that I too resent the intrusion of calculus into zoology and have always sworn by your famous exhortation to the visiting "younger, more critical herpetologist" concerning his substitution of a slide-rule for a species sense.[93]

Although he had already substantively proved his point regarding snapping turtles, Carr aligned himself with the work of naturalists while recognizing a transition within university biology departments. That he directed his response to Barbour, and not to Stejneger, reflects the strong aversion to confrontation that Carr maintained throughout his life.

Barbour recognized the validity of Carr's argument but diplomatically deferred to his senior collaborator: "I think you have set forth your views with remarkable lucidity and I am convinced that you are right; however Dr. S is old and set in his ways and I must give him the last word."[94] Stejneger's death on February 28, 1943 (reported to Carr by Barbour via telegram) ended the debate. Upon receiving the sad news, a disheartened Carr immediately wrote a crestfallen letter to Barbour:

> The news of Dr. Stejneger's death was somehow as much of a shock as though he had been fifty instead of ninety-three. My mind had been on him considerably of late and I had frequently thought of how much fun

it would be to go up to see him and try to stir up an argument about turtles if I should go to Washington. It's trite but as true as anything can be to say that an epoch in Zoology has closed. I hope his annoyance at me over that cocky letter wasn't one of his last emotions. I'm sure you must feel that you've suffered a terrible loss.[95]

Barbour tried to put Carr's mind to rest:

I am afraid dear old L. S. never saw your last letter and I am making a large number of changes in the check list and taking the responsibility therefore. I am heading this edition with a little postscript as follows: "A message received but an hour ago tells me that since February 28th Leonhard Stejneger has gone and left but the shadow of his great name. He was the knightly heir of a hardy and virile race. A viking in the great tradition. This is no place to set forth his magnificent record of achievement, the simplicity of character nor the versatility of his agile mind. It is but to say that for me a new and sadder day has dawned which will only close when I close my own eyes for the last time."[96]

Carr was maturing as a scientist. As his confidence grew, he could challenge one of the giants in American herpetology, but only in a nonconfrontational way through an intermediary. Above all, this episode indicates the development of Carr's intellectual commitments and confidence as an independent scientist.

Joining the War Effort

As more and more friends and colleagues enlisted to serve in World War II, Carr indicated his growing frustration with his inability to join the war effort: "I haven't heard anything more from the Engineering Corps. That doesn't necessarily mean anything, however, for the ways of the army are devious. The head of the Spanish Department here has just been appointed to what seems like a swell job with Naval Intelligence and thinks they might be able to use me."[97] A few months later, after receiving a rejection letter from Naval Intelligence, Carr wrote Barbour, tongue in cheek, "I'm going to try now for the Specialists Corps, if there really is such a thing, and after that—something else, maybe the FBI. . . . As well as I know the lower West Coast I should think I could be of some help even without jiu jitsu. It's a devilish Jap ritual anyway. I'm also going to write the Rockefeller Inst. InterAmer. Affairs again. I read that some of them were jailed for sedition the other day, so maybe there's an opening."[98] Barbour lobbied on Carr's behalf by contacting other scientists who had landed commissions.

By October 1942, Carr began to think he might be drafted after all:

I think I may be drafted before long. Frank Young tells me of several one-legged draftees and they've taken our last graduate student who was terribly crippled with polio. The ironic part of this is that the only officer candidate school open to physically defective selectees is the kind which has taken over the third floor of our building and which trains administrative officers. That directs me definitely toward two lovely stripes as my goal in this struggle. I'd rather be dead than be a bookkeeping shavetail.[99]

Carr's hopes were dashed five weeks later when the local draft board categorized him for "occupational deferment;" this meant he would have to enlist as a private to join the war, which would not receive the approval of the university administration.

Nevertheless, an opportunity to serve arose when the University of Florida joined the war effort. Finally, it seemed that Carr would support the war by training young recruits. By the time a commission became available for Carr, Marjorie was expecting their first child, and Archie was loathe to leave Gainesville for Washington, D.C. He expressed his ambivalence in a letter to Barbour. Along with the rest of the biology faculty, Carr was teaching physics (a subject he hated) to Air Corps cadets. Moreover, it was unclear if he could receive a leave of absence from UF, and his salary had not been determined. At least the physics teaching position offered the security of a job after the war. Still, Carr clung to the hope that he might be dispatched to South America.[100]

Barbour was adamant that Carr should stay in Gainesville and take care of Marjorie.[101] When the President of UF, John J. Tigert, refused to grant Carr a leave of absence, Barbour wrote to add some perspective: "Of course that is a nice salary but I suppose you will be perfectly furious with me that I think old Tigert did exactly right and that ten years from now you will be glad. Washington is a hell hole. Your chances of finding a decent place to live are just a little less than zero. That compared to living in your own comfortable house—well I have said too much."[102] Carr's next letter, a week later, suggested that he had accepted his fate, perhaps begrudgingly: "I haven't had an opportunity to do anything about the hammock dope that you requested. I am busy all day, Monday through Saturday, until 5 PM; teaching or attending classes in Physics; then I study math until midnight. It's a new experience and I'm not sure I like it, but I don't have time to get very morbid."[103]

Barbour spent more than a month with the Carrs in March and April of 1943, and this must have come as a welcome distraction for Archie from the demands of physics. After Barbour returned to Cambridge, he wrote Carr with good news:

I have some interesting news as I had a long visit with Wetmore and Miss Cochran yesterday. The long and the short of it is that beyond some notes on the soft-shelled turtles, which can probably be polished up and published, dear old L. S. left no turtle manuscript. Apparently the thing had him bogged down and he got into the frame of mind that Merriam got into with the Brown Bears. I think Miss Cochran will bring out the *Amyda* paper under his name and I think the rest of the notes will ultimately go to you. Just keep this tightly under your hat.[104]

Carr thanked Barbour profusely in a letter: "I was delighted to hear of the possibilities at the Nat. Mus. and am enthusiastically grateful for the kind words you must have spoken in my behalf. I don't know when I'll ever have the opportunity of doing anything along such lines again, but, whatever the shape of Dr. Stejenger's stuff may be, he had a tremendous amount of material and I'm sure I should perish from frustration if Burt or some such ass took it over."[105]

During the course of Barbour's visit to Gainesville, the Carrs heard of an opportunity to buy another home, and Barbour tentatively offered to assist them with the mortgage after he returned to Cambridge and assessed his financial state. To his dismay, his finances would not allow him to make a loan to the Carrs.[106] Despite Barbour's inability to help, the Carrs planned to sell their house for $8500, which would give them greater equity in the new property. They also considered renting out their house. Within the month, all was resolved as the Carrs moved out of their old house, and Barbour was able to provide a mortgage of $2500 with a delay on payments until November 1943.

In the meantime, Barbour's popular writing career was flourishing. He sent the manuscript of a book regarding the nature of Florida to the Carrs in early May. Archie wrote an enthusiastic prepublication review: "There are few people alive who know Florida as Dr. Barbour does and none who are so well qualified for the difficult task of revealing to the layman the fascination which this land holds for the naturalist. His book fulfils all the high promise innate in the character of the material and in the talent of the author."[107] In Late May 1943, it appears that Carr found himself overwhelmed by the demands of moving and teaching. Like a parent desperate for news from a child, Barbour peppered his letters with questions regarding the Carrs' well being. On June 13, Marjorie sent a letter filled with the details of their move and new home that Barbour so craved to hear. Nine days later, Archie sent a telegram to Barbour informing him of the birth of their daughter, Marjorie. Barbour regretted that the war prevented him from responding with a telegram, as telegraphed congratulations were prohibited as the war intensified.

An Infant and the Challenges of Wartime

With the added burden of their newborn baby, Barbour received only the sketchiest details of the Carrs' life for the next month. Finally, he resorted to that time-honored parental attempt to reestablish communication—the questionnaire—which included ten questions regarding the baby, the new house, Zep, and wildlife. Archie quickly answered with a response to each question and then added a lengthy note: "Have been busier than ever before in my life. It's almost impossible to get any help, and trying to keep Physics going and do carpentry, plumbing, horticulture + animal husbandry needed at the place really keeps me in high gear. Besides Rhoda, Zep (and the baby) we have 10 ducks (muscovies and mallards), around 35 chickens, 4 geese and two pigs."[108] Moving to a farm had insulated the Carrs from some of the deprivations of the war, but the move (and the baby) dramatically increased their domestic responsibilities. Carr included with the letter a young Kemp's ridley sea turtle captured at St. Marks on July 4 with the dismissive note: "Coals to Newcastle!" Barbour responded with customary verve: "The questionnaire arrived yesterday. The answers and the addenda were noted with breathless interest.... Coals to Newcastle, nothing! The British Museum wants a ridley. I'll bet they have one but as long as they don't know it no harm is done and I shall pry something high grade out of them with luck."[109] Barbour's comment suggests that sea turtles were still relatively unknown and that specimens represented valuable commodities in the barter system of major natural history museums.

The one-sided pattern of correspondence continued for several more months as the Carrs adjusted to their new life. Despite the lack of letters, Barbour was clearly on Carr's mind when, in October 1943, he nominated Barbour for an honorary doctorate at the University of Florida and wrote to Barbour's secretary for the details of his illustrious career. The following spring, Carr presented Barbour for an honorary doctorate of science at UF's commencement on May 29:

> Mr. President: It is a privilege to present for the degree of Doctor of Science, Thomas Barbour, naturalist, teacher, author, world traveler and able and generous leader of one of the world's great museums . . .
>
> His love of Florida and interest in her fauna and flora date from 1896, when as a boy of 12, he first visited the enchanting wilderness now the metropolitan lower East Coast. Since that time he has spent part of nearly every year in the state. His efforts toward conservation of our wildlife have been tireless. To graduate students at this University he has given encouragement and support for investigations in natural history. It was solely through his efforts that excavation for fossils in Gilchrist County was undertaken. This site, now famous as the most important terrestrial Miocene vertebrate locality in the Eastern States, he has presented to this University.
>
> Mr. President, I present a great Floridian.[110]

Despite the many demands on his time, Carr should have been dumbfounded by the August 1943 *National Geographic*. Buried among several articles regarding the war, there was an article titled "Capturing giant turtles in the Caribbean." It followed Allie Ebanks, a Cayman turtle captain on a sailing cruise to the Miskito Cays of Nicaragua and then to Key West, Florida. Though the article provided little in the way of data on the biology of green turtles, it was filled with details about the economics and culture of turtle fishing. More than a decade would pass before Carr was able to visit Grand Cayman and meet Allie Ebanks himself, but the *National Geographic* article would have supplied daydreams for the rare idle moment.[111]

We have already seen how Carr benefited from productive interactions with Helen Gaige, Thomas Barbour, and Leonhard Stejneger. In early December 1943, Carr received a letter from another of his original contacts in herpetology. Albert Hazen Wright of Cornell University offered the maturing herpetologist an opportunity of a lifetime. In his letter to Carr, Wright first noted how much he had enjoyed visiting Carr in Florida many years before. After a few more pleasantries, Wright asked Carr to write a handbook of the turtles of the United States for a reptile series published by Comstock Press (at Cornell).[112] Having laid the foundation for such a book in his work on the herpetology of Florida and his taxonomic studies of *Pseudemys*, Carr accepted this task almost immediately. Once again, a contact that Carr had cultivated at the beginning of his career offered him a major opportunity. Nevertheless, Wright's offer hardly represented patronage, as it demanded a significant commitment on Carr's part. Despite his extensive preparation in herpetology, the *Handbook of Turtles* consumed much of Carr's effort over the next eight years.

A week or so after Carr received Wright's letter, Barbour asked Carr to resume efforts to visit the Thomas Farm dig in Gilchrist County. After approaching the rationing board for gasoline and once again being denied, Carr relayed the disappointing news to Barbour.[113] Carr's letter then went from bad to worse: "This is a dismal letter. Little Zep was killed Thursday. He went hunting with Rhoda and failed to return with her. We hunted him nearly all night and the next day put an ad in the paper. That evening a man brought by Zep's collar in to the [Gainesville] *Sun* office. The poor little fellow had tried to crawl through a woven wire fence about a mile down the creek from us."[114] Ever sympathetic, Barbour recounted his family's heartfelt response to the news of Zep's demise in a letter of condolence to Marjorie.[115] On a brighter note, Carr reported that he was serious about writing a handbook of turtles for Wright's series and he was only awaiting the approval from the publishers.

Sea Turtle Systematics

The problem of sea turtle systematics continued to irritate Carr. In November 1944, he sent a letter to Barbour to explore how he might address this issue:

> In going over the sea turtle literature in connection with the handbook I have become progressively more exasperated at the shambles that sea turtle taxonomy presents. It seems to me a shameful blot on the herpetological escutcheon that so little is known of them. As intriguing a zoological picture as they present, and as important economically as they are—I suppose the green is the world's most valuable reptile—it is a disgrace that (among many other defections) no one has gone to the trouble to make an adequate comparison of Atlantic and Pacific forms of *Lepidochelys*, *Chelonia*, *Dermochelys*, and *Caretta*.[116]

Carr admitted that he was obsessed with this issue and said that he hoped to address it before turning in the manuscript of the turtle handbook ("some months hence," he predicted optimistically). To fully explore sea turtle systematics, he needed to conduct a detailed examination of a range of specimens from both the Atlantic or Caribbean and the Pacific. Possible sites for this research included commercial fisheries on the Mosquito Coast, the Cayman Islands, and Key West (the cheapest option) and in the Caribbean and Baja, California, Guatemala, or

Costa Rica on the Pacific coast. Carr's hope (and reason for writing to Barbour) was to explore the possibility of funding either from a federal agency such as the National Research Council or a foundation such as the Guggenheim. No one in his department had applied for such a grant. Carr added: "To summarize the above: I want to apply for a grant of $300 minimum for a 2–4 weeks' trip to turtle fisheries on Pacific Coast of Central America."[117]

Though this correspondence was ostensibly a query letter regarding funding opportunities, Carr described a well-conceived research agenda with realistic objectives. From his survey of turtles of North America for the *Handbook*, it was clear that sea turtles represented the most significant gap in knowledge. By comparing sea turtles in the Pacific as well as the Atlantic, Carr hoped to make a significant contribution to the literature. Barbour replied to Carr's letter with his customary speed and resolve. First, he wondered whether the United Fruit Company could accommodate Carr while he pursued his research. Next, Barbour offered to fund a trip to Key West and one to the west coast of Central America with MCZ funds. Other possible sources of funding included the American Philosophical Society or the National Academy of Sciences. Apparently, in the course of writing this letter to Carr, Barbour heard from his contact at the United Fruit Company. He added the following postscript:

> P.S. Miss Duggan, Secretary of A. A. Pollan, who is the Executive Vice President of the United Fruit Company, says your best bet is Costa Rica, Taca plane from Miami to Tegucigalpa, and from there on to Costa Rica. Wilson Popenoe can easily put you up at the Agriculture School in Honduras, and you can discuss with him there (as he is very familiar with the situation) where will be the best place on the Pacific Coast to find a turtle fishery established and then proceed accordingly. He knows all the local ropes! Let me know if you want any detailed introductions and so forth.[118]

Barbour was combining all his roles by providing useful advice regarding contacts in Central America and additional granting bodies. Moreover, he offered to fund the preliminary trips himself with MCZ funds and made suggestions regarding where Carr might find sea turtles. Once again, Carr had turned to his friend Thomas Barbour, and Barbour promised minor financial backing and a potential contact. This time, however, Barbour's assistance would lead to the opportunity of a lifetime for Archie and Marjorie Carr, as they would spend nearly five years in Honduras as a result of this exchange.

Conclusion

Thomas Barbour occupied many roles in the Carrs' life: mentor, parent figure, colleague, collaborator, benefactor, role model, and most of all, friend. Over the course of their friendship with Barbour, the Carrs matured as scientists. Archie in particular published extensively as a result of his taxonomic research on the turtle collections at the MCZ. Later, when his interests shifted from systematics to ecology and conservation, Carr continued to draw on taxonomy to determine which species most needed protection. In fact, throughout his long career, he cited taxonomy as one of the critical components of conservation in general and sea turtles in particular. Marjorie benefited from Barbour's encouragement to publish her master's thesis. No true friendship is one sided, and the Carrs reciprocated Barbour's kindness in many ways, from sending oranges to arranging for an honorary doctorate to heaping lavish praise on his popular books. It is possible to identify in the long friendship with Barbour the seeds of the characteristics that would make Carr a renowned scientist, conservationist, and writer. No other person had a greater influence on the trajectory of Carr's career.

CHAPTER 4

Exploring Tropical Ecology in Honduras

For the first five years or so after he received his doctorate, Carr was able to pursue his passion: the natural history of reptiles, particularly turtles. With Thomas Barbour's friendship and support, Carr developed as a naturalist. He revised taxonomies of turtles and other groups. He also described several species new to science. Forays into Mexico broadened Carr's view of natural history beyond U.S. borders. Nevertheless, like millions of other Americans, World War II forced him to put his career on hold as he and Marjorie supported the war effort in several ways. Many aspects of their life were curtailed by rationing or by other wartime demands. Carr was required to teach physics to young cadets. Few subjects within the sciences could have been further from his area of interest. Marjorie worked as a nurse's aide. Their ability to travel was constrained by gas rationing. The war hung over their lives like heavy fog. Just as the fog began to lift, an opportunity to teach at the Escuela Agricola Panamericana (EAP) in Tegucigalpa, Honduras, arose (facilitated by Barbour). Without the slightest hesitation, the Carrs grabbed this opportunity and ultimately spent nearly five years exploring the tropics from their base at the EAP.

The desire to travel and study the flora and fauna of a new region has flowed through the veins of many naturalists going all the way back to Linnaeus. So typical is this tendency among naturalists that the exceptions, such as Gilbert White, the country parson content to catalog the natural history of Selbourne, stand in stark contrast to Linnaeus, von Humbolt, Darwin, Wallace, and scores of lesser known naturalists who risked honor, fortune, and even life and limb to explore terra incognita. Those who traveled in search of nature generally found it, but they turned a blind eye to culture except in pursuit of anthropology. In Honduras, Carr investigated culture alongside nature, just as he had in Florida. Unlike other naturalists,

Carr acknowledged the dynamic between nature and culture. He spoke with Hondurans walking 30 miles or more to reach a market, and he respectfully listened to their stories about poisonous snakes, butt-headed fish, and sea turtle eggs. Even before his first encounter with a nesting sea turtle, Carr discussed the subject at length with the men who harvested the eggs. Moreover, he savored the taste of turtle eggs and lemon, slurped from a hole in the shell. From all accounts (and especially his own notebooks and publications), Carr relished such cultural experiences with the same gusto as documenting a sea turtle making its nest.

An Initial Foray to Honduras

On December 29, 1944, Carr shared his initial impressions of Wilson Popenoe and the remarkable agricultural school he had founded: "Dr. Popenoe has a burning sort of missionary drive, the effects of which are evident throughout the place. He has the sons of Guatemaltecan Indians and of Costa Rican millionaires living and working together in what seems to me complete harmony. I really had no idea such a place existed anywhere in the world."[1] Carr later elaborated on his initial impressions of Popenoe and his school: " 'Pop' had just finished building the Escuela Agricola Panamericana for the United Fruit Company in the beautiful high valley of the Yeguare River in the mountains of southern Honduras. The first crop of graduates had just been turned out, the budded mangoes were about to bear, you could get a good horse for twenty-five dollars, and the sun flooded the valley through the most exciting air I had ever breathed."[2] Equally important for Carr, however, was news of sea turtles, although he planned to divide his time between the coast and the tropical forest. He wrote to Barbour: "I have learned that there are lots of sea turtles in Guanacaste and that they lay in January on the beaches near a little place called Murcielago. I'm going down to Fonseca soon to see if there's anything there and if not I'll head for Nicoya. I expect to get a mule and go up to the high pinewoods this afternoon. At the 6000 foot level there is cloud forest on one of the peaks that looks intriguing—I want to get up there too before I leave. I surely wish you were here to see all this."[3]

Nearly two weeks passed before Carr found the time to write to Barbour again, this time with the wonderful news that Popenoe had offered Carr a position at the school to survey the vertebrate fauna and help with science teaching. For Carr, the lifelong naturalist, it seemed that an opportunity of a lifetime had fallen into his hands. He later reflected on its significance: "I had just wound up a two-year tour in the Army Air Force Pre-Flight Program at the University of Florida, where I taught elementary physics to impatient cadets and dreamed of the time when I could be a naturalist again. I was ripe for Dr. Popenoe's offer. I took it and went home to tell my wife. She was ready to go before I finished the story."[4] His enthusiasm was barely diminished by the discovery that nesting sea turtles would be difficult to reach:

> It appears that the only point on the Pacific where turtles are regularly taken is a little town in Guanacaste in Costa Rica. The fisheries at

Amapola have been temporarily abandoned in favor of iguana-hunting which has become much more profitable. Since the Guanacaste trip is pretty complicated, involving travel by plane, boat and horse, and since I feel fairly confident of returning in the near future, I've decided to content myself with short jaunts out from the school.[5]

For such trips, Carr was guided by Wilson Popenoe's son, Hugh, whom he described as "an extremely intelligent and energetic 15-year-old, with a flair for natural history."[6]

The Escuela Agricola Panamericana

In developing the Escuela Agricola Panamericana, Wilson Popenoe, a long-time United Fruit Company employee, realized the philanthropic dream of the company's president, Samuel Zemurray.[7] In his extensive history of Zamorano (as the EAP became known), a former director, Simón E. Malo, explored the unique set of circumstances through which the school emerged. Zemurray was a successful businessman with a profound sense of philanthropy. He willingly committed some of his own resources to establish an agricultural school in Central America that would train the region's young men in agriculture through direct, hands-on experience to the greatest extent possible. Students would receive training at no cost.[8]

Zemurray's vision was underwritten by the United Fruit Company's remarkable windfall of $200 million that could not be reinvested in the aftermath of World War II. Tom D. Cabot, long-time board member of the school, noted that such a large amount of capital made it difficult to reject Zemurray's proposal to create a practical school. In addition, Honduras at that time was still an undeveloped nation and one of the few countries in Central America without an agricultural education program. The Honduran president, Tiburcio Carias, and agricultural minister, Juan M. Galvez, were dedicated to improving the state of agriculture in Honduras. In fact, Carias and Galvez facilitated the establishment of a foreign-owned school despite local opposition. The decisive factor in the development of the school was its first director, Popenoe, who was widely known as a plant explorer for the U.S. Department of Agriculture before he joined the United Fruit Company. In describing Popenoe's influence, Malo wrote: "Without his practical vigor, experience in the region, great conviction regarding what was needed, stubborn dedication, and yes, lack of formal academic training, the EAP might today be another of the hundreds of theoretical colleges in the region, or perhaps simply a copy of a 'white-coat' U.S. university."[9]

When the Carrs arrived at the EAP, the school was fully functional, and the inaugural class had graduated. The courses that Archie Carr taught do not appear to have been demanding, and this left ample time for exploring the hills around Zamorano on horseback. Both Archie and Marjorie developed their knowledge of tropical ecology. Their previous trips to Mexico had taken them into the tropics only on a limited basis, but Honduras placed them solidly in the heart of the tropics for nearly five years. Certainly, Archie continued to work on his *Handbook of Turtles*, but he and Marjorie took every opportunity to get to know the tropics: the ecosystems and the flora and fauna, but also the people barely eking out a living in the forests and farms of the region. For years, Barbour had regaled the Carrs with stories of tropical exploration. In Honduras, they reveled in explorations of their own.

Planning and Packing

January 1945 was a busy time for the Carrs: they sold Marjorie's mother's house and their own. Archie took four consecutive trips to the hospital in Ocala for gastrointestinal diagnostic examinations for an undiagnosed stomach condition that had been bothering him for two years and seemed to be intensifying. Unable to find lesions, doctors dismissed the problem as a spastic duodenum attributable to nerves. Ironically, Carr began taking flying lessons in anticipation of flying in Central America. Marjorie shared her reaction to the move in a March letter to Barbour: "I am certainly anticipating our move. You know how I have always wanted to spend at least a few years in South or Central America but I never dared hope for such an opportunity as this. . . . At the rate time is slipping by I'll have a baby tonight and we'll leave for Honduras day after tomorrow."[10] Marjorie was expecting their second child while their first had unexpectedly developed a dislike of herpetology (or at least the subjects thereof): "Much to his surprise and chagrin Mimi doesn't like reptiles or amphibians. Isn't that queer? Her one exception is turtles. They have bitten her several times, however."[11] Marjorie also asked Barbour if he wanted specimens (birds or mammals) from Honduras and wondered if he could recommend books about the region. Barbour recommended Thomas Belt's *Naturalist in Nicaragua* as the "best thing ever written on Central America" as well as Wallace's *Malay Archipelago*, Bates's *Naturalist on the Amazon*, and Hudson's *Naturalist in La Plata*. He enthusiastically accepted the promise of skins: "You bet your life we'd like skins from that region. Anything from Honduras would come in mighty handy."[12] He prioritized possibilities in a handwritten note on the letter: "Mammals! Birds. Freshwater Fish!! Myriapods!!! Spiders!!!"[13] Assuming that the number of exclamation points suggests desirability, Barbour was much more interested in invertebrates than vertebrates.

Discovering Zamorano

Though the move to Honduras hardly went as fast as Marjorie's facetious timetable, Archie was established at the school by June 3, 1945, and he finally had a moment to write to Barbour (on EAP stationery): "I'm finally ensconced in these Elysial fields after what was easily the most hectic month of my wicked career. The job of packing one's worldly goods for over-water shipment is at best a chore, and when the chief executive is away with the job of unpacking a second-born, things really get thick. At least I found it so."[14] Archie worried about how

Marjorie would find a home for their dog Rhoda. The other significant challenge facing Marjorie was the long flight to Honduras with two small children, but their friend Lucy Dickinson had agreed to travel with her and help. Any collecting Carr hoped to do was delayed by the demands of preparing a new course in agricultural chemistry, but he had conscripted the students to help: "However, the students have brought in some stuff, including a beautiful big coral snake and several other things that I missed last winter. It's going to be a great racket having 160 boys looking out for varmints."[15] Meanwhile, new challenges confronted the Florida naturalist: "The rainy season has begun here and the frogs are singing—lots of them, and I don't know a single one!"[16]

Unfortunately, Marjorie's trip to Honduras was delayed by the processing of her passport. She wrote to Barbour about contacts in the State Department in hopes of expediting the process, but she also provided the first details of her newborn child: "The babies are fine. Mimi is browning slowly—very slowly, but at last she doesn't burn. We have named the little boy—A.F.C. III but we call him Chuck."[17] With Barbour's assistance, Marjorie finally traveled to Honduras at the beginning of July 1945. The flight proved to be an ordeal. Mimi experienced mild airsickness, while Lucy's husband, J. C. Dickinson, Jr., an ornithologist at the Florida Museum of Natural History, became violently ill.

The Carrs took their first collecting trip on July 23. They were joined by their friends the Dickinsons and several others. The trip was a quest for the "four-eyed" fish (*Anableps*), a species named for its eyes, which seem subdivided. Due to an asymmetrical lens, four-eyed fish can see clearly above and below the water simultaneously. Carr seemed to revel in his return to field natural history (his work on the turtle handbook had been limited to a review of the literature):

> The behaviour of the fish is about what you would expect from a look at the eyes. They are appropriately almost exclusively surface-swimmers and are quite as interested in the atmosphere above as in the water beneath—if not more so. . . . Apparently nearly all the activities of the fish are carried on at the surface. When mildly alarmed, they cruise away with the eyes above water. When badly frightened, they make off like flushed quail in a series of prolonged jumps which remind one of the saltations of hound fish and ballyhoos.[18]

Carr's reference to Florida fish conformed to a pattern that he maintained throughout his life. Wherever he traveled, Carr drew on his background in the nature of Florida for his basis of comparison. This fish proved to be difficult to catch, so the scientists resorted to shotguns as the collecting tool as opposed to nets or even rod and reel. Carr recounted the strengths and weaknesses of this device to Barbour: "If damage to the specimens were no drawback I think a shotgun would be the ideal collecting instrument. Certainly it is the most sporting, yielding nothing to bird shooting if the fish are in full aerial flight."[19] Barbour relished Carr's tale and promised to use the information in one of his books.

Science and Passion

A s was often the case, it was Marjorie who provided Barbour with the details of their new life in Honduras. Clearly, the change of scenery was profound, and yet the greatest changes arrived in the form of lifestyle. Marjorie breathlessly described for Barbour their house (provided by the EAP): "I'm very pleased with the house—its high rooms are such a change—they must be 15 feet high with great beams."[20] Moreover, during the war, the Carrs had struggled with the demands of their home and small farm. In Honduras, they had help, as Marjorie relayed to Barbour: "We have four servants including a niñera for Mimi. I am learning Spanish rapidly. Our cook is pretty good."[21] Marjorie went on to describe a particularly delicious meal: "Seafood is practically non-existent now but we have had something delicious in that line: Cascas de burros—a big thick shelled *Arca* that lives in fresh water streams. I've never heard of a fresh H_2O *Arca*. This one resembles *Arca ponderosa* but has a more ponderous shell. They are tough so to be eaten raw we chopped them and left them in lime juice + pepper sauce with plenty of onions. They are elegant."[22] There was still more magic to life at Zamorano: "The most exotic thing that has happened to us is a serenade the other night. One of the Moyo's has the most velvety sensual voice imaginable."[23]

In *High Jungles and Low*, Carr attempted to capture the romance of their life in Honduras:

> It was nearly always springtime there in the valley. There were seasons, but mostly they were just different kinds of springtime. The days were golden, and there was a special kind of night that came very often, when the cool air drifted down from the mountains and across the day-warm chaparral, gathering spice all the way, stirring the fireflies in the pastures, and bringing the thin yammer of distant coyotes through the rustling palm crowns. The slow wind raised the curtains at the tall windows and kept you too aware of the world to sleep; but it didn't matter because on nights like that two carpenters got restless, and when the valley lights began to wink out, they came down from the creek on their hillside to serenade—Paco with his guitar and Julio with his velvet voice; and when they sang *El Arriero* under the palms in the yard you heard your wife reflexively clawing the pins from her hair and fumbling for a lipstick in the dark.[24]

It is difficult to imagine a stronger contrast with the Carrs' experience during the war. At Zamorano, they had found a tropical paradise.

Though Marjorie was impressed by the domestic arrangements and the cultural differences, it was the land that held her attention. In her descriptions of the natural landscapes around her new home, Marjorie captured the idyllic nature of the area around Zamorano. After offering her perspective on the collecting trip for four eyes, during which she also skinned and ate a Macaw that someone else had shot, Marjorie described Honduras:

> We ride nearly every day. There are so many good trails and trips to take around here. I am reminded of New England by the countryside.

The mountains are more rounded than craggy. The farms are small and enclosed in stonewalls. The streams, cold, swift and rocky and the lanes are tree shaded, only on closer inspection do you realize that trees are pines, the meadows filled with Cochonal (?) and the big trees by the bank are Cebus. The bird and frog calls are dramatically strange.[25]

She went on to describe other things that impressed her: "No mosquitoes. The coolness and yet burning power of the sun—the impudence of the vultures—the

Figure 11. Archie and Marjorie Carr on a collecting expedition with several EAP students (ca. 1947). Photograph courtesy of Mimi Carr.

lack of native handcrafts—the fairly high standard of living for the natives—the legs of the women (mountain climbing does it, I guess)."[26]

Identification of the many new species of birds and other animals proved to be a challenge for both Archie and Marjorie. Archie queried Barbour about guides that might touch on the region. After discovering that the *Fieldbook of Birds of the Panama Canal Zone* (the only source for the birds of Central America) was out of print, Barbour devised a different solution, as relayed by his secretary: "Tell Archie that the best way for him to get birds identified is to send two or three specimens up here from each kind, and we will send him back a named set of everything that comes up, and he can keep them and use them for identification material. This refers to anything else that may be a conundrum—frogs, lizards, snakes, etc."[27] Eventually, Barbour also sent a copy of Witmer Stone's report on birds of Honduras, which Marjorie studied carefully.

Barbour encouraged Carr to collect coyotes because virtually nothing was known of the species's range in Honduras. As was often the case with Barbour's requests, it was no simple matter to collect coyotes, especially since Barbour warned Carr that poisoned carcasses would not make good skins.[28] Nevertheless, Carr was optimistic: "Strangely enough getting specimens of coyotes is much more difficult than I had anticipated. Partly because of a local *creencis* that shooting a coyote ruins one's aim for deer and partly due to the sagacity of the beasts themselves. I still haven't rounded up any specimens. However, my latest offer of 16 *Lampiros* each stirred up considerable enthusiasm and I hope to have some good news before long."[29] Despite the proffered reward, coyotes remained elusive: "The coyote quest is exasperating me beyond endurance. The entire population of the valley knows that I crave coyotes and will pay what they regard as an insanely high price for them and yet none come in."[30] Though Archie's Spanish skills were near fluent and improving, Marjorie struggled with the language. Meanwhile, their young daughter Mimi had no difficulty assimilating the new language, and Archie found himself in a strange position: "Mimi is learning Spanish at least as rapidly as English. I frequently have to translate her stuff for Margie!"[31]

Passage of a Mentor

During November and December 1945, the Carrs' correspondence with Barbour languished, save for a hurried exchange regarding Mrs. Barbour's illness and hospitalization. On January 4, 1946, Barbour wrote to Archie for the last time, mentioning a colleague who was considering a position at the EAP, his daughter's imminent wedding, and his wife's recovery from illness. Life at the MCZ was returning to normal, and the Navy had moved out of the buildings they occupied during the war. Barbour closed his letter with another prompt for news: "Best wishes for the New Year for you all and do drop me a line." And he handwrote a note: "Love all Carrs. TB"[32] Thomas Barbour died four days later on January 8. No doubt the Carrs had many fond memories to share about their old friend. Marjorie probably recalled the time she found an aging Barbour sitting on their back porch with tears streaming down his face. When she asked him what was the matter, Barbour said he could no longer hear the high-pitched choruses of cricket and tree frogs.[33] The Carrs expressed their heartfelt sympathy in letters and a cable to Barbour's widow. She responded: "Your cable and your sweet letter touched me deeply. I need hardly tell you that Tom loved you and Margie dearly, and considered you two almost as his own children. Your friendship has meant much to us both for many years and I find it difficult to express my appreciation and gratitude."[34] The passing of Thomas Barbour left a void in the Carrs' life that they filled with fieldwork and a growing family.

Exploring the Cloud Forest

"As one wanders about the highlands of Honduras and asks people the names of their highest local peaks, the answer comes back, again and again, 'La Llorana', that is to say, *La Montaña Llorana*, which means 'The Weeping Woods.' The people thus allude, with characteristic imagery, to the tearlike fall of water that condenses on the trees of the cloud forest."[35] This is the evocative description of cloud forest with which Carr opens *High Jungles and Low*, his book about the time he and his family spent in Honduras. Published in 1953, this work introduces the layman to the ecology and culture of the tropics in and around Honduras. In 1950, Carr published a formal description of the ecosystems of Honduras: "Outline for a Classification of Animal Habitats in Honduras," in the *Bulletin of the American Museum of Natural History*. In his *Bulletin* article, Carr wrote of cloud forest: "This often loosely applied term is used in the present case in its broader sense to include any upland woods that owes its character to lower temperatures and cloud condensation, or perhaps more directly to the low evaporation rates . . . rather than to direct precipitation."[36] Comparison of the book with the journal article show how Carr easily shifted from lyrical, romantic prose to formal scientific writing. Yet, both convey a sense of the essence of cloud forest in utterly different ways.

While in Honduras, Carr came to know Charles M. Bogert, who was chairman of the Department of Amphibians and Reptiles at the American Museum of Natural History. Bogert visited Carr in Honduras. Carr was appointed to the position of research associate at the museum and listed this as his affiliation, along with the Department of Zoology at UF.

The Carrs discovered the cloud forest through a series of day trips on horseback or in a pickup. At first, they both struggled with the sheer diversity of novel plants and animals: "It is next to impossible to look up and determine with certainty which leaves belong to a particular tree. The confusion of interlacing branches is complete, and a tree which bears, say, three tons of leaves may, according to my reckoning, support five tons of epiphytes ranging in bulk from microscopic algae, tiny mosses, and half-inch orchids to enormous, thick-leaved, woody parasites."[37]

Yet, like so many tropical naturalists before him, Carr was struck by the relative scarcity of animals, especially within cloud forests. He cautioned: "It is entirely possible to wander about for hours, however, and even for days amid this floral splendor and see only a little more in the way of animals than might be found in a well-kept greenhouse."[38] Carr went on to illustrate this point by noting that in many regions of the tropics, bromeliads provided homes to a wide range of frogs, salamanders, lizards, snakes, insects, and even mollusks. Carr and a companion visited a cloud forest known as *Portillo de los Arados*. There were more bromeliads than Carr had ever seen in one place. But after dumping the water reservoirs of dozens of large bromeliads, Carr was cold, wet, and utterly frustrated: "We hauled them down one after another, dumping upon our shivering persons the quart or gallon of cold water that each contained and finding not one single vertebrate animal. Of invertebrates there were only some sow bugs, an occasional centipede, several scorpions (one of which stung me), and swarms of ants (nearly all of which stung me)." This discouraging and painful mishap drove Carr out of the cloud forest for two weeks before he returned to find evidence of animal life, in this case the "ethereal songs of scores of *jilgueros*, the incomparable notes of which express so precisely in fluid sound the spirit of the high forest."[39] *Jilguero* (most likely the brown-backed solitaire [*Myadestes occidentalis*]) is one of many Spanish names for common fauna that Carr saw no need to translate or offer a scientific name, thus allowing the lyrical and onomonopoetic name to reflect the beauty of the bird's voice.

It was also in *Portillo de los Arados* that the Carrs first observed one of the most spectacular birds in all of the tropics: the resplendent quetzal (*Pharomachrus mocinno*). Marjorie recalled the sighting in her notes from November 23, 1948 (the Carrs kept a joint field notebook to record their observations): "At 11 o'clock, as we entered the lower trail above the waterfall we heard a bird calling ahead of us not unlike a big pileated woodpecker. Archie led the way and as he approached a turn in the path the most outrageous calling broke out—loud and complaining, woodpeckerish. Archie turned, shaken, to call, "It's a Quetzal right here!" In a small tree about 30 ft tall a male quetzal sat not 20 ft in front of Archie. He changed his loud complaint to one softer and more whining and about one a second."[40] Archie's account read differently:

> Much later on this same mountain, I had the supreme reward of seeing my first male quetzal. It was five-thirty in the afternoon of a day spent in fruitless search of quetzals. We had scaled the dripping peak of El Volcán, descending it on the opposite side and laboriously working our way back around the base to the homeward trail. As the sun was setting we crossed a little clearing bounded by the towering silvery trunks of the primeval forest. I sat down to spend the few remaining minutes of daylight watching the forest border for anything that might emerge while my companion went to fetch the horses. The brief sunset spread flame through the clouds behind the western ranges, and desultory shreds of mist began to spiral down in to the milpa. For once there was almost no wind, and the only sounds were occasional incredibly sweet passages from the *jilgueros* and the low, duotone chant of the Mexican trogon in the depths of the forest. Suddenly a harsh, crackling call came from a tall tree at the edge of the woods. I rose and walked toward the

tree while the call continued. As I approached, several green toucanets emerged from the tree and flew off over the milpa, pushing their banana-like beaks before them. The raucous cackle continued. Then three quetzals, a female and two gorgeous males, rose above the crown of the tree. One of the males and the female flew directly into the woods, but the other flew and hopped from one tree to another, dashing out the tree and back, making vertical sallies into the air above and descending again in a wholly uncalled-for series of swoops and dips and pirouettes to display his crimson breast, the blue-green fire of his wings, and the grace of his yard-long tail. This was reward enough, and indeed if the forest grew for no other end than to support this bird it were no waste.[41]

Both accounts reveal the potential of the naturalist's experience in the cloud forest.

Later, the Carrs returned to the cloud forest and managed to collect quetzals. Some readers may detect a note of cognitive dissonance here. How can someone with so much appreciation of nature as clearly revealed by the Carrs destroy what

Figure 12. Archie Carr with collected resplendent quetzal (ca. 1947). Photograph courtesy of Mimi Carr.

they love? As we have seen through the many expeditions to Mexico and elsewhere, the practice of natural history consisted of collecting specimens, more often than not very large or "complete" ranges of specimens. For generations of naturalists stretching back to Linnaeus and before, the best evidence of sighting in the field was the specimen (i.e., the actual object sighted). Carr was an avid hunter in his youth, and these skills could be put to good use on collecting trips. Although this practice began to change in the United States during the early part of the twentieth century, the nature of tropical forests and the limitations of optical and photographic equipment necessitated collection.[42]

Studying Nature and Culture

There is no question that Carr found the tropical forests to be wondrous places. He and Marjorie devoted days and weeks at a time to learning everything they could about these ecosystems. Yet, on many trips they encountered people who were trying to eke out a living in or near the forests. It seems that Archie rarely missed an opportunity to engage anyone they encountered. Sometimes they would share a meal or a place to stay. Despite a lack of any formal training in science, many people had insights into the flora and fauna of the region. For example, Carr was amazed at the endurance and patience of the poor mountain people as they traveled: "The sheer doggedness exhibited on these trips always moves me peculiarly to a point between pity and anger. Whole barefoot families, including children of four or five years and even infants in arms, will strike out through rain and mud or merciless sun to walk halfway across the republic."[43] He went on to describe incredible trips of 30 miles or more across mountain ranges that rose to 6,000 feet undertaken by Hondurans on a daily basis. Late one night, Carr encountered a man driving pigs from Jamastram to a market Tegucigalpa, a distance of nearly 70 miles.

Sometimes local Hondurans offered insights into native flora and fauna. Carr's ears were especially well attuned to these vignettes. His barber, for example, kept a small snake in a jar that Carr knew to be a viper in the genus *Bothrops*. Carr relished the barber's story of the snake's predilection for the breast milk of nursing mothers and the risk that posed to nursing infants: "If the child seeks comfort at the occupied breast, the viper coils at once into a tight ball. The doomed infant, fumbling for the nipple, takes instead the coiled snake into its mouth. Thereupon the *casera* shows its true character. It crawls down into the child's stomach and strikes repeatedly at all vitals within reach."[44] The barber's fanciful tale bespoke a tenuous relationship with nature.

Finding Sea Turtles

Although myths of nature offered insights into the values and beliefs of Hondurans, Carr also encountered Central Americans with knowledge of natural history. Carr met a Salvadoran egg collector and gleaned from him knowledge of the nature of sea turtle eggs inside the "contrabandist's" hut: "At least half the floor area, including all that under the hammock, was smoothly paved with a layer of turtle eggs half embedded in the sand and so closely spaced that they looked like curious hemispherical tiles on bathroom floor. Titco noticed my astonishment and said, 'Turtle eggs last a long time that way.' "[45] Having discovered that turtle eggs did not spoil at room temperature, Carr joined in the business at hand, which was to consume turtle eggs with salt and lemon. He described the technique:"He reached for another turtle egg, pinched a hole in it, gave it a quick squeeze of lemon and noisily sucked it out of the leathery shell; and then he lay back in the hammock with his head on his hands."[46]

Carr's personal field notes present a rather different image of his first encounter with the nests of sea turtles. On October 10, 1947, Carr drove for eight hours to San Lorenzo. The next day, he sailed to Isla de Ratones in Fonseria, a trip that took about four hours. On the island, Carr quickly located three camps of Salvadoran egg hunters. All the egg hunters stayed in thatched shacks with turtle eggs neatly laid out on the floor. Carr convinced the egg hunters to let him have the beach for the night. After watching the sun set, Carr and his companions waited. At 10:04, the first turtle climbed out of the water, resting three times, before stopping in flood tide wrack about 52 yards from the water's edge. Carr meticulously recorded every detail of the nesting process:

> Began throwing sand with fore flippers and then hind til shallow hole for her body. This made deeper behind than in front. When the hind part of carapace about 5" below general level of the sand she began excavation of actual nest cavity. A half teacup of sand was pushed up by the left hind flip. Under rear margin of shell carried out behind + to the side and dropped; as this fell the other hind flipper kicked sand straight back. Then the process was repeated in reverse. As the hole deepened by alternate digging it also was condensed, but much more on the side toward the turtle's head than on the others; it was not circular in cross section, but squared to some extent, with definite ?? rounded corners. As flippers had to reach deeper into the nest excavation digging became slower and hind part of shell gradually sank into the upper basin which also gradually deepened. When hole as deep as flipper could reach digging stopped. Turtle slapped at tail and cloaca twice with each flipper, knocking away adhering sand. Tail lowered into hole [extending vertically downward about 4 inches, and after a few seconds 1st egg fell into nest at 10:25 PM. Eggs came every 4–10 seconds in bunches of 2,3, or 4 [few times 1], usually 2–3. Turtle's eyes closed and plaster with tears and sand. This must be of some function. Before extrusion of each pair or trio of eggs turtle raised head to 45°, then lowered it slowly as the contractions began and as the eggs emerged pushed chin against sand. Last eggs [2] laid at 10:37 PM after few somewhat lengthened intervals between extrusions. She immediately began filling the hole raked in by hind flippers from

surrounding ramp. There appeared to be some selection of quality of sand as anus was pressed against each section of the ramp to be dragged into the hole and sometimes the lot . . . was rejected. [We poured a small amount of dry sand on the ramp behind the hole; this was touched with the partly everted anus and not used. Packing soon began, the turtle lifting herself on fore flippers while sand pulled in by hind and letting plastron fall upon mound with a wobbling motion. This was continued for 4 minutes and was done very thoroughly the pounding noise produced being audible at a distance of several yards.

At 10:45 she began flipping sand into the shallow upper hole with front flippers, meanwhile rotating body to bring in sand evenly from all sides. We especially noticed that she did not urinate in or on the nest at any time. She crawled across the nest spot twice and started back to the water at 10:52. Halfway down the beach we turned several lights on her and she seemed completely confused as to orientation and started back up the beach! The lights extinguished she turned back toward sea. At 10:58 back in water.[47]

In these initial observations, Carr noted several aspects of the breeding biology of sea turtles that would form a basis for further study: site selection, light sensitivity, and the condition of the eyes while out of the water.

Fish of the Nicaraguan Great Lakes

Locals also shared insights regarding the remarkable fish diversity of the Nicaraguan Great Lakes, Managua and Nicaragua, known collectively as the Sweet Sea. According to Carr, several landlocked saltwater species (a shark, a sawfish, and a tarpon), a red fish, and a "butt-headed fish" were unique to the region. To a biologist steeped in evolution, the lake shark was of particular interest in the way it had diverged from its marine counterpart: "When the lake shark lost contact with its marine relatives, four things happened to it: the position of its eyes shifted slightly, its gill openings enlarged by a trifle, the free tip of its second dorsal fin grew longer, and its disposition got worse."[48] Unlike marine species, even small lake sharks would attack humans, making them more dangerous to bathers and fishermen.

Beyond his interest in the landlocked species, Carr was fascinated by the cichlids of the Great Lakes. Cichlids represented one of the most diverse families of fish in the world. By the time Carr was writing *High Jungles and Low* in 1952, he had visited Lake Nyasa in Nyasaland (now Malawi) in Africa, where the diversity of cichlids is the greatest in the world. But he had been impressed with the diversity of cichlids in Nicaragua: eighteen species, seventeen of which were endemic (unique to the Sweet Sea). Carr acknowledged that the Nyasa cichlids were far more diverse than the ones in Nicaragua, but the Nicaraguan cichlids were interesting in their own right: "It is a very different sort of thing that sets the Sweet

Sea cichlids apart. It is a tendency for some of the perfectly distinct and evidently only distantly related species there to show two of the same bizarre variations—a gross high hump on the forehead and a color phase of golden red."[49] Although Carr had collected cichlids in Mexico and Honduras, he was caught off guard when a friend returned from the market in Managua with fifteen pickled cichlids: four were red while others had huge, swollen foreheads (butt-headed). Neither trait seemed to be linked to species.

To clarify the problem, Carr visited the market in Managua, where he found that some fishwives segregated the red cichlids while others left them in with the others. He interviewed several fishwives, but he found them to be almost as confused by the traits of the various cichlids as himself. He doubted one fishwife's tale that a local *guapote* (one of the cichlids) showed both variations (redness and butt-headedness). Though skeptical, Carr met the woman on the lakeshore before the *Barrio de los Pescadores* to await the arrival of fishing boats. As fishing boats returned, the fishwife waded out to each one in search of the bizarre fish until at last she located one and showed it to Carr, to the dismay of the boatman. Carr recounted: "The solemn boatman seemed depressed at the eccentric fervor of his shorecoming welcome and looked at me mistrustfully, sure that there was gringo *sonbichismo* somewhere back of the fishwife's behavior. I hastened to reassure him by saying heartily that this was just the kind of fish I'd been looking for all over the market and all along the beach."[50] (The term *sonbichismo* was most likely Carr's own neologism [a la Hemingway] juxtaposing the American phrase "son of a bitch" with the Spanish suffix "ismo" to produce a rough approximation [and far more lyrical] form of "son-of-a-bitchedness." Alternatively, Carr's ear was particularly attuned to Creole constructions, and it is possible that he heard this term in use.)

Once again, local knowledge had revealed new research pathways to Carr, and he lamented that he would not have time to develop the studies to explain the variation, whether genetic or environmental, among the Lake Nicaragua cichlids. Instead, he reviewed the biological literature to determine whether any other scientist had contemplated the biology of the Sweet Sea cichlids. He found only one article on the fishes of the Nicaraguan lakes. Although the author described several red forms and commented on their relative abundance, he failed to account for the "rubrism" (red coloration). Like Carr, he visited the fish market in Managua where the red forms appeared to be the best sellers, but he found no reason for this. Carr left the matter for other researchers to solve in the future. His conclusion linked the biology of the fishes to the history of the lake and the colonization of Nicaragua: "The red fishes are still there, swimming the Sweet Sea, like fish image carved from copper-tainted gold—like trinkets the hidalgos took from old Nicaro, bartered for a hawk bell and a lie. They still run like sunbeams before the dugouts, as they fled the bows of the brigantines and frigates standing out across the lake to shoot the white water and thread the forests down to the northern sea, and flock with the *flota* home to Spain."[51]

Rainforests of the Caribbean Slope

Near the end of his tenure in Honduras, Carr spent a month in the Caribbean rainforest. This opportunity arose when Paul Shank, master forester for the United Fruit Company, and another member of the staff at the EAP invited Carr to join a timber cruise in the wildlands near Pearl Lagoon on the Caribbean slope of Nicaragua. Carr's official role in the expedition was to serve as the only meat-hunter (thereby keeping firearms to a minimum). In this capacity, Carr would supplement the standard daily ration of *gallo pinto* (red beans and rice). It was the ideal arrangement for Carr, as he enjoyed hunting game, and it gave him a privileged status: "I took the job, and as it turned out none could have been more to my liking, for it kept me out in front of the cruising parties where I was all alone in the still forest and could skulk along over a hundred miles and more of new-cut trails and soak up sights and sounds and smells."[52]

The expedition began in Bluefields, a quiet Nicaraguan port, which reminded Carr of the Yamacraw district of Savannah as it was during his childhood. Here Carr encountered the Mosquito people and tried to unravel their lineage:

> The Mosquito people still feel keen nostalgia for their century-old heyday and still hold England in most extraordinary reverence. They are dark folk in whom a hereditary background of mosquito [Miskito] and Sumo Indian, diluted by the genes of buccaneers and British traders, is to the eye almost completely hidden by African blood, the first injection of which occurred in 1641 when a slave ship was wrecked at the Cape and two hundred Negroes were released. They are poor, pleasant people who call themselves Creoles if they are black and Indians if brown, speak English with an astonishing accent, and live in a state of strained mutual tolerance with the Latin minority.[53]

From Bluefields, the party traveled by water through estuaries and creeks and finally through a canal all lined by mangroves to Pearl Lagoon. The canal traveled a wet savanna that reminded Carr of the Florida Everglades: "This shallow ditch traverses a wet savanna set with scattered little islands (we call them hammocks back in Florida) of Caribbean pine and skinny, broad-leaved palms, producing a landscape strongly reminiscent of the Florida Everglades."[54]

One of Carr's significant ambitions for the expedition was to see a tapir in the wild. On the tenth day of the trip, Carr did in fact spot the beast (called the "mountain cow" by locals). This first sighting of the remarkable animal filled Carr with conflicted feelings. He wrote: "With respect to my attitude toward varmints, I have always been afflicted with a Dr.-Jekyll-and-Mr.-Hyde complex—an altogether unresolved conflict between the instincts of a naturalist and the urge to shoot things. These dual drives have given me a lot of trouble from time to time, and today they did their worst."[55] Carr's reference to Jekyll and Hyde carried greater meaning in that he was notorious at Zamorano for attending costume parties dressed as Mr. Hyde. Once he even appeared with a live black vulture on his shoulder! His complex came into play when he surprised a male tapir bathing in the flood plain of a small creek. Later, Carr realized that his hunter's instinct had cost him a unique opportunity:

> I could have stood who knows how long savoring the fulfillment of my desire to see a tapir, watching this one in his most private moments as he rolled and snorted and steamed himself in his oxbow bathtub; but I muffed the chance. Through inexorable reflex I raised my puny rifle and started shooting as fast as I could pull the trigger. I don't suppose the stream of hollow-point bullets that rattled slantwise against his flinty thorax did the tapir any harm; I hope not. But they put him in a great sweat to get out of the water and away from the neighborhood. He floundered out of the mudhole with most incongruous agility. Ignoring in his haste the usual exit, he burst easily through the solid palisade of lianas and vine-strung saplings that enclosed his retreat. Heading straight for a bamboo thicket, he again sought no trail, but drove squarely into the dense stand of fishing poles, bending the live canes and snapping the dead ones that opposed him with the racket of a dozen rifles. He charged away through the brake with unabating fury, and I stood and listened in complete dejection as the noise of the popping canes grew feeble with distance.[56]

Twelve days later, Carr encountered his second tapir. It was bigger than the one he had shot at. The second encounter lasted longer and was far more satisfying for Carr, as it confirmed his fanciful notion that tapirs resulted from a cross between an elephant and a donkey! After a few moments, however, the tapir became wary and retreated, but not without turning and producing a drumming noise with its front feet. Carr surmised that the tapir was most likely a female with a calf in the vicinity, a possibility that tempered his desire to shoot it: "I lean to the theory that this was a female with a calf nearby, and it was mostly this feeling that kept me from shooting it, which I could easily have done. I also cite the abstention as atonement for my irresponsible attack on the first tapir of the ravine mudhole, and claim double credit because I could already feel the sting of the tongue-lashing in store for me when Arnold and the Lagoon boys should hear of my improvidence."[57] Several years later, while on an African safari to hunt elephants, Carr completely lost the urge to hunt.

Florida and Honduras

Florida was never far from Carr's mind while he was on the expedition, as a near encounter with a jaguar suggests. Carr had stopped to catch frogs when he was overtaken by another member of the party When Carr resumed his travels on the same path, he noticed that a jaguar (*tigre*) had taken up the same path, and it occurred to him that the jaguar was actually following his colleague. It seemed to Carr that the jaguar had chosen its path not out of malicious intent or hunger.

Figure 13. Archie Carr dressed as Mr. Hyde for an EAP costume party (ca. 1948). Carr carried the black vulture on his arm until it bit him. Photograph courtesy of Mimi Carr.

Rather the jaguar trailed his friend out of "feline caprice": "The situation took me back to an afternoon in the Florida scrub when I sat on a deer stand on a low hill overlooking a recent burn and watched two figures move off toward the horizon—one of them an ant-sized deerhunter who believed himself all alone, and the other a trailing bobcat."[58] Carr continued on the jaguar's trail until he lost it a few hundred yards from camp. Unfortunately (or perhaps fortunately), the large cat was not to be seen.

Occasionally, local myths about nature reminded Carr of similar stories he had heard in Florida. One night, he was wading in the river when he heard a small group of toads calling. Carr knew the toads to be *Bufo marinus*; people referred to them as "marine toads" as a result of bastardizing the scientific name of the species. But the inappropriate common name did not irk Carr. As he waded, one of his friends warned him about wading near *Terciopelos*. Carr explained that the word, which meant "velvet," generally referred to the bushmaster but also the fer-de-lance (one of the most poisonous snakes in Latin America). Carr pointed out that the singing came from toads, not snakes, and that there was little danger. His friend was adamant, so Carr waded upstream to where the toads were and caught one (of course, it and all the other toads stopped calling immediately). Nevertheless, Carr showed the toad to his friend and invited him to walk where he might see the toads singing. His friend dismissed the animal as the vocalist and again warned Carr not to wade when the snakes were singing: "I've been living around here all my life, and I know better than to go fooling around in the river when the *terciopelos* are singing!"[59]

Carr wrote of this experience, "For a time I was saddened by this intrusion of mysticism into the woodsmanship of my new friends and respected mentors. But then I began to realize that this was just another demonstration of the dogged consistency with which non-zoologists everywhere exclude from their woodslore even the most elementary grasp of the natural history of reptiles, and accept any cock-and-bull story concerning them."[60] The episode reminded Carr of a similar one in Florida:

> I know a dozen Florida farmers—intelligent and successful men with deep freezers at home and children in college, who will slow down as they drive past a rain-flooded flatwoods, in which a thousand tiny oak toads cheep, to call your attention to the "scorpions whistling." By "scorpions" they mean lizards in the genus *Eumeces*, and since no one of their set has ever been so shiftless as to slog out in the wet to check the identification, they have no cause to doubt it.[61]

Here Carr drew a distinction between nonscientific and scientific ways of knowing. Nonscientists, whether Mosquitoes, Nicaraguans, or Floridians, viewed nature through a particular lens, in which myth might override empirical evidence. It was because Carr held his new friends in such high regard that he found their beliefs surprising, as this statement suggests: "And thus Charlie, who can name every tree along a rain-forest trail, and Sosa, who has laid out and worked rubber lines ten days walk from human habitation, feel no responsibility for testing the tradition, handed down from who know what remote origin, that vipers sing on the rocks in the river!"[62]

Carr's time in the rainforest drew to a close after just one month, and he had to return to the Escuela Agrícola. He wrangled with the pilot of the plane for a route that would in some measure retrace at least part of the route that he had traveled. Finally, they settled on a course over the agricultural experiment station at El Recreo and the United Fruit Company's oil palm plantation along the Escondido. Moreover, they would see some excellent rainforest from the air. Secretly, Carr hoped to see the Rio Huahuashan from the air. As they flew above the jungle, Carr noticed flashes of a brilliant, blue light emanating from somewhere above the trees. At 2,000 feet in altitude, he had to wonder what caused the flashes. "Butterflies," came his pilot's reply. Immediately, Carr agreed with this designation while marveling that the small area of a butterfly's wing (even one as large as a blue morpho [*Morpho menelaus*]) could throw light so intense as to reach the plane. His pilot scoffed that he had noticed similar flashes from 8,000 feet. For Carr, the flight was also a time for reflection on the previous month: "This was where I had been. Out there were the halls of the mountain cow, where the *Montaña* was even now stealing back into our clearings and hiding the cold ashes of our camp fires; and I watched till the last glint of silver sank into the sea of green and the circling plane turned its tail toward Huahuashan—River of Racing Water, and river of many a dream to come."[63]

The Naturalist as Political Activist

The stories that Carr heard as he and Marjorie explored the rainforests and cloud forests of Central America left him with a deep respect of the people: Hondurans, Nicaraguans, Salvadorans, Mosquitoes, and others. Carr recalled:

> I saw Central America from two quite divergent points of view. My interest in natural history took me away from the cities and out into the backcountry, where public opinion is born but not often canvassed. I wandered about among the country people, eating their *tortillas* and beans, contracting their parasites, seeing their despair—moving among them with no axe to grind, predisposed to like them, and in my interest in their land and its varmints predisposing them to like me. Time and again my wife and I stopped at some lonely *aldea* to tie our horses before a mud-floored inn, and over coffee and *pan dulce* talk with the *dueño* about Honduras and its chances for a better future.[64]

At the EAP, Carr met Central America's future as he taught boys from Central American families, distinguished and unremarkable alike. Like his students, his colleagues and the staff represented a cross-section of Central America. He wrote: "Back at the school we were at a crossroads of inter-Americanism. The staff and student body were a cross section of Caribbean America and there was a steady stream of visitors, most of them in the field for some international agency, or fired with private schemes for saving or seducing Indo-America."[65] More often than not, such hopes floundered in the treacherous waters that rushed at the intersection of ideal, actual, and practical. Carr kept a laundry list of the most egregious examples of failed plans:

> The war was on, and funds for buying the favor of our neighbors flowed freely. Projects to court the people with the promise of technical or agricultural enlightenment sprouted like mushrooms. Sixteen-hand Missouri mules were brought in to bog down in fields where only oxen could plow, or to pull shiny new red-and-green farm wagons and smash

their wheels on roads that only oxcarts could travel. New corn that yielded triply but made bitter *tortillas* was urged upon the Indians, and blooded livestock of all kinds was brought in to languish and waste away in the face of unforeseen factors like the drought or the rainy season, or the dog days or the altitude, or from peculiar localized agents such as screw worms, *tórsolas*, vampire bats, spider bites, abnormally dense tick populations, or the *comelengua*.[66]

In a footnote, Carr noted that the *comelengua* (tongue-eater), a mythical beast part snake, part buzzard, and part puma that preyed on cattle by eating their tongues never appeared in agricultural reports but that it should have. Once again, Carr incorporated local myths into a serious critique of American policy in Honduras and Central America.

When Carr returned to the United States, he developed seven recommendations for future relations with Latin America. He elaborated on each of these in an itemized list in *High Jungles and Low*.

1. Improve the quality of the personnel we send to the tropics.

2. Stop underestimating the intelligence and sensitivity of the rural Latin American people.

3. Avoid irresponsible spending for unnecessary technical assistance.

4. Increase opportunities for Latin American students to go to the United States for training.

5. Support expansion of public health programs.

6. Accelerate and expand agricultural research and education.

7. Encourage and help the Latin American countries in establishing and maintaining campaigns of conservation and restoration of renewable resources.[67]

According to Carr, the quality of personnel sent to the tropics could be improved dramatically by requiring fluency in Spanish and by requiring knowledge of the geography and history of the country of interest. Carr felt that U.S. bureaucrats consistently underestimated the intelligence of Central Americans: "The distinction between ignorance and stupidity, between lack of advantage and lack of capacity, has been repeatedly overlooked in our dealings with peons, not only in our commercial contacts but in our often abortive or misunderstood attempts to help them."[68]

Not all of Carr's recommendations were negative. He supported additional educational opportunities for Central Americans at U.S. schools and universities (this is one area in which he continued to work after returning to the University of Florida). He also recommended the expansion of public health programs and agricultural education and research. As a model agricultural school, Carr cited the EAP: "A dozen schools like Escuela Agricola Panamericana would not be too many, but we should keep them simple and practical—dedicated to the training of farmers and possibly of the county agent type of demonstration worker but not of research men or engineers, or of office staff for the ministries."[69] In recommending campaigns for the conservation and restoration of renewable resources, Carr recognized that of nutrition, public health, and conservation, conservation was the most difficult. He mentioned William Vogt's *Road to Survival* and the criticism that it inspired fear in those who read it, as did Thomas Malthus's *Essay on Population* some 150 years earlier. In response, Carr wrote: "He was trying to scare people. Whether his tactics were good or bad I won't try to say, but I used to get scared before I ever read Vogt. Whether or not you approve of Vogt or know anything about world demography, you can't ignore the tragic dependence of the peons of Latin America on ruined or wasting environments."[70]

How did Carr the political activist emerge from Carr the biologist? Like other naturalists, he spent much of his time in the tropics exploring the cloud forests and rainforests while he and Marjorie collected amphibians, fish, birds, and mammals. During these trips, he frequently encountered people with whom he discussed the future of Honduras. The view of Honduras that developed out of these conversations diverged from those of the various Latin Americans and Americans at the agricultural school. Carr also encountered visitors with plans for Honduran development.

But the roots of Carr's social awareness as a biologist developed before his conversations with local Hondurans. In the Department of Biology at the

Figure 14. Archie Carr and the view from the Escuela Agricola Panamericana (ca. 1948). Photo courtesy of Simón Malo.

University of Florida, there was an explicit concern with the relationships among all animals, including humans. The authors of Carr's introductory biology text examined the organism as a unit in a social-economic complex, asking, What role does the organism play in the economy of nature? How does it affect the other members of its society? What other organism affects its welfare? Ecological relationships were at the core of human social systems as well as animal societies. In *High Jungles and Low*, Carr integrated these principles. He recognized that people were an integral part of the landscapes he studied in Central America. Thus, when he wrote his narrative of tropical nature, it was only natural that he included descriptions of humans and their struggles. In the meantime, because of the departure of Rogers and Hubbell (Carr's former advisors) from the University of Florida, Carr found the Department of Biology in a state of crisis when he returned from Honduras.

Conclusion

What Carr learned or at least confirmed many times in Honduras was that the people had valuable stories to tell about their lives and nature. Even their myths contained interesting bits of knowledge and perspective. As Carr's seven recommendations suggested, well-meaning U.S. bureaucrats consistently underestimated the people of Honduras and discounted or ignored their claims. It would be fair to say that most scientists, particularly northern scientists, found little or nothing of value in the stories of Latin American peasants. Without the scientific method, without scientific names, without extensive training, what could locals tell scientists? Carr recognized the value of stories, from a barber's viper to a smuggler's sea turtle eggs. Everywhere he went Carr sought out local stories and the wisdom they contained. Because he spoke Spanish fluently, Carr had access to information that few northern scientists could capture and even fewer could fully understand. Appreciation of local people and their stories served Carr for the rest of his long career, but most especially as he embarked on a detailed investigation of the ecology and migrations of sea turtles. His conservation programs consistently acknowledged and incorporated the interests and needs of local culture.

Like his students in Florida, many of his students at the EAP never forgot him. As we saw at the beginning of this chapter, the EAP students came from throughout Central America. Most returned to their home countries after their time at the EAP. Some went on to careers in government, while others managed their families' holdings. Years later, as Carr worked to develop conservation programs for sea turtles, he had personal connections with some of the most powerful individuals, families, and government officials in Central America.

Honduras held even greater significance to the Carrs' marriage in that the Escuela Agricola Panamericana enabled them to live at a very high standard of living. It was the only time in their married lives that neither childcare nor cooking and cleaning were necessary demands of day-to-day life. Both Archie and Marjorie could focus exclusively on the study of tropical biology for at least part of most weeks. They had arrived with two small children (Mimi and Archie III, or "Chuck"); they left with two more additions to the family (Tom and Steven). Their fifth child, David, was born after they returned to Gainesville. For the next decade and a half, Marjorie ran the Carr household and kept the children busy while their father wrote, taught, and conducted research. If there were hard times, and it's hard to imagine that there were not, memories of Honduras and the EAP surely carried them forward.

CHAPTER 5

Study and Conservation of Sea Turtles

With few exceptions, successful individuals acknowledge the role of serendipity in their achievements, no matter how remarkable. Skill, intellect, and determination set the stage, but good fortune often plays a central role in success. The trajectory of the first twelve years of Archie Carr's career (1937–1949) resulted to a significant degree from his close friendship with Thomas Barbour. There is no question that Carr worked hard throughout this period, but fortune was critical too. Nearly five years had passed since Archie and Marjorie Carr had left Gainesville for Zamorano. Shortly after their return, Carr was promoted to full professor. Before he could devote himself wholly to the study of sea turtles, however, Carr was dedicated to the completion of his magnum opus, the *Handbook of Turtles* (1952). He also wrote his first popular travel narrative, *High Jungles and Low* (1953), about his time in Honduras. With these two works in print, Carr applied for several research grants to conduct pilot studies of sea turtle colonies throughout Florida and the Caribbean.

As in Honduras, Carr relished the cultures that he encountered in his study of nature in the Carribean, and for this reason Caribbean culture served as the central theme of Carr's second popular work, *The Windward Road* (1956). Yet it was the plight of sea turtles (and their role in culture) that captured the attention several philanthropists, including Joshua Powers and John H. "Ben" Phipps, who facilitated Carr's conservation efforts by starting the Caribbean Conservation Corporation (CCC). The development of the CCC was one of the most serendipitous events in Carr's life.

Changes in the Biology Department
at the University of Florida

While the Carrs were studying tropical ecology and raising their children in Honduras, the Department of Biology at the University of Florida (UF) was undergoing significant changes. In 1946, J. Speed Rogers returned to the Museum of Zoology at the University of Michigan, and with him went Theodore Hubbell. The department thus lost its chairman and another strong member of the department in the same year. Given the number of World War II vets using the G.I. bill for college tuition, the university had to provide courses for many more students. To cope with rising enrollment, the biology department granted faculty positions to several graduate students. During this time, the biology faculty included more and more UF graduates, which left the department open to charges of academic inbreeding. With the loss of two of its strongest senior faculty members, the department floundered without a leader. When Carr returned from Honduras, an earnest search for a department chair was underway. Given the department's emphasis on organismal (whole animal) biology, experimental orientation, and social concern, Warder Clyde Allee of the University of Chicago was an ideal candidate to chair the department.[1]

Warder Clyde Allee (1885–1955) led the ecology group at the University of Chicago during the early part of the twentieth century.[2] In the zoology department at Chicago, Allee studied under Victor Shelford, a pioneer ecologist, and Charles Manning Child, an animal physiologist. Allee's upbringing as the son of a Methodist minister and his experiences in World War I contributed to his development as a socially aware scientist. In 1938, he wrote *The Social Life of Animals*, which he based on a series of lectures he presented at Northwestern University the previous year. The book explored cooperation among animals, with human implications.[3] The ecology group (W. C. Allee, Alfred E. Emerson, Orlando Park, Thomas Park, and Karl P. Schmidt) produced *Principles of Animal Ecology* (1949), and it became a standard textbook at the University of Florida, where it became known as "AEPPS," in reference to the initials of its contributors.[4] In 1950, Allee retired from the University of Chicago. Most biologists know Allee for the concept of a positive relationship between population density and per capita growth rate. This idea has become important in conservation because certain species (e.g., passenger pigeons) seem to require a critical mass for long-term population growth, without which the population crashes.

Personal tragedy dogged Allee throughout his life. He witnessed the death of his son to a horse-drawn streetcar and endured the death of his wife. Worse, a mass of differentiated cells of an undeveloped twin grew near his spine and paralyzed him below the waist. He was also severely injured when backing his wheelchair into an open elevator shaft. Allee did not let his personal tragedies or his considerable weight limit his activities: he called on graduate students to transport him on fieldtrips, even if that meant carrying him in his wheelchair.

When Allee was entertaining the possibility of heading the UF Department of Biology, he wrote each member of the biology faculty and asked them to assess the program's strengths and weaknesses. Before he made the decision to accept, he again wrote to the faculty soliciting their reactions. Carr responded enthusiastically: "In answer to your inquiry concerning my reaction to the possibility of your acceptance of the headship here, I believe I add my voice to many (probably to all) when I say that I can think of no more promising solution of our problems here.... I believe that your influence would integrate efforts and smooth out personal wrinkles which are at present serious obstacles to progress."[5]

As head of the Department of Biology at UF, Warder Clyde Allee would continue the efforts of J. Speed Rogers and Theodore Hubbell to produce socially aware biologists. In addition, he demanded that graduates of the doctoral program seek experience elsewhere before joining the department as faculty members (in the hopes of limiting the effects of academic inbreeding in the department). As Allee's research emphasized social interactions among animals in nature, he called for cooperation among the faculty at UF. Allee's policies assured continuity with the past and awareness of other approaches in biology. To understand Carr as a naturalist and ecologist, it is necessary to look beyond the Department of Biology at the University of Florida.

Handbook of Turtles

During his years in Honduras, Carr's work on the *Handbook of Turtles* had progressed slowly in the face of numerous teaching commitments as well as other projects, but he was able to finish the project after he returned to UF. Gradually, descriptive natural history was being restricted to the introductory notes of experimental ecology studies and, of course, popular accounts, where natural history still commanded a large audience. Much of the *Handbook of Turtles*, however, suggested classically descriptive natural history, but Carr integrated much of the known and hypothetical or anecdotal biology (that is, life histories) for every North American turtle.

The *Handbook of Turtles* was published in 1952. Although the book is divided between two potentially dry subjects (taxonomy and life histories), Carr sustained a lively tone throughout. Like Carr's other books, the *Handbook* opens with a literary flourish:

> Two hundred million years ago the reptiles, newly arisen from an uncommonly doughty set of amphibians, were on the verge of great adventures. They bore the mark of destiny in the shape of impervious scales and the new cunning to lay shelled eggs, and these devices insured them against the age-old disaster of drying out, both before birth and after, and let them gratify their own curiosity about the vast and almost empty land. Along with the new equipment they had imagination and no end of notions for novel body designs. Today we call these old beasts cotylosaurs, or stem reptiles, because all the lines of vertebrate life above the amphibian level lead back to them as branches converge in the trunk of a tree.[6]

Figure 15. Archie Carr and a Florida soft-shelled turtle examining the *Handbook of Turtles*, 1952. Courtesy of the Department of Special and Area Studies Collections, George A. Smathers Libraries, University of Florida.

Rarely has evolutionary history and descent from a common ancestor received such elegant prose. Carr continued to relate that, eventually, a new animal emerged: "The new animal was a turtle. Having once performed the spectacular feat of getting its girdles inside its ribs, it lapsed into a state of complacent conservatism that has been the chief mark of the breed ever since."[7] By this Carr meant that turtles have retained their basic body plan since their emergence more than 200 hundred million years ago.

With characteristic panache (unexpected in a formal biological monograph), Carr explained the dramatic biological transformations that occurred in the millions of years since turtles appeared on earth, including the ascent of winged insects and archosaurs, the dawn of dinosaurs and birds, and finally the arrival of mammals (and a sense of irony):

> Turtles went with them, as tortoises now, with high shells and columnar, elephantine feet, but always making as few compromises as possible with the new environment, for by now their architecture and their philosophy had been proved by the eons; and there is no wonder that they just kept on watching as *Eohippus* begat Man o' War and a mob of irresponsible

and shifty-eyed little shrews swarmed down out of the trees to chip at stones, and fidget around fires, and build atomic bombs.[8]

Here Carr's narrative skills are on full display. While the body plan of turtles remained essentially unchanged, life as we know it developed. Dinosaurs disappeared, while other groups came and went. Humans descended from the trees. Carr's reference to atomic bombs came at a time when many Americans and scientists in particular were reeling from the implications of the development and use of nuclear weapons and the commencement of the Cold War.

Having summarized the long paleontological history of turtles, Carr went on to review the basic biology of turtles. Turtle respiration was particularly interesting in that aquatic turtles had the ability to maintain respiration under water through various means:

> To augment their oxygen supply, many aquatic turtles use the highly vascular pharyngeal cavity as a sort of gill, sucking in and expelling water and obtaining by this means sufficient oxygen to increase materially their capacity for remaining submerged. In a similar way, additional underwater respiration is effected by some species that augment the work of the pharynx by filling and emptying, through the anus, two thin-walled sacs that communicate with the cloaca. The currents set up by these pumping movements may be easily demonstrated if a small amount of dye or suspended silt is placed near the anal or nasal openings of a live turtle.[9]

Though circulation, excretion, and digestion were for the most part unremarkable, reproduction attracted the interest of both scientists and nonscientists. Carr noted that two genera of aquatic turtles (*Chrysemys* and *Pseudemys*) performed the most elaborate courtship rituals among North American turtles. In both groups, the male swims backward before the female as he strokes her face with his greatly elongated claws. As for duration of copulation, Carr acknowledged a general lack of data. But he could not resist the temptation to relay a popular legend regarding sea turtles that was transcribed in 1708:

> Among the many bizarre amatory feats that popular legend ascribes to sea turtles, I find one that I believe will escape editorial censorship. While it is of course apocryphal, it is entertainingly so, and it also serves to emphasize the remark that we really know almost nothing about the subject: "All the turtles from the Charibbeas [Caribbean] to the Bay of Mexico, repair in Summer to the Cayman Islands. . . . They coot for 14 days together, then lay in one Night three Hundred Eggs, with White and Yolk but no shells. Then they coot again, and lay in the Sand; and so thrice; when the Male is reduced to a kind of Gelly within, and blind; and is so carry'd home by the Female."[10]

Although Carr regularly incorporated local wisdom into his scientific research and particularly into his writing, occasionally he acknowledged cases in which scientific knowledge contradicted folk beliefs. He wrote, "Anyone who has watched a dozen turtles slide off a log in concerted response to a slight noise a half mile down-river will be loath to accept the pronouncement that turtles do not hear well, but such appears to be the case."[11] Despite a complete auditory anatomy, turtles do not receive atmospheric vibrations but are instead highly attuned to vibrations in the water or the substratum. It surprised Carr to discover that numerous authors had ascribed to turtles a "voice." Carr explained these sounds as the result of the exhalation of breath or frictional contact between parts of the body. Cases of barking or grunting during copulation did, however, seem to constitute a "voice" or at least a vocalization, despite turtles' lack of vocal cords. To convey the state of confusion regarding turtle vocalizations, Carr recounted popular myths regarding scorpions singing in northern Florida (*Eumeces*) and the fer-de-lance calling in Nicaragua. Popular myths and physical explanations notwithstanding, there was a need for more research: "When such intrenched misconceptions as these, and all instances of merely incidental mechanical noise, have been discounted in the case for the voice of the turtle, there still remain a few examples of genuine vocalization that would well repay further investigation."[12]

It seemed to Carr that improved refrigeration techniques would exact heavy tolls on turtle populations, despite the shifts in turtle markets that resulted from World War I and II. International demand for both diamondback terrapins and green turtles was down as far as Carr could tell. Nevertheless, new markets were emerging, even in Gainesville, as Carr recounted:

> I know a number of people who within the past few months have made their first enthusiastic acquaintance with green-turtle meat through a local establishment that features it on a popular carry-away lunch. The deep-fried steaks here are a far cry from Key West chicken turtle with chines and calipees, or from an authentic curry, or green-turtle soup in New Orleans; or for that matter from Carr's broiled ridley filets with lime butter; but people like them, and this bodes no good for the green turtle.[13]

"Carr's broiled ridley filets" undoubtedly refers to his own successful attempts to prepare turtles for consumption. Recall that Carr skeletonized a great many turtle specimens in a soup pot during World War II. Such experiences enabled him to appreciate the economic value of turtles on a deeper level. Another turtle that was becoming popular with diners was the snapping turtle (*Chelydra*), although Carr noted that Philadelphians had appreciated the gastronomic qualities of snappers for many decades. Soft-shell turtles also seemed to Carr to be candidates for an expanded market in turtles. Based on his extensive experience and catholic tastes, Carr noted the qualities of two lesser known species: "Better than either of them (in my private judgment) are the gopher and the 'Suwannee chicken,' but like most of their near relatives these are too locally distributed to support a commercial market."[14] "Gopher" referred to the gopher tortoise, which had not yet declined, and the "Suwannee chicken" was one of the *Pseudemys* species. Both were restricted to the Southeast.

Of the sea turtles, Carr recognized in the green turtle the greatest potential for market development:

It seems to me that in any plan to extend human food resources by more intelligent exploitation of the sea, the green turtle should receive careful consideration. It not only furnished meat of unsurpassed quality but, being herbivorous, it is able to utilize huge volumes of forage provided by the submarine pastures of turtle grass and manatee grass that cover immense circumtropical areas. While at the present time the green turtle is not abundant, we have good evidence that it once grazed the pastures in numbers incomparably greater than now, and that the depletion was the result of short-sighted exploitation by man. To me the fact that green turtles have survived at all indicates that they are an uncommonly tough breed.[15]

Successful management of this extraordinary resource was limited by lack of information on the biology of green turtles, among other factors. Nevertheless, it seemed to Carr that the problems were surmountable: "When painstaking investigations have furnished a sound basis for practical schemes of protection and control, and international agreements have implemented these, it seems highly probable that the green turtle hordes could be restored to their three-fathom meadows to harvest for us the almost inexhaustible stores of energy held there."[16]

Carr's interest in turtle stories extended to folklore, and he addressed this topic in the *Handbook* under a section titled "The Inscrutable Turtle." Ancient Hindus believed that the earth rested on the backs of four elephants that in turn stood on the back of a gigantic tortoise, while American Indian myth held that there was a great turtle floating in the sea before there was anything else and all animals lived on its shell. Notwithstanding the exalted role of turtles within these two belief systems, Carr found that turtle stories paled in comparison with the snake tales he encountered all over rural areas in the United States. A notable exception to this rule was the richness of stories regarding sea turtles. The shear diversity of stories and languages in which they were told impressed Carr: "I have heard the same gripping yarns in Spanish and in English of every Caribbean shade, and they are told every day in Carib, Mosquito, French, and Danish, and who knows in what other recondite tongues elsewhere."[17] To Carr's considerable regret, convention dictated that he omit the most salacious tales regarding the sexual activities of the sea turtles.

Given the contested territory of turtle classification and taxonomy (recall Carr's extended debate with Leonhard Stejneger), Carr used the nomenclature from *A Check List of North American Amphibians and Reptiles* by Stejneger and Barbour (1943), with a number of changes that Carr deemed trivial. As Carr attempted to rationalize turtle names, he also endeavored to establish a consistent descriptive terminology of reptiles in general and turtles specifically. He criticized other scientists for their indiscriminate and confusing use of technical terms.

As an antidote to such a state of affairs, Carr drew upon Stejneger for a more logical system of descriptive terms, thus honoring his old rival:

> This is a melancholy situation, and there is no real reason for its existence. The late Dr. Leonhard Stejneger was well aware of this awkward weakness in terminology and many years ago adopted in his own writings a more rational scheme, by which the horny and bony parts of the carapace are designated by two wholly distinct sets of names. His system was a consistent adherence to the policy of some early writers to use Greek names for the bony pieces and Latin names for the scales of the carapace. Unfortunately Dr. Stejneger published no full explanation of his revised nomenclature, and consequently it was not extensively adopted by other herpetologists.[18]

Carr proceeded to describe each of the 79 species and subspecies of turtles known to occur within the United States, Canada, and Baja California. Each entry in the *Handbook of Turtles* included range, distinguishing features, description, habitat, habits, breeding, feeding, and economic importance. Where there were few or no data, Carr stressed the need for more research. Beyond describing each turtle and its natural history, Carr identified the value of each species to humans in a section called "Economic Uses." It was in this section that he addressed the need for conservation and further study of sea turtles. In the chapter on the Atlantic green turtle (*Chelonia mydas*), Carr noted the value of the turtle as a dietary staple in the Caribbean and as a delicacy in France and England. After locating the center of the turtle industry in Florida and the Caribbean, he cautioned against the overexploitation of the turtle:

> Although the green turtle is in no immediate danger of extinction, it will support no resurgence of the industry. It seems almost certain that with modern methods of refrigeration and food preservation to enlighten the inland public concerning the gastronomic properties of this succulent reptile, the pathetic remnant of the once-teeming hordes will be pursued with harpoon and stop net.[19]

Carr believed that the only way to reverse the trend was through further study of the ecology of the sea turtles.

Carr called for investigations into the Atlantic green turtle's breeding sites and season, breeding biology, location and behavior of young, migrations, and volume of annual egg collections. Concluding that such studies could save the green turtle, he wrote: "If adequate solutions to these problems could be obtained—and they await only a proper investigation—it seems probable that the green turtle could not only be saved from virtual extermination but might even be encouraged to regain something approaching its primitive range and abundance."[20] Thus, Carr managed to suffuse a classically taxonomic and descriptive text with both biology and a call for conservation and ecological study—that is, natural history. Ironically, many of the original photographs for the book included a ruler (to indicate length), but the production director noticed that the ruler contained Louis Agassiz's oft-quoted exhortation: "Study nature not books." As such advice seemed inconsistent with the publisher's goals, each photo was cropped or airbrushed.[21]

The *Handbook* suggests several trends in natural history. First, Carr's individual life histories of turtles range beyond simple description to analysis of aspects of behavior and ecology. Second, in the *Handbook*, Carr expressed an explicit concern with conservation, which was becoming increasingly important to biologists, naturalists, and ecologists as the twentieth century progressed. The book won the Daniel Giraud Elliott Medal of the National Academy of Sciences for meritorious work in zoology. After the completion of the *Handbook of Turtles*, Carr's emphasis narrowed as he began to unravel the mysteries of the ecology and migrations of sea turtles. By studying the life histories of sea turtles, Carr hoped to establish the scientific basis and method for their conservation.

Preliminary Sea Turtle Research: *The Windward Road*

In his fateful letter to Barbour that led to five years in Honduras, Carr raised the question of funding for a project devoted to sea turtles. This problem continued to occupy Carr's mind while he was in Honduras. When he returned in 1949, Carr resumed his campaign for a grant supporting sea turtle research. In 1952 and 1953, he successfully petitioned the American Philosophical Society for two small grants that would provide the basis for further research. With a $500 grant for each year, Carr obtained data from Yucatán, Honduras, Costa Rica, Panama, Colombia, Venezuela, Tobago, Barbados, Antigua, and Puerto Rico.

Just as Carr had drawn on the stories of locals from his graduate student days on, he consulted fishermen, turtle hunters, and fisheries officers for information regarding sea turtle nesting and migration. In an article summarizing all this information, "The Zoogeography and Migrations of Sea Turtles," Carr established several key facts that would form the basis of support for future grant proposals. First, he identified breeding areas for three species of sea turtles and isolated a 15–20 mile stretch of beach east of Tortuguero in Costa Rica as potentially the most important single breeding site in the Caribbean. The report also established that turtles could be counted from the air in small planes. With reasons for further study delineated, Carr concluded by emphasizing the project's motivations: "The data gathered during the reconnaissance supported by these grants show clearly the American sea turtles present problems of exceptional interest both from the standpoint of pure natural history and as potential subjects for the application of conservation practices."[22] This statement is one of the first indications that Carr hoped to apply his efforts to the continued survival of sea turtles. During the preliminary study, it appears that Carr had realized the urgency of the plight of sea turtles:

> The green turtle seems to the grantee in a dangerous state of depletion in American waters; and yet it would seem to be at the same time most peculiarly amenable to conservation manipulations. It could almost

surely be restored as an abundant source of protein in tropical seaboards where protein is scarce. Under present conditions, however, it seems probable that the green turtle will be extirpated from the Caribbean in twenty years.[23]

Carr's projection was based on extensive surveys and interviews throughout the Caribbean. In many areas, sea turtles had already disappeared, and Carr was reasonably confident that he had found one of the largest remaining breeding grounds. Clearly, sea turtles were endangered. But the reasons to protect them extended beyond the mere scientific or aesthetic; sea turtles provided a significant source of protein for people in a protein-starved part of the world. Just as Carr found the jungles of Honduras teeming with people eking out a living, he found the Caribbean filled with fishermen and turtle egg collectors. It was through these people that Carr discovered the beaches where sea turtles came to nest. Just as the sea turtles were threatened with extinction, Carr found that the cultural resources of the people of the Caribbean was likewise vanishing. It was their story that Carr captured in his second book for a popular audience, *The Windward Road*.

The Windward Road developed out of the grants from the American Philosophical Society. Each chapter evokes a different part of the Caribbean and the people who Carr encountered in his travels. In the opening lines of the book, Carr addressed his compelling title:

Down in the Caribbean the trade wind blows so honestly that in some of the islands you rarely hear the cardinal directions used, but people speak instead of living to leeward or of going to windward to visit an aunt. On all but the smallest or most rugged or least populated islands there are roads that lead to or run along the upwind coasts, and these are known as windward roads. I got to thinking in these terms and liked it, and then it occurred to me that the book that I was writing grew mostly out of the hundreds of miles I had walked along the beaches of the Caribbean, where the good beaches are the windward ones built up high and clean by the driven surf. These beaches were the roads I walked, and it is a good road. If you are in the tropics and have trouble seeing the good in where you are, work your way to windward where the trade comes in to land.[24]

Though Carr reserved the first two and last two chapters for his thoughts on sea turtles, the rest of the book is about Caribbean places and peoples. *The Windward Road* captures the atmospheric essence of the Caribbean—the intense heat and the correspondingly measured pace of life.

Carr had arrived in Puerto Limón eager to charter a plane to fly him along the coast to Tortuguero. From a low-flying plane, it would be possible to count the number of turtle tracks as the females dragged themselves up the beach to dig their nests and lay their eggs. But the attendant at Aerovías Costaricienses (the only place to charter a plane in Puerto Limón) informed Carr that their plane was "discomposed." There was a possibility that Carr could fly to Tortuguero on the flight to deliver *guaro* (locally made sugarcane rum). Carr waited five days for the flight to disembark, and during that time he visited Parque Vargas each day at noon, where he seized the opportunity to study sloths. Nine Gray's three-toed sloths (*Bradypus variegatus*) resided in the twenty-eight *laurel de India* trees (genus: *Ficus*). What riveted Carr's attention and imagination was the sheer slowness of sloths: "For example, a sloth may initiate some simple, straightforward move—like reaching for another handhold, say—and you may find that you must wait many minutes before it is clear whether he is carrying out the act or has stopped to reconsider the whole plan."[25]

As Carr sat, shaded from the midday sun, he contemplated the cultural and evolutionary implications of sloths and slowness. He recalled two other arboreal animals that moved with a ponderous deliberation: lorises (one of the lower primates) and chameleons. Given the taxonomic distance between these animals and the existence of quicker terrestrial relatives in each case, it seemed to Carr that the move to the trees had slowed down the muscles of the animals. The animal and the pace of its movements had become so conflated that in both Spanish and English the animal's name meant "slowness." But Carr was quick to note that in Central American Spanish, the preferred term was *"perico ligero"* or "lively Pete," a vernacular irony that Carr relished. Having watched the sloths for five days, Carr noticed one approaching another, and his mind wandered to the reproductive biology of the sloths: "It could be that I was at last to be allowed to witness what I had hoped for five days to see—the love-making of the lively Pete. Slow as the animal is, and upside down . . . I was intensely, perhaps even morbidly curious, and this approach was the first hint of sex that had crept into the activities of the sloths."[26] At the moment when contact might occur, Carr's reverie was interrupted by the arrival of a boy who carried the message that the *guaro* plane was ready to travel to Tortuguero at long last. The biologist was supremely frustrated: "I was not content. I was cheated. Five day's waiting and hardly a foot between two sloths and their sex rites; and suddenly the wretched plane was composed."[27] Having made the most of a typical delay in travel, so characteristic of the Caribbean, it was difficult for Carr to see the object of his rumination slip away.

The trip to Tortuguero was revelatory in several ways. First, Carr developed an abiding respect for his pilot, a feeling that was sharpened by Carr's own efforts to learn to pilot a small aircraft. After suggesting that they fly a quarter mile out from the shore, Carr asked how low they could fly, in case they saw a turtle. He liked the pilot's answer: "It's better not to splash salt water on the engine."[28]

Carr also made several discoveries about sea turtles. He soon spotted sea turtles, including a pair in the process of mating. Carr noticed another male turtle, quietly floating 20 feet from the pair, and like so many of Carr's observations, this one transported him back to Florida and the turtles that mated in his pond: The pattern of two males competing over one female was common in aquatic turtles also. Carr's lascivious tone, so plain in his study of lively Petes, had disappeared, to be replaced with the authoritative voice of a biologist who had been studying turtles for most of his life. Of course, the subject was the same (i.e., breeding biology), but Carr was scratching for every available piece of information on the turtles, while the sloths provided a diversion from a seemingly interminable delay. Carr concluded: "I was impressed when, on the occasion of my first good view of the *Liebespiel* of sea turtles, these huge greens, courting on the swells off a tropical shore, turned up in the same familiar triangle."[29]

Carr noted that sea turtles mated at the same time that the females lay the eggs. While Carr's biological findings were interesting in their own right, perhaps his most significant discovery involved practical matters, specifically that it was in fact possible to conduct surveys from a light plane: "Besides this I had proved to myself that sea turtles floating in clear water could be easily identified from a light plane flying at safe altitudes, and that nesting trails on the beach could be readily seen. This meant that small planes could be used to make turtle censuses and surveys of breeding grounds as they are used in studying waterfowl migrations. This was a good thing to know."[30] Aerial surveys would become a critical part of assessing populations of sea turtles around the world. In the 1970s and 1980s, Carr directed a program to conduct extensive aerial surveys throughout Florida and the Caribbean see chapter 9.

After one harrowing attempt to land on the beach, in which Carr's pilot narrowly avoided a pack of feral dogs feeding on turtle eggs, and another equally harrowing landing, which involved following an unseen curve through the trees on the beach, Carr's plane arrived at Tortuguero. Without ceremony, his Costa Rican pilot turned him over to his Afro-Caribbean host, George. Ever the taxonomist, Carr was struck by the transition:

> An immense black man of great age stood under the shelter. There seems to me to be a strain of West African Negro that produces a high proportion of almost perfectly preserved septuagenarians—dynamic, thigh-slapping women and vast, silver-bearded patriarchs to go with them and keep them content and do pretty much the same day's work they did at twenty-five. The man under the shelter was one of these. He was a Carib, but it takes more than the blood of salt-water Indians to dim the strong West African blood.[31]

George had met the plane to transport the *guaro* back to the village at Tortuguero. With a casual ease, he hoisted the 15-gallon container on to his shoulder and balanced the hundred pounds with one hand. With the other, he collected a few of Carr's belongings. Carr asked if George (a man in his seventies) was sure he could manage the load himself: "George thought this was funny. He rumbled with humor and then said with a nice touch of sarcasm; 'I think so, sah.' "[32] From the moment that George launched his cayuca or canoe, Carr's mind filled with images of all the tropical rivers he had visited in his life. Interrupting Carr's meditation, George asked about a lizard standing on a nearby tree: "The Sponish people call this Jesucristo. Jesus Christ, they call it. You know why, sah?" Though Carr had seen the animal and the behavior from which it derived its name, he asked why and George responded: "Because this onny-mul walk on the wa-teah. Now you watch him closely." George's efforts to dislodge the basilisk from its branch were stymied until Carr pulled out his slingshot (a decent substitute for a pistol where permits could be difficult to obtain). The first shot (fashioned from wet sand) knocked the lizard into the river from which it escaped by running on its hind feet on the surface of the water! This demonstration satisfied George, who said: "You see, sah? You see why this onny-mul get the name of Jesus Christ?"[33]

Eventually, George left Carr at the dock of the Atlantic Trading Company, where he was greeted by the manager of the logging and banana depot, Don Yoyo Quiroz, who agreed to take him the rest of the way to the village, but not before a detour upriver to see banana plantations. Carr fumed over the noise produced by the large outboard motor that powered Quiroz's cayuca. Arriving at the loading area, Carr's introduction to the cultural diversity of Tortuguero continued when he noticed that the workers were Miskito Indians. Quiroz delivered him to Tortuguero, where he showed Carr where he could sleep. Having been in transit for much of the day, Carr encountered Creoles (as Jamaican immigrants called themselves), Miskito Indians, and Caribs as he sought something to eat. Unfortunately, no one knew where he might find someone to sell him a meal, so Carr continued to walk until a young woman (a "Nicaraguan mestiza") led him to a house where he might purchase a meal. There he met Sibella (a "refined looking mulatto woman"), whose accent was not Jamaican, Miskito, or Creole. As she set to preparing a meal of dolphin and green turtle for Carr, he asked her about her origins. Sibella wanted to know why Carr's speech was so queer, and when he revealed that he was from the United States, she said that she had always wanted to visit.

With the meal underway, Sibella went in search of tortillas, leaving Carr to his notes, the aromas of the dolphin and turtle, and musings about a gigantic sow that lived beneath the house. It was the sow that eventually drew him to the porch, to ascertain its size ("the tallest pig I ever saw"). From the porch, Carr's attention turned to the amorous efforts of a young Miskito couple:

> An adolescent Mosquito couple was courting in the lee of a log, and this reminded me how you almost never see Hondurans even holding hands in public—anyway, none but the very highest castes. Of course, calling Tortuguero beach public is stretching a point, but the difference in the people is there all the same. I went on to wonder if the reticence of the mountain Hondurans had anything to do with the alleged loss of interest of the Mayan in sex, which one anthropologist holds responsible for the race's dying out. From this standpoint the future of the Mosquito race is sound as a dollar.[34]

In general, Carr did not ask why one group of people was more likely to reproduce rapidly, but he was keenly aware that there were cultures in the tropics whose reproductive rate was rapidly outstripping its resource base. Given that natural areas accounted for a significant part of the available resources, overpopulation troubled Carr. By the time he contemplated overpopulation in Central America, Carr had five children of his own, but there is no indication that he was struck by the irony of his own contribution to overpopulation.

Carr went back inside Sibella's house and devoured his supper with abandon. So delicious were the fish and the turtle that initially he could not decide between them, but the turtle eventually won out, and Carr finished all of it. By the time he had finished his meal, night had fallen. As Carr sat and considered the food left on his plate, he was shocked to hear a "hideous shriek" from beneath his feet. Something had upset the sow. Sibella rushed out to determine the cause of the pig's distress. Carr quickly joined her only to find it too dark to see. He asked Sibella what was the matter and was stunned to discover a turtle, trying to dig a nest, had disrupted the sow. Carr had gone to the Caribbean to study sea turtles or at least to conduct the preliminary research to begin studying sea turtles. But the search for turtles and their nests was never simply a "natural" experience completely divorced from its cultural context. For Carr, the natural and the cultural were linked. At Tortuguero, cultural diversity matched natural diversity, and Carr immersed himself in both.

While at Tortuguero, Carr spent the majority of his waking hours walking the beach in search of turtle nests. Like many beaches of volcanic origin in the Caribbean, the beach was black. Turtles nested on the beach at night, but Carr knew that he could follow their prominent tracks to the site of the nest. Unfortunately, the word around Tortuguero was that the van or fleet of sea turtles was late and turtle tracks were few and far between. The object of Carr's quest was the nest of a trunkback (leatherback) sea turtle. By the time he had walked five miles, the temperature was climbing, and it was getting hot on the beach. Driftwood and flotsam blocked Carr's path, and he found himself traveling down to the surf and returning to the area above the high tide line (where turtles nested). So taxing were these exertions that Carr considered abandoning the project for a nap in the shade.

Then he saw it: "a short, broad-limbed V, deeply engraved in the beach above the tide zone." The trail was so wide and the nest so deep that it reminded Carr of the work of a tractor, but there was no question what had disturbed the beach:

> It was the nest of a trunkback. It was the first I had ever seen but there was no mistaking it. It was the first ever recorded for Central America, but its significance to me far transcended that statistic. To me it was the long-sought land-sign of a sea creature I had looked for since child-hood—a monster of the deep ocean guided ashore one time in each year by the primal reptile drive to dig a hole in earth and drop in it the seeds of trunkbacks of tomorrow, and cover the hole with toeless flat feet, and pound back down to the sea never looking behind. It was the work of a water reptile pelagic as a whale or a plesiosaur and at home in the oceans of the world—the last vestige of landcraft left to a bloodline seabound for a hundred million years, and left then but to one sex for one hour on one night in the year.[35]

Even standing at what was clearly the nest of a leatherback turtle, Carr was perplexed as to the actual location of the nest, a situation that he attributed to the deceptive efforts of the female leatherback. Carr determined that the turtle had flung sand across an area 15 feet in diameter: "Since it offered no evidence, at least to my eye, by which the field for search might be narrowed down, I had to cover every square foot of it; and since the clutch of eggs might lie waist-deep beneath the sand, the job ahead was imposing."[36] Once again, culture intervened. First, a dog startled Carr and then Carr startled a woman on horseback. After regaining control of the horse, the woman endeavored to continue on her way, saluting Carr with a terse *adios*. Carr fully appreciated the meaning of this greeting: "*Adios* said that way means you are going on by. In a matrix of circumstance such as this it becomes a bivalent greeting, a salutation with connotation that a parting will follow immediately. It is hello-goodby, and a word that, as far as I know, has no counterpart in English or North American. Spanish can be shaded delicately. It is nowhere near so simple as my textbooks and teachers made out."[37] Having seen turtle eggs in her baskets, Carr stopped the woman with a hearty *buenas tardes*. Carr tried to derive the woman's origins from her appearance, but for once he was stymied, though it was clear that she was not native to Tortuguero. She reminded him of women he had encountered in the mountains of Honduras.

After brief introductions, Carr asked if she knew the type of turtle that had made the nest. Though the woman was in a hurry to collect debts from several Mosquito Indians, Carr convinced her to help him locate the eggs by offering a few colones more than the total of the debts. Mrs. Ybarra (as Carr referred to her) proceeded to tie her horse to a tree (Carr knew the tree to be mangineel, a highly poisonous species, and suggested that the horse might fare better if tied to a different tree, but Mrs. Ybarra was not concerned). After finding a more suitable stick to probe the nest, Mrs. Ybarra drove a dozen more holes into the area of the turtle nest. When this too failed to yield the site of the eggs, the unlikely couple joined forces in an attempt to drive the stick still further into the sand, but it broke, which was just as well since Carr had begun to suspect that the disturbance was a "false nest" dug by the turtle but empty. When he offered to pay Mrs. Ybarra for her trouble, she declined, noting that she had agreed to *find* the eggs. Carr insisted and walked up the beach to place the money in the horse's saddlebags. But the horse had broken the branch that had held its reins and was rolling on top of the two baskets of turtle eggs. The horse's contortions had destroyed both the turtle eggs and a chicken. To make matters worse, the horse was covered in eggs. Carr was overcome with guilt: "A feeling of despondency spread over me. This poor woman—what misery I had brought her! How utterly my stubbornness had wrecked her hopes and her day! I turned to her, in my shame ready to crawl, or to force on her every last colon I could claw out of my pockets."[38] Mrs. Ybarra was laughing uncontrollably at the sheer hilarity of the situation. When Carr looked again at the horse, this time through Mrs. Ybarra's eyes, he too saw the humor in the considerable mess it had created. He generously joined in as his new friend bathed the horse in an attempt to remove the layer of crushed eggs from its coat.

In Trinidad, Carr's search for turtle tracks and nests led him down a beach for more data, when he realized that he did not know the local name of the beach to include with his data. Music wafted across from the coconut grove, so he decided to find the source of the music and ask whomever was around the name of the

Figure 16. Archie Carr on the beach at Tortuguero (ca. 1954). The dogs to the left of Carr are the feral dogs he noted when his plane was landing at Tortuguero. Courtesy of Mimi Carr.

beach. As Carr drew closer, he saw no one, but he could make out the lyrics of the music, a familiar song about love and seduction called "Kitch." The sound of the music enabled Carr to find its source: an old Victrola. Less clear, however, was who was playing the music, until Carr spotted two pairs of feet sticking out from behind a coco palm. Initially, Carr thought he would slip quietly back to the beach. On closer examination, the orientation of the pairs of feet suggested to Carr that the couple could be interrupted, so he stomped around on palm fronds to make his presence known. The couple now disturbed sat up, and Carr saw that they were an African-Caribbean boy and an East Indian girl. He asked them the name of the beach and apologized for the interruption. Nevertheless, the encounter made him uneasy:

> I should have been sorry. It was not light matter, disturbing that pair— not like walking up on Swedes or Sioux or Georgia crackers. Their

preoccupation was no casual dalliance. It was the unwinding of the future. Those two were the stuff of the most fecund strain in all the exploding populations of the Caribbean, where people are spreading faster than anywhere else in the world. The Caribbean Afro-Asian is one to keep your eye on—your descendants may be a lot like him. They could be a lot worse, too; and if they get the African tolerance and humor out of the deal, the world may be saved after all. But meantime the thing that makes me nervous is, there are just getting to be too many people; and it is obviously going to get worse before anybody finds a solution. The minute you give that strong new hybrid stock medical care and a decent diet, you had better stand aside—or be ready with a brand-new and very handy contraceptive.[39]

There is a temptation when reading a statement like this one to label Carr as a racist. It would be simple to explain Carr's reference to "the most fecund strain in the Caribbean," as a reflection of the attitudes and prejudices developed over a lifetime in the South, and we should not ignore the context in which Carr lived most of his life. Nevertheless, as a taxonomist it would be difficult for Carr not to view the races of man in the same light he viewed sea turtles. Taken in this light, a reference to a "new hybrid stock" is entirely consistent with Carr's philosophy of biology. Few things are more basic to an evolutionary biologist's outlook than the notion that species reproduce at different rates. Otherwise, there would be nothing for selection to work on. In the fifty-odd years since Carr wrote his thoughts regarding the "Caribbean Afro-Asian," however, scientists have developed a more nuanced view of human races, most notably that the variations within races are generally greater than between races.

In Bocas del Toro, Panama, Carr once again came nearly face to face with Caribbean sexuality. His lodging for a few days was a jook. (Carr bemoaned the mispronunciation of the word "jook." Having grown up hearing the term, Carr felt the authority to expound on the preferred pronunciation [rhymes with "took"], but he recognized that the original, pure pronunciation was being swamped by a new pronunciation [rhyming with fluke].) Just as Carr had been stuck observing sloths in Limón, Costa Rica, he was similarly constrained in his visit to the place he referred to as "Mouths of the Bull." But rather than occupying his time in idle conjecture regarding the mating habits of sloths, Carr was kept awake nights by the mating habits of other humans. By day, he was trying to determine a mode of transportation that would deliver him to Chiriquí Beach, 40 miles along the coast towards Colón. He had just a week to reach his destination. Since only sailing cayucas made this trip, the midsummer doldrums and the nightly activities had laid Carr low. This was a case where Carr offered a disclaimer: "It may seem to you that I have gone to odd lengths to explain my position. The thing is, I want to get all the facts before you. I was traveling on a grant from the American Philosophical Society, and to a sponsor like that my sitting in a jook with a beer at midmorning would sound like pretty irresponsible behavior. I want to make it clear that I was at the mercy of elemental forces. I was a pawn in the hands of capricious air masses—weak perhaps, but not overtly dissolute."[40]

When the moment arrived to sail for Chiriquí, Carr noticed a cayuca arriving with a load of hawksbill turtles. As his primary interest in the trip was the search for green turtle nesting sites, he asked if they also nested at Chiriquí. Everyone laughed at the suggestion, but Carr hoped for a more sober response. A Mr. Peterson (the "most reliable man" Carr had seen in Bocas) assured Carr that only hawksbills nested at Chiriquí, but he knew of another authority on hawksbills. The captain and crew of the sail cayuca agreed to wait while Carr and Mr. Peterson went in search of an authoritative statement. As it turned out, the new authority was the contractor for the north beaches from Bocas to Sixaola, and when the market was good, he would keep up to fifteen men out fishing hawksbills. Carr asked where green turtles nested, and he commented that green turtles nested all along the Panamanian coast, but only hawksbills nested in the Chiriquí rookery. He also mentioned that the only green turtle nesting site of any significance was Turtle Bogue (Tortuguero). Having returned to the dock to pay the captain, Carr remained there to continue the conversation with Mr. Peterson. As he inspected the day's catch, Carr noticed that except for the one boat of hawksbills, all of the turtles were greens and they were all males. This observation reminded Carr of his observations on the flight to Tortuguero: "I recalled how the *Liebesspiel* of the greens I saw from Paco's plane usually involved extra males; and an old saying of the turtle men came back to me too—one that says that in cooting time you strike the female and you get two turtles, maybe three."[41] Carr noted that rutting male green turtles could be threatening, particularly if one was unlucky enough to capture the object of its affection (a female): "Foiled in such an attempt, the loving turtle, aslosh with hormones, crackling with short circuits in his vagus control, may thrash about your boat in a wild, trial-and-error quest for an embraceable substitute. He flaps, scrapes, and bites at the planking, he chews your paddle blade or hugs your oars till they snap off. Throw out a board or a buoy and he assaults in frantically. Fall over yourself and you're out of luck."[42] Carr had heard stories that the much larger leatherback sea turtles could be just as aggressive and far more imposing.

In keeping with his grant from the American Philosophical Society on the migrations of sea turtles, Carr "migrated" as well, and everywhere he went he encountered cultures of the Caribbean. What other scientists may have seen as a distraction, Carr embraced. He acknowledged the Caribbean people and their stories as an essential part of his study. At Tortuguero, Bocas del Toro, and Trinidad, Carr found compelling human stories that were closely linked to the stories of turtles. Often, he located the established authority on sea turtles by word of mouth. Carr sought out one group from which he knew he would obtain anecdotal evidence of the migration of sea turtles: the turtle captains of the Cayman Islands. It seemed to Carr that every story of homing ability in sea turtles traced back to the Cayman turtle captains:

> Every time I dug into the homing rumors I found the Cayman turtle men back of them. All the people who told me the stories turned out to be quoting, or misquoting, the stanch professionals of the Cayman

Islands. The reason for this was that Caymanians not only catch more turtles than everybody else put together, but they haul them all over creation, all the way from Nicaragua to Florida, for instance; and besides this, they brand every particular ship or owner.[43]

In 1952, it was no simple matter to travel to the Cayman Islands; flights were intermittent and unreliable, and boats were small and time-consuming. Carr opted to fly from Jamaica. Carr trusted the views of the turtle captains because they came from generations of turtle fishermen, and their livelihood depended on knowing where to find sea turtles: "These men are specialists in an exacting fishery, and they learn things no zoologist knows because it is the only way they can succeed in their calling. To a man the Cayman captains believe that green turtles make long-distance migrations at breeding time."[44]

Caribbean culture and the natural history of sea turtles intersected in the turtle captains. Carr learned that the Caymanian captains caught green turtles at Mosquito Cays, located off the Mosquito Coast in Nicaragua, roughly 350 miles from the Cayman Islands. The turtle captains informed Carr that by night sea turtles fed on turtle grass (*Thalassia*) and slept on or under submerged rocks. Turtles might travel four or five miles from feeding to resting areas, and it was these short-distance trips that convince the turtle captains that turtles could make longer ones: "This is not a thing you can read about in the zoology books. Seeing the turtles thus casually commuting predisposes the turtle captains to believe that they are capable of purposeful, controlled migration journeys."[45] Experience taught the captains that they could not fish for sea turtles year round. From late May to August, the sea turtles disappeared from the Mosquito Cays. The captains interpreted the absence of turtles as a sign that they were away on a migration to breed. Again, Carr noted that textbooks would not include this information: "Here again you cannot corroborate the assumption in the literature of science; but the operations of the ancient and successful Caymanian turtle fleet are geared to it, and the captains don't worry much over the lack of support their ideas get from professional naturalists."[46]

The turtle captains claimed that the turtles migrated from the Mosquito Cays to Tortuguero, which they called Turtle Bogue, to nest. Corroboration for this view came from the Costa Ricans, who believed that a significant number of green turtles arrived at Tortuguero from the Mosquito Cays. Carr wrote: "The Caymanians see their turtles leave the banks and at the same time the Costa Ricans see hosts of greens gathering off their shore and coming up on the Black Beach to lay their eggs."[47] Together, the Caymanians and Costa Ricans realized that they saw turtles at different nodes in the same cycle: "The Cayman captains and the Costa Ricans have figured it out between them that the June exodus at Mosquito Cay and the massing at the Bogue are parts of the same phenomenon."[48] Such a conclusion was perfectly reasonable as far as Carr was concerned, but science required a higher standard of proof:

> Their case seems so clear that mentioning the lack of experimental proof sounds like quibbling. Nevertheless, proof is lacking, and getting it is one

of the things I am determined to do. I want to go down to the Bogue again and meet the fleet and mark hundreds of nesting greens with monel metal tags, stamped, in Spanish and English, with my address and with the offer of a reward to whoever returns a tag to me with full details of its recovery. I am as sure as you can be in such cases that tags will be picked up at Mosquito Cay.[49]

Carr (and the turtle captains) suspected that tags would be recovered from places spanning the Caribbean.

Through aerial surveys of the beaches at and near Tortuguero, Carr determined the extent of nesting from Limón to Tortuguero. During a second trip to the Black Beach, nesting was sparse, so he chartered a plane back to Limón at an altitude of 100 feet. Over the span of the 24-mile section of the Bogue, only twenty-five turtles had come up during the three preceding nights. The next 8 miles to the south showed very few tracks at all. From that point on, however, there were numerous tracks. On one 6-mile stretch of beach, there were too many tracks to count, but Carr guessed that there were hundreds if not thousands of tracks layered upon one another. Most if not all of these showed no sign of actual nesting: the turtle had crawled up above the high tide line and then returned to the water without excavating a nest, let alone laying eggs (later Carr called such tracks "half moons"). Carr interpreted these findings as follows:

> Because I had talked to turtle captains and walked the Bogue and waited with the people there for the *flota del sur*—the southern fleet—to arrive, I interpreted what I saw as sign of a vast shoal of greens just in from the south and on its way to Tortuguero. The trial strandings seemed to mean that the school was prospecting—was somehow aware that the voyage was near an end, but could be sure only by testing the sand for whatever mystic properties they are that call the fleets to Turtle Bogue.[50]

Again, a part of Carr wanted to accept as fact what so many Costa Ricans believed. But the scientist demanded proof:

> So my own snatched observations fit in perfectly with the folk-beliefs, and I was tempted to call the matter established fact. But a little thought showed that even then nothing had been really proved. It was a good thing to have seen, what I saw from the airplane, and it certainly added weight to the deductions of the people on the Bogue. But attributing to a reptile an ability to carry out long-distance, open-water migrations is a serious thing. It is serious because you are attributing to him powers of orientation, which are not prevalent among his kind and about which we understand next to nothing where they do occur. It is true that a lot of different kinds of animals have such powers, but adding a turtle to that gifted list is not a thing to do lightly or without impeccable grounds.[51]

So Carr had a specific goal in mind when he went to the Cayman Islands to interview the turtle captains. He sought stories of long-distance migration in sea turtles. Everywhere he had traveled in the Caribbean, he encountered stories about turtles, and most of these he was able to trace back to Caymanian turtle captains. If anyone could tell him stories about homing in sea turtles, the turtle captains could. In fact, each of the captains he interviewed offered stories of long-distance homing behavior. Carr saved the most compelling story of turtle homing for last. It was told by Captain Allie Ebanks, one of the youngest experienced captains. An expert at judging and grading green turtles, Allie was asked to select the five best turtles from a large group on a catboat. Once selected, a man carved his initials in their carapaces and loaded them on to a schooner, *The Wilson*, heading for Grand Cayman. Unbeknownst to Allie, *The Wilson* made good time and deposited the turtles in a small rock crawl. A major storm appeared, and the water rose high enough in the crawl for the turtles to escape. Twelve days later, a catboat captain caught one of the marked turtles near the Mosquito Cays. The turtle had traveled 350 miles in twelve days. Carr was amazed that the turtle must have averaged nearly 30 miles per day, even if it had taken the most direct course possible. If the turtle had wandered from its course slightly, its speed became even more incredible. For Carr, the conclusion was obvious: "So, in spite of the lack of final experimental proof, there can remain little serious doubt that the fleet that comes to Tortuguero in June is made up of migrants from Mosquito Bank, and perhaps from many other points in the Caribbean as well."[52]

As Carr searched for turtle nests at Tortuguero, Bocas del Toro, Trinidad, the Cayman Islands, and elsewhere, he found people scratching out an existence. Carr developed a secondary passion for the stories of Caribbean peoples. After traveling throughout the Caribbean in search of sea turtles, Carr realized that the populations of most species were declining rapidly. To make this point, he surveyed the history of Caribbean exploration and revealed that sea turtles and particularly the green turtle had played a profoundly important role in the exploration and colonization of the region. In 1503, Columbus had witnessed vast numbers of sea turtles between three fairly small islands, so he christened them *Las Tortugas*, later renamed The Cayman Islands. By way of contrast, Carr and his friend Coleman Goin walked the beach at Grand Cayman at the height of the season until they were tired and found just one turtle track (and Carr doubted that it was a green). Carr contrasted the demise of the green turtle to that of the American bison:

> Perhaps the story behind this change is not as dramatic as the story of the bison on our western plains. The bison was in the public eye from the start. It cluttered land now Illinois real estate. It gave comfort to difficult red Indians and blocked the scant traffic on proud new railroads. The bison passed in a blaze, watched by everybody—not without lamentation here and there, but with little interference. It had to go in the mind of the day, because it hindered progress. The green turtle on the other hand, hindered nothing. The turtle fleets passed secretly and without commotion. They were just too good to last.[53]

To the trade winds that aided the exploration of the Caribbean, Carr added the green turtle. No other animal provided such a rich and lasting source of protein for the return trip to Europe. Carr wrote: "It was only the green turtle that could take the place of spoiled kegs of beef and send a ship on for a second year of wandering or marauding. All early activity in the New World tropics— exploration, colonization, buccaneering, and even the maneuverings of naval squadrons—was in some way or degree dependent on turtle."[54] And throughout the Caribbean, the turtle colonies had disappeared, save for a few that clung to existence in the remotest areas. Carr traced the decline of green turtles in the Caribbean across time and space:

> One by one the famous rookeries were destroyed. The first to go was Bermuda and next the shores of the Greater Antilles. The Bahamas were blanked out not long after, and boats from there began to cross the Gulf Stream to abet the decimation in Florida, where the crawl was once more common than the hen coop—where Charles Peake caught 2,500 greens about Sebastian in 1886 and in 1895 could take only 60; where vast herds foraged in the east-coast estuaries and on the Gulf flats of the upper peninsula and a great breeding school came each year to Dry Tortugas.[55]

As the Florida colonies waned, turtlers sailed to the Cayman Islands and then on to Cuba. As these colonies succumbed to the slaughter, the Cayman turtle captains set their sights on more distant colonies, and the Mosquito Cays became the preferred hunting grounds.

The prospects for long-term survival of the green turtle worried Carr. The ongoing slaughter of adults and collection of eggs was but one among several concerns. Loss of habitat suitable for nesting and the rapid proliferation of the human population seemed to present long-term problems as well:

> The Caribbean people are among the most prolific strains on earth, and they are breeding fast. There are outboard motors on the dugouts, and little airplanes will set you down nearly anywhere. Where twenty years ago most Caribbean shore was wilderness or lonesome cocal, aluminum roofing now shines in new clearings in the seaside scrub. The people are breeding too fast for the turtles. The drain on nesting grounds is increasing by jumps. It is this drain that is hard to control, and it is this that will finish *Chelonia*.[56]

Thinking about the impact of humans on turtles, Carr focused on Tortuguero. He listed the remarkable diversity of predators on the beach: dogs, buzzards, and fish. Nevertheless, Tortuguero was the best nesting beach in the western Caribbean: "But the fact remains that, in spite of the dogs and the natural predators, the Bogue is the best nesting beach in its half of the Caribbean, and this suggests its potential role in a plan to save the green turtle. The Cayman captains believe their industry is wholly dependent on this breeding ground. They are certainly at least partly right; and . . . the importance of the rookery may be far greater than that."[57] If Tortuguero was the best beach for the nesting of sea turtles, and Carr's extensive research throughout the Caribbean certainly substantiated that claim, then it played a vital role in the continued survival of green turtles. Carr asked, "How then, it is natural to ask, is *Chelonia* faring there; and if the Bogue is in the future, what does that faring hold?" The answer was bleak: "As near as I can make out, it holds extinction."[58] This impending extinction was due to a significant degree to a remarkably efficient commercial system with the capacity to capture and remove virtually every female turtle from the beach before she had the chance to lay her eggs.

Nevertheless, Carr concluded on an optimistic note:

> Viewed from the standpoint of the opportunities for intervention, the situation looks different. The very migratory habit that makes that green turtle vulnerable could be the basis for its rescue, and even conceivable for bringing it back to the abundance of former times. We are not at deadlock with the green turtle as we were with the buffalo. We are killing it out idly, aimlessly, with no conviction of any sort, with most of us not even aware that it is going. We have no need for its habitat for any of our own schemes. Territorially, its interests and ours overlap only on the sea beach, and even there *Chelonia* comes when we are asleep. And because the creature congregates each year to mate and lay its eggs, real protection for a few beaches not only would help save for the future a species now threatened with extinction, but might even bring back the fleets Columbus found.[59]

There is no indication that Carr suspected that this call for the conservation of sea turtles would start a movement with exactly that aim.

Grants: National Science Foundation and Office of Naval Research

After completing his initial surveys with the American Philosophical Society grants, Carr applied to the National Science Foundation (NSF) and the Office of Naval Research (ONR) for more significant funds to further examine the natural history and ecology of sea turtles. He wisely sent a preliminary proposal to George Sprugel, Jr., who was the acting program director of environmental biology at NSF. Sprugel responded as to the likelihood of the proposal receiving support from NSF: "You know, of course, that the Foundation supports only basic research. The preliminary proposal, as it now stands, would face two probable hurdles while being evaluated by our advisory committee. First, the panel would likely say that much of the research outlined is of an applied nature and, second, it would be almost sure to rule that the proposed research is the type which might be more suitable for support by the Fish and Wildlife Service."[60] What Carr had identified as a significant motivation behind his study (the nutritional value of sea turtles to the people of the Caribbean) shifted the emphasis of the study from basic to applied research and thus out of the realm of NSF funding.[61] It is clear from the final proposal that Carr heeded Sprugel's advice; there was no mention of the nutritional value or the need for conservation.

Late in 1954, Carr submitted his NSF and ONR proposals under the title, "A study of the ecology, migrations, and population levels of sea turtles in the Atlantic and Caribbean with special reference to the Atlantic green turtle, *Chelonia mydas mydas* [Linné]." The first sentence of the proposal revealed Carr's intent to focus on basic research: "Within recent years it has become evident that sea turtles, and especially the Atlantic green turtle, present some fundamental and arresting biological problems."[62] After describing the dependence of the green turtle on marine seed plants, Carr posed a lengthy series of unsolved problems, most of which focused on the little-known natural history and biology of sea turtles. The description of the research concluded with an explanation of the choice of the Atlantic green turtle as subject: "The green turtle will be used as the experimental and observational subject because its herd grazing and mass nesting make it more easily available for tagging and other manipulation and study."[63]

What was the nature of Carr's experiment? There were three parts to the procedure of the NSF grant: a netting and tagging program along the Gulf Coast in Florida, a tagging program at Tortuguero in Costa Rica, and trips to suspected nesting areas. In Florida and Costa Rica, Carr's assistants would place metal tags offering a reward for the return of any tag to the University of Florida. Each returned tag would support the anecdotal evidence of Caribbean fisherman regarding the extensive postbreeding migrations of sea turtles. Carr's proposal was elegant in its simplicity. Moreover, Carr proposed a low budget.

In the initial proposal, Carr suggested a budget of just under $24,000 for three years of research. More than $13,000 would go to the first year and set-up. Remarkably, that total included all travel, summer salaries for Carr and several assistants, and expenses. Nonetheless, after several months of consideration, NSF indicated that it would fund the project on a reduced budget of $18,000. Carr revised his budget estimates mainly by eliminating much of the most expensive travel. Official approval came through on June 14, 1955. In the meantime, Carr withdrew his proposal to the ONR from consideration, but he would later reapply to the ONR for additional support (specifically for the extensive travel dropped from the NSF grant). With NSF approval, Carr could continue the research he began in Honduras and had continued with American Philosophical Society grants. That research had formed the basis of the NSF proposal. In addition, Carr wrote *The Windward Road*, his second popular book, based on his initial study. This book quickly attracted popular support for sea turtle conservation.

The public was a crucial component of the conservation movement in America. The various state-level Audubon societies founded during the 1880s were among the first organizations to promote popular environmental activism. The first Audubon Society was founded to stop the slaughter of herons and egrets for their plumes. Bird feathers (and sometimes the birds themselves) were popular with high-society ladies for use on the most stylish hats of the day.[64] While members of Audubon societies recognized the need for federal, state, and local legislation, they initially focused efforts on informing the public with pamphlets. Through their efforts, the fashionable ladies' hat became the symbol of the destruction of wild birds. Besides the pamphlets, Audubon societies petitioned millineries where the hats were made and lobbied for legislation protecting birds. In addition, they hired wardens to enforce existing legislation and identified and bought areas that were in the greatest need of protection. The success of the Audubon societies in the protection of birds indicated a well of popular support that scientists and conservationists could access on the behalf of ecological causes.[65] Such popular support coalesced around sea turtles in the development of the Caribbean Conservation Corporation.

The Caribbean Conservation Corporation

On March 11, 1958, Carr received what he might have dismissed as yet another letter from a well-meaning fan of his books. The letter was from S. G. Fletcher, who happened to be the managing director of The Gleaner Co., Ltd, (publisher of The Daily Gleaner, a Jamaican newspaper). Fletcher wrote:

> I have read your delightful book THE WINDWARD ROAD with a great deal of interest. A copy was sent to me by Mr. Joshua B. Powers of New York who seems to be equally interested in turtles.
>
> If I understand you correctly, it would appear to a layman's mind less tuned to the mysteries of turtles that the practical thing to do is to protect the beach of Turtle Bogue and that of Chiriqui and for this purpose someone should enlist a corporation of the governments of Costa Rica and Panama. If the turtles from all over the Caribbean go there to breed it seems obvious that many of your points are met if these two beaches can be protected. If not all of them can be protected then maybe it would help if a good slice were. To protect from the dogs would be merely a matter of a good fence. To protect from man is another matter but it is obvious that the two governments could do it if they were really interested.
>
> How does one go about starting a movement?[66]

Carr did not dismiss Fletcher's letter. In fact, he sent a reply March 18 thanking Fletcher for the comments, and he agreed that protecting Turtle Bogue would be the most important step short of protecting all nesting beaches. Carr also mentioned that his NSF-sponsored tagging program had recovered a tag from Morant Cay, not far from Jamaica. In closing, Carr welcomed Fletcher's support: "The support of people in your position will be indispensable when the time comes for the conservation program to be developed, because it will not be easy to put over."[67] Unbeknownst to Carr, the conservation program was already underway.

Joshua B. Powers (1892–1989) was a successful publishers representative.[68] Although Powers represented newspapers throughout the world, the Caribbean and Latin America had been long-term interests for him. He had been considering how he could help the people of those regions since World War II. On reading *The Windward Road*, Powers found his cause, and he decided to alert his friends to the book and the plight of the sea turtles.[69] On January 31, 1958, Powers sent the following letter and Carr's book to a list of nearly two dozen friends, most of whom worked in the publishing industry:

Dear _____:

This is to advise you that you have been elected to membership in the Brotherhood of the Turtle.

As a token of membership, you are endowed with one copy of "The Windward Road" by Archie Carr, who is Grand Admiral of the Fleet.

The purpose of the Brotherhood is to make sure that Winston Churchill continues to have his cup of Green Turtle soup every night before retiring.

Members of the Brotherhood pledge themselves to save the Green Turtle from passing the way of the wood pigeon, and to cooperate with the friendly people of the Caribbean lands in keeping the good things they already have and helping them to find more.

Sincerely Yours,
J.B. Powers
Chief Patrolman of the Beaches[70]

Like early Audubon Society members, Powers believed that he and his colleagues and friends could organize an effective conservation movement. Though humorous, Powers's letter (and the inclusion of Carr's book) suggests a sincerity of purpose. Moreover, Powers shared Carr's sympathy with the people of the Caribbean. On March 21, 1958, he wrote to Carr.

Dear Grand Admiral of the Fleet:

You may be surprised to know that you are the spiritual head for the Brotherhood of the Turtle.

This, of course, all came about through your two books, but especially through your "Windward Road". Last December I visited the University of Florida and discovered your earlier book, upon which Bill Pepper told me about "The Windward Road" and insisted that I should get a copy before I left town.

The book is a delight, and you deserve, shall I say, "adhesions".

In any event, I sent out letters accompanied by copies of the book to the following…[71]

Powers then listed the names and addresses of recipients of the letter.[72] The list indicated that Powers had contacts at many of the major Central American publishing organizations and newspapers, as well as at major newspapers and magazines in the United States. Powers provided the full text of the original letter for Carr's benefit. This brought him to the reason for writing to Carr:

It seems to me that the time has now come to draw the Grand Admiral into this picture for the purpose of saving the Green Turtle. I have had answers more or less enthusiastic from most of these brothers, including the following letter from the Military Aide to the Governor of Puerto Rico:

Before leaving on his trip to Washington, the Governor asked me to tell you how very much he appreciates your cordial letter of January 31st, and your courtesy in sending him the book entitled "The Windward Road". The Governor has placed this book on his priority reading list and he sent you his very best regards.

I suggested to the newspaper supplement HABLEMOS, which circulates as a supplement with one newspaper in each of the countries of the Caribbean area, that an article be written about you and your book.

The next step ought logically to be for us to get the Chief of State of each one of these countries behind us and get some legal action initiated for the purpose of patrolling the beaches. Or do you have a better idea? In any event, we haven't got long, or else Winston Churchill won't any longer be with us to have his cup of soup and to know that other men and true can have a cup also when they retire, a bit overladen with brandy.[73]

As in his other letters, Powers struck the delicate balance between light-hearted fun and ponderous sincerity. Listing the editors of major North and Central American newspapers and magazines and quoting from the military aide to the Puerto Rican governor served to legitimize Powers, who needed to assure Carr that despite the levity of the letter, he was not a crank. While the references to Churchill's turtle soup seem eccentric, they suggest that Powers appreciated one of the values of turtles. By casually suggesting that they contact "chiefs of states" (and making pointed reference to one), Powers indicated that he could gain access to government officials in Latin America.

In his response, Carr first joined in the fun and then laid out his plans:

Dear Chief Patrolman of the Beaches:

It would be hard to tell you how pleased I was to learn that the green turtle had suddenly acquired such unthought-of sympathy and support. In one flutter of letters you have done more to bring the case to the attention of people who could help insure turtle soup for the future than I could have done in years.

I have held up any concrete proposals for protection of sea turtles in the Caribbean, partly for lack of knowledge as to how the negotiations should best be started, and partly for lack of the airtight scientific proof that the green turtle makes mass journeys to distant nesting grounds and thus lends itself to effective protection in preserves of limited extent. What I'd really like to see is pan-Caribbean protection for the nesting turtles, wherever they lay; but the complexity of the international maneuvering necessary to bring this about makes it a plan not to be tackled without heavy reserves of influence.

If I can prove that making a few places like Tortuguero preserves will materially improve chances for the creatures' survival, this would be a more practicable move—possible to sell to the governments affected, and not likely to abort once it got under way.

Shortly after *The Windward Road* was published I started a program of research on the problem of sea turtle migrations, under sponsorship of the National Science Foundation. We tagged around a thousand green turtles down at Tortuguero and have had international returns from both northward and southward of Costa Rica, and one from clear across the western Caribbean at Morant Cay near Jamaica. While I can't say we have final, positive proof that the fishermen's stories about the movements of turtles are right, everything seems to point that way and I think our work this summer is just about going to tie it up.

That is one reason why I haven't stirred up an uproar over the need for an international conservation program. We need at least this one more summer's data to build up the case. The other reason is that I haven't had the vaguest notion how to swing the scheme politically. It is wonderful to know that your help can be counted on in getting a decent hearing.

I had a letter three days ago from a representative of ICA in San Jose, wanting to know what I thought about the possibility of promoting the exploitation of green turtles as a way of aiding Costa Rica economically. If this man, or anybody else, succeeds in developing even a fraction of the potential market for green turtles (that is, gets them out of the luxury vittles class) before a management program has been worked out *Chelonia* surely will last only a short while in American waters.

So I am very grateful indeed for your willingness to accept the post of Chief Patrolman of the Beaches, for the interest you have shown in the poor old green turtles, and for the very generous things you have said about *The Windward Road*.[74]

With the introductions made, Powers set to work by arranging with Latin American newspapers like *Hablemos* (a newspaper supplement that circulated along with 17 newspapers to 360,000 readers) and a wide-reaching radio station in Latin America to carry reports about Carr's research and cause. On January 7, 1959, Powers met with Carr in Gainesville, and the two men discussed the new organization and the possibility of financial support for a sea turtle restoration project. Carr followed up the meeting with a letter in which he wrote: "I was very glad to hear that you believe it may be possible to find financial support for a restoration project. I enclose a tentative project plan and budget for the first two years of such a program. Naturally nobody can guarantee results in any conservation project, but I see no reason why five years of the kind of manipulations proposed in this plan shouldn't bring worthwhile advance in the nutrition level of the Caribbean seaboard."[75] Carr went on to discuss the budget and how they could use graduate students in biology as a source of inexpensive yet proficient labor. Carr next directed Powers's attention to a section of the proposal that he had deliberately left vague: "You will notice that in the project plan I go into no detail as to how the protection of nesting beaches in various areas named can be achieved. This is where the membership of the Brotherhood will come in, and I suppose they'll have to play it by ear, since their problem will range all the way from asking that active commercial turtle turning be prohibited to simply asking that existing protective laws be enforced."[76] After years of watching the plans of U.S. coordinators fail in Honduras, Carr was convinced that conservation programs required local participation and support. Members of the Brotherhood could best determine the most appropriate course of action in their home countries, thereby maintaining their sovereignty.

In the hopes of avoiding any suggestion of foreign imperialist aims (even of the most benevolent kind), Carr suggested a separate name for the restoration plan:

> What do you think of the possibility of our giving the actual management project some label other than "Brotherhood of the Green Turtle," which may have an up-beat ring pleasing to the influential people necessary to get anywhere with the plan, but which may fall oddly on the ear of others whose blessings we will need. We'll have to identify ourselves in dickering with boards of conservation and ministers of agriculture, and maybe in putting up a sign here and there. Mightn't it be good to keep "Brotherhood" for the organization itself and think up another name for the organization's program? Something dull and reassuring like, "Florida-Caribbean Restocking Project," or just "Caribbean Restocking Project," or maybe "Green Turtle Restoration Plan." What do you think?[77]

It seems that Powers accepted Carr's suggestion as he distinguished between the Brotherhood and the restoration program in a letter inviting John H. "Ben" Phipps of WCTV Television Station in Tallahassee, Florida, to join the Brotherhood on April 14. Powers wrote:

> Plans are well advanced for the incorporation of what may be called: "CARIBBEAN GREEN TURTLE REPRODUCTION PROJECT INC." Archie Carr will be the unpaid technical director and the officers and members of the board will include those who now form the Brotherhood of the Green Turtle. We have members in the United States, Bermuda, Cuba, Jamaica, Trinidad, Costa Rica and Mexico, and we expect to add others, paying particular attention to those areas where there are beaches that formerly were frequented by green turtles.[78]

In one brief paragraph, Powers conveyed that the Brotherhood had a concrete plan, expert guidance, and broad representation in the countries of the Caribbean. Powers elaborated on the details:

> We have a concession of four miles at Tortuguero in the Caribbean, where we expect this summer to establish a turtle hatchery. We estimate that we may be able to give 10,000 baby turtles a chance to get into the sea this season. Some of these will be shipped like day-old chicks to different places in the Caribbean, some will be shipped to the University of Florida where they will be raised until they are a year old in the Marine Laboratory, and

some will be released directly into the waters off the beach where they are born. In other words, they will be saved from destruction by the various predators that hold the turtle population down.[79]

No doubt moved by Powers's invitation, Phipps joined the Brotherhood, became its president, and was one of the strongest supporters of the as-yet unnamed restoration project. Phipps would soon emerge as the organization's most significant benefactor. For years, his contributions accounted for the lion's share of the operating budget. Later on, Phipps's family assumed his role. Clay Frick also provided key support.

On May 4, Powers sent a brief memorandum under the title "What happened so far" to all the members of the Brotherhood (on Brotherhood stationery). In addition to the concession of a 4-mile stretch of beach for the turtle hatchery at Tortuguero, Powers noted that the Costa Rican Minister of Agriculture had prohibited taking turtle eggs for sale and limited the catch of adult turtles for export, and the equivalent official in Nicaragua had prohibited turtle fishing along parts of the coast. Incorporation of a nonprofit organization, the Caribbean Conservation Association, under the leadership of Carr, would be in Florida, and Powers solicited donations towards the $5,000 budget that would cover the construction of a permanent camp at Tortuguero, expenses of two graduate students and a helper at Tortuguero, and contingencies.

The Brotherhood of the Green Turtle had its first official meeting at the Metropolitan Club in New York City on May 19. Though he was not able to attend, it became clear to Powers that Phipps was the best choice for president. By June 23, Powers could report to the Brotherhood that Larry Ogren and Harry Hirth had arrived at Tortuguero to tag adult turtles and rear hatchlings for distribution around the Caribbean. Moreover, a turtle tagged at Tortuguero was captured between Isle of Pines and Cuba, which meant that the turtle had traveled roughly a thousand miles. Since the Brotherhood had officially decided to create the Caribbean Conservation Association, they had raised $5,150 in tax-deductible contributions to cover the expenses of the first season.

By the time of the first official meeting, the Brotherhood had agreed to call the public side of the group the "Caribbean Conservation Association" and to organize it as a nonprofit, tax-exempt Florida corporation. Ultimately, the name "Caribbean Conservation Corporation" became the official title of the organization. No doubt the initials CCC, reminiscent of the Civilian Conservation Corps of Roosevelt's New Deal, suggested goodwill and constructive efforts to Americans. Phipps formally accepted the presidency.

The CCC provided Carr with three things that were vital to his conservation efforts. First, it established a small fund that supported expenses that official granting agencies such as the National Science Foundation could not cover. Second, it validated the efforts of Carr's long-time friend Guillermo "Billy" Cruz, who became the organization's Costa Rican representative. One of the first people to describe Tortuguero to Carr, Cruz was particularly active in its protection. Third, it brought publicity to the conservation of sea turtles in the United States and throughout Central America, which would attract sympathy and support. Literally dozens of articles and news briefs about Carr's efforts with turtles began to appear in newspapers throughout the Americas. The CCC and Powers raised popular concern for the sea turtles across a broad spectrum. All of these efforts enabled Carr to focus on developing life histories from scientific studies while guiding conservation efforts in his capacity as technical advisor to the CCC.

Conclusion

The seven years since the Carrs returned from Honduras had proved to be highly productive and successful. Carr published *High Jungles and Low* and the *Handbook of Turtles*. He also inaugurated his study of the ecology and migration of sea turtles with funding from the American Philosophical Society, the National Science Foundation, and the Office of Naval Research. Carr's experiences during preliminary research trips provided fodder for another popular travel narrative, *The Windward Road*. Fortuitously, *The Windward Road* came to the attention of Joshua Powers, who rallied the support of his friends to form the Brotherhood of the Green Turtle and the Caribbean Conservation Corporation, an organization dedicated to the conservation of sea turtles. Thus, even as Carr's study of sea turtles was gaining momentum, he secured additional and unanticipated support for related conservation efforts.

In 1959, as the CCC relieved Carr from the politics of organizing a conservation movement, the University of Florida reduced his teaching obligations by promoting him to graduate research professor (by choice, Carr continued to teach his favorite courses such as community ecology). He was the first professor at UF to achieve this rank. His activity in other areas continued or accelerated. In scientific journals, he analyzed various aspects of the ecology and migrations of all sea turtle species. Later, he wrote another book, *So Excellent a Fishe* (1967), in which he interpreted findings and reported the story of the science behind the migrations of sea turtles for the public. Most important, Carr noted the implications of his findings for continued efforts in sea turtle conservation.

CHAPTER 6

The Ecology and Migrations of Sea Turtles

The thing is—as you may know, but I had not discovered then—adventure is just a state of mind, and a very pleasant one, and no harm to anybody, and a great asset if you use it right."[1] Anyone who ever encountered Carr in the field, whether in Florida, Costa Rica, or Africa, knows that he believed his words with every fiber of his being. Larry Ogren, one of Carr's first research assistants and a long-time friend, recalled traveling to Tortuguero for the first time with Carr in 1956. When they reached San Jose, Carr eschewed the two hotels as "too fancified," and chose a small inn where they could leave their "travel clothes." They took the train to Limón to see the rainforest as it passed by. While waiting to charter a plane, they spent an afternoon watching sloths in the park (and Carr shared his desire to observe mating sloths *in flagrante delicto*). Finally the time arrived to board a small Cessna to Tortuguero. Shortly after arrival, Carr left his assistant with these parting words: "Now you know it's hot as hell down there; you get sand in your ass walking the beach; you get bored to death. If you do, please get on the 'Miss Bessy'—that's a converted WWII landing craft that comes up and down the coast every so often—get on the Bessy, go into town, and get a beer. Take a break, because it's going to be a long, wet summer. See ya!"[2] And with that valedictory, Carr told the pilot to take off (he had to get back to San Jose to catch a flight to Ecuador) and left Ogren to his own adventure.

Ogren's experience as Carr's research assistant mirrored that of many of Carr's students. Invariably, they found their time with Carr to be exhilarating; the professor's exuberance for all things natural inspired and thrilled them. Doing science with Carr was fun. After the publication of the *Handbook of Turtles* in 1952, Carr devoted his research almost exclusively to sea turtles, and the nature of the work required him to employ students. Yet, the demands on Carr's time were such that

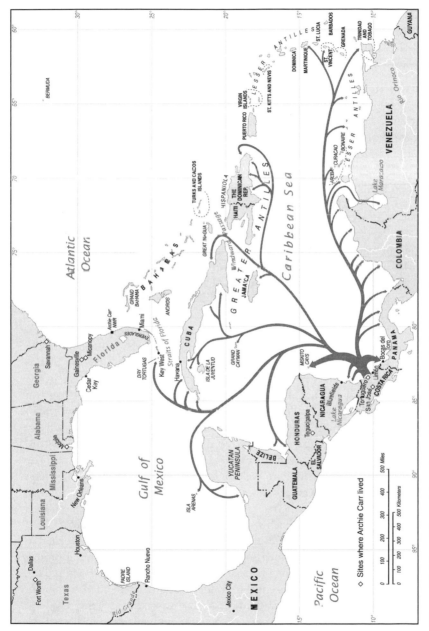

Figure 17. Map of Central America and the Caribbean, including southern United States. Diamonds indicate sites where Archie Carr lived. Arrows indicate green turtle distribution according to tag return. Adapted from map in *The Sea Turtle. So Excellent a Fishe*, (University of Texas Press, 1984).

he could only afford brief periods of direct involvement. Carr demonstrated great faith in his assistants by trusting them to complete their work independently and well. At the same time, he typically credited his assistants as second author in his scientific papers. As Carr delved deeper into the ecology of sea turtles and more students joined him, he was able to explore a broad range of questions that touched on behavior, reproduction, migration, taxonomy, and biogeography. Naturally, he initiated his research locally with the study of sea turtles in Florida.

Sea Turtles in Florida

To advance knowledge of sea turtles, Carr had to clearly delineate what was known about them. At the outset, Carr acknowledged that the study of the green turtle in Florida was limited by the fact that the local population was for the most part composed of nonbreeding juvenile individuals. He estimated the breeding weight of sea turtles to be approximately 130 pounds and noted that the heaviest Florida green he had seen was only 115 pounds. Six sources of data supported Carr's claim: observations of nonscientists (some of whom were highly qualified in Carr's assessment), the absence of pregnant or egg-carrying females, the absence of hatchlings on the beaches or in the water, the length of the nesting season (May to October), the disappearance of turtles en mass in the fall and their reappearance during the spring, and historical accounts and observations.[3]

The research on Florida turtles essentially functioned as a pilot program for the larger program in the Caribbean. One of the critical issues was to find a suitable tag with which to mark the mature turtles. Carr and his research assistant experimented with three tags. The first two tags both needed to be connected to the turtle's carapace (shell) with wire, but Carr soon abandoned these tags for a two-inch cow-ear tag of monel metal that could be fastened to the hind edge of the fore flipper, where it did not disrupt the turtle's swimming ability. This tag had several other advantages. It could be fastened in one-third the time that it took to fasten the other tags, and it was extremely resilient, meaning that it was not likely to fall off or to snag as the turtle swam. To test the effectiveness of the various tags, Carr attached them to several of the turtles held in captivity at the "Gulfarium, The Living Sea" (an aquarium in Fort Walton Beach, Florida). Carr worried that the bent tag (it had to be wrapped around the flipper) might confuse the people he was hoping would return the tag to the University of Florida, but he hoped common sense would prevail. Each tag was inscribed with a simple abbreviated message in Spanish and English offering a reward (five dollars) for tags returned to the University of Florida. Any hope of acquiring data from the tagging project depended upon the good will (and financial state) of the individual who found the tagged turtle. Still, five dollars was equal to if not greater than the market value of an individual turtle in most areas at the time, so it was savvy to set the reward at that level.[4] Remarkably, tags arrived in Carr's department mailbox from all over the world, and the messages that accompanied the tags displayed a stunning variety.

During the first season of tagging, Carr and his research assistant tagged and released forty-three green turtles from April to October, the regular fishing season, on the Withlacoochee-Crystal River Grounds. Six tags were returned, and they recaptured two tagged individuals with tag holes clearly evident. All but one of the recoveries involved a return to the original site of capture, and one turtle made the return trip twice. Carr felt that even these limited data suggested homing behavior in sea turtles, a view widely held by turtle fishermen. Even in his technical reports, Carr acknowledged local knowledge. Ironically, if the only behavior on display was local homing, these preliminary data would undercut Carr's greater objective to establish that sea turtles migrated over long distances, so he hinted that future studies would distinguish between these two types of movement.

One of the problems with the Cedar Key pilot-tagging program was the expense associated with tagging turtles near a commercial turtle fishing area. Carr lamented these additional costs: "Tagging on any commercially exploited green turtle feeding ground is expensive, because you have to buy the turtles by the pound, then pay a reward if they are retaken, buy them again, and so on."[5] Though Carr and his students managed to tag just fifty-three turtles, the cost was considerable because they had to pay by the pound for each turtle. Yet, despite the expense and the limited sample size, Carr gleaned valuable data on the sea turtle aggregation in western Florida. Using the data from tagging and returns, Carr developed a rough estimate of the population of turtles on the west coast of Florida (approximately 5,600) based on the number of recaptures and total number of marked turtles released and on an estimate of the total number of turtles captured. Similarly, Carr was able to comment on growth rates in green turtles, and his sketchy notes were the only ones published for forty years.

Carr's first season of data also provided important information on the shells of turtles. From his taxonomic studies of turtles, Carr knew that taxonomists had used the proportions and shape of the shell in attempts to define species and races of sea turtles. Such ratios were highly variable in Carr's data, which was the first adequate sample ever appraised in print. Carr argued that a more useful character

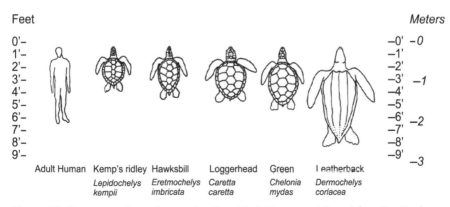

Figure 18. Size comparison of sea turtles found in U.S. waters. Adapted from Sea Turtle Survival League. Courtesy of the Caribbean Conservation Corporation.

for evaluation was the length–weight relationship.[6] Though he directed the bulk of his research effort at green turtles, Carr began to consider the life cycles of other species.

The Riddle of the Ridley

Although Carr's limited data on green turtles laid the groundwork for further study, his data on the Atlantic (now Kemp's) ridley turtle (*lepidochelys kempii*) represented a major contribution to knowledge of this elusive species. The Atlantic ridley had frustrated Carr since his days as a graduate student. Even his first published account of sea turtles gave only the barest sketch of the details of the life history of the Atlantic ridley, focusing instead on the taxonomy of sea turtles (see chapter 3). Carr drew attention to two aspects of the Atlantic ridley: it was inclined to react violently when captured (particularly when placed on its back), and its breeding biology was completely unknown. It was the latter observation that Carr expanded upon when he wrote about the Atlantic ridley in his *Handbook of Turtles*. He noted that Richard Kemp (for whom *kempii* was named), when submitting a specimen for classification, noted that "We know that they come out on the beach to lay [in the Florida Keys] in the months of December, January and February, but cannot tell how often or how many eggs."[7] Carr was surprised to be able to offer little additional information regarding the species, and he had to speculate on the possible breeding sites of the Atlantic ridley: "It seems likely that breeding occurs all along the Florida coast, where suitable beaches exist, and that the lack of data means merely that the ridley is confused with the loggerhead by most people who find turtles nesting."[8] For the time being, the ridley left Carr perplexed, but he relished the challenge.

As usual, Carr reviewed the lore of the Atlantic ridley as shared by fishermen. For once, he expressed skepticism regarding their claims. There were three stories that fishermen told about the ridley. The first suggested that the ridley did not in fact breed on its own but resulted when two species hybridized (hence the popular name, "bastard turtle"). Other stories acknowledged that turtles did in fact breed, but at an unknown location at a great distance from where the given fisherman was speaking. Before he had devoted considerable time and effort to the exploration of Caribbean coastlines, Carr found such claims reassuring. By the time he wrote about the ridley, he was fairly confident that the species did not breed in the Caribbean, or if it did no one knew where. Still another story circulating among fishermen theorized that the ridley nested in very small numbers along with the other species in the same places and at the same time. Further examination of such claims revealed little or no evidence to support them.

In addition to the fishermen's explanations, Carr considered and dismissed in turn the possibility that the ridley did not in fact breed or that it bore live young (like some snakes but no known turtles) or that it bred at a strange time (mid–winter, for example), but no one had recorded a pregnant female. As for the possibility that ridleys were in fact nesting along with other species on Florida's beaches, Carr wondered how he and many of his friends and correspondents had spent hundreds of hours on the beaches of Florida without witnessing nesting ridleys. Not even the illegal east-coast turtle-hunters (poachers), whom Carr numbered among his consultants, had ever seen a ridley climb onto a beach, and if one did, Carr expected to hear about it: "By not moralizing on their ways, I have made friends among these poachers, and if a ridley comes up on one of the good mainland beaches in the turtle season, I bet I hear about it within hours."[9] But Carr's riddle of the ridley remained unsolved.

Against the backdrop of the mystery of the ridley, Carr's analysis of the first year's data collected under the auspices of the grant from National Science Foundation was striking. In his first *American Museum Novitates* contribution, Carr noted that the Atlantic ridley was not known to breed anywhere and that it was not known to nest in areas where other sea turtles nested and that his search of the Caribbean had failed to produce ridleys or any knowledge of the turtle. Yet, the research project at Cedar Key had yielded more tantalizing stories. In 1954, Carr repeatedly heard rumors of a "ridley with eggs" that had been butchered locally. Even more intriguing, on October 27, 1955, he heard that a ridley of unprecedented size (93 pounds) had been found with 100–150 yellow, ovarian eggs the "size of marbles."[10] Meanwhile, the green turtle tagging program at Tortuguero began to yield results.

The First Season at Tortuguero

As the Florida tagging project evolved, Carr reported on the first season of sea turtle research in Costa Rica as well, which also occurred in 1955. Carr's research assistant in Tortuguero was Leonard Giovannoli, who spent the months of July and August tagging turtles at a makeshift station on the beach. Giovannoli had been one of Carr's professors and colleagues in the Department of Zoology at the University of Florida, but had recently retired. Having spent the previous three summers engaged in reconnaissance in Tortuguero and elsewhere in the Caribbean, Carr and Giovannoli established a small camp near the beach. Giovannoli lived there from July 2, 1955 to August 29, 1955, and Carr spent a few days at the beginning and end of the summer opening and closing the station. At this time, Tortuguero was still an active site of egg collection for commercial sale, so Carr and Giovannoli were fortunate to obtain a 2-mile stretch of the beach that was not disturbed by commercial interference (they acknowledged the Costa Rican Ministry of Agriculture for this protected stretch of beach). Shrewdly, Carr hired the men who usually turned turtles on this part of the beach for export interests in Limón. Unlike ridley turtles, green turtles could survive days and weeks lying on their backs, so turtle hunters would flip them over before butchering them. This was also useful to Giovannoli, who could tag and release the turned turtles the next morning.[11] In fact, despite having lost a leg, Giovannoli set a long-standing record by tagging fifty turtles in a single night using this method. As in Florida, the scientists experimented with two kinds of tags and found that

Figure 19. Harold Hirth, Archie Carr, and Larry Ogren measure a green sea turtle at Tortuguero (ca. 1961). Courtesy of the Department of Special and area Studies Collections, George A. Smathers Libraries, University of Florida.

the flipper tags remained in place longer than the shell tags. Giovannoli measured the length and width of the shell of each turtle, counted the postocular scales of each side, and documented the site of capture as accurately as possible.[12]

Over the course of the summer, Giovannoli caught 644 turtles (149 of these were recaptures that had returned to the beach, thus diminishing somewhat the total number of tagged turtles). As in Florida, the data were preliminary, but suggestive. One of the common pieces of turtle lore suggested that turtles migrated in large groups or schools. Carr believed that the data from the single season of tagging provided a small measure of scientific support to this claim. His example was four turtles that were tagged on August 11, 1955, within a quarter of a mile of each other. On August 24, Carr and Giovannoli recaptured all four turtles within 2 miles of the original site. Another pair of turtles demonstrated a similar pattern of return. Because offshore currents and environmental conditions precluded congregations immediately off the beach (confirmed by reconnaissance via airplane), Carr concluded that the turtles were roaming the coast searching for protection and food. Thus, the return of turtles together to the same region of beach after a two-week interval suggested some level of at least local homing ability.[13]

Carr perceived a serious threat to nesting turtles at Tortuguero in that the females were extremely vulnerable to hunting pressure as they hauled themselves up the beach. It seemed to Carr that such one-sided hunting of females would lead to an unbalanced ratio of males to females. Of course, the data Carr and

Giovannoli were collecting at Tortuguero involved only the mature females that used that beach, but Carr knew of a turtle fishery at Mosquito Cays 300 miles to the northwest of Tortuguero. At the Mosquito Cays, the turtle catch was indiscriminate, so Carr felt that it would reflect the rough ratio of males to females. Carr was attempting to verify what he considered the "surprising opinion" that the sex ratio of turtles varied from year to year.[14] Carr determined the numbers of male and female turtles captured by two vessels between February and April 1956. One caught 27 males and 66 females, and the other caught 105 males and 271 females. This was clear evidence of an unbalanced sex ratio.[15] Like the riddle of the ridley, the variable sex ratio among sea turtles would not be explained for years (see chapter 9).

In establishing the tagging program, Carr hoped to shed light on the long-distance migrations of sea turtles. By the time he published the report on the first season of tagging, he had received ten tag returns. Seven came from Mosquito Cays or the Nicaraguan coast, and three came from Panama. The greatest distance traveled was 300 miles (both to the north and the south by different turtles). Carr reiterated the problem of lost tags due to the initial tag and wire combination. One of the turtles (#213) traveled 300 miles in ten days, which according to Carr corresponded with the rate of travel recorded by one of the captains whom Carr met during reconnaissance in the Cayman Islands: "So the story changed quickly from tragedy to just a marvel of nature. Looking at the map, you see that when the turtle got out of the pen in Georgetown there was no coastwise shallow-water course home that he possibly could have traveled in twelve days. Even if he went straight across the western end of the Caribbean he made the trip at a rate of thirty miles a day."[16]

Carr recognized the accumulated years of knowledge of the natural history of sea turtles in the stories of the turtle captains. It must have been gratifying to obtain scientific evidence to verify the long-distance traveling abilities of green turtles. In a sense, the rigorous scientist in Carr managed to confirm what the dedicated naturalist in Carr already knew from his long discussions with turtle captains. A key shift occurred here from collected anecdotes regarding natural wonders to data points that contributed to a clearer knowledge of the natural history of sea turtles. In addition to questioning turtle captains, Carr scoured the zoological literature for references to sea turtle studies. At the time of the first season at Tortuguero, only one other researcher had published on the biology of sea turtles, and to find him Carr had to look on the other side of the world. It was Tom Harrisson who studied green turtles in Borneo.[17] Turtle researchers were even more widely dispersed than their subjects.

Nesting Behavior in Green Turtles

Although Carr's preliminary efforts in Florida were directed toward establishing the tagging program, during the first season he was also able to contribute to an understanding of the reproductive biology of green turtles. Carr's objective was to elaborate on general observations with meticulously collected and timed descriptions of each element of the natural history of reproduction in green turtles. There was more at stake than knowledge for its own sake. Carr had already made timed, move-by-move records of the Pacific ridley (*Lepidochelys olivacea*), and he considered the collected data on the loggerhead turtle (*Caretta caretta caretta*) to be adequate. By documenting the nesting behavior of sea turtles, Carr hoped to shed light on their evolution.[18]

With this goal in mind, Carr set out on August 25, 1955, to document the nesting process for a green turtle. After rejecting several candidates that had already reached an advanced stage of the nesting process, Carr found no. 227, a turtle that had been tagged the night before, at 9:15. Fortunately, Carr was able to find the turtle without a flashlight; turtles are highly reactive to light before nesting. He had learned of the green turtle's light sensitivity through experiments and anecdotes: "We have, experimentally, often sent an emerging turtle scurrying back into the water by one flick of a flashlight beam across her eyes. The veladores say lighting a cigarette at the coco-plum line sometimes scares away a turtle coming out of the surf."[19] Over the course of the next twenty minutes, the turtle dragged herself out of the water and to the edge of beach vegetation (55 paces by Carr's estimate up the beach). It took another hour or so to dig a suitable hole for the eggs, and twenty minutes after that the turtle had laid her eggs and had begun to cover them with sand with her back flippers. Carr described the entire process in meticulous detail. At 11:32, the turtle had finished covering up the eggs and quickly returned to the sea. She disappeared into the surf at 11:43. Two and a half hours had elapsed since the turtle first left the water.[20] As Carr was fleshing out the life history of the green turtle, other species remained enigmatic, but an opportunity to spend two years at the University of Costa Rica developing its biology program enabled Carr to study leatherbacks.

University of Costa Rica, 1956–1957

In August 1956, Carr began a contract with the University of Costa Rica located in the capital, San Jose. According to his contract, Carr was to serve as a visiting professor and technical advisor to help the university organize its Faculty of Sciences and Letters and contribute to the development of the Department of General Education. Before 1956, the University of Costa Rica (UCR) did not have a formal school of biology, and Carr's main responsibility during his sixteen months at the university was to develop such a program. Carr worked with Rafael Lucas Rodriguez, who completed his doctorate at the University of California at Berkeley and was invited to start the program in biology at the UCR.[21] To that end, Carr developed a lecture course in biology for groups of fifty students selected by the dean. In addition, Carr offered a seminar in biology to students in the Department of General Studies. These courses would serve as models for other professors associated with the department. To be effective Carr would have to deliver all his lectures and seminars in Spanish. He presented a series of lectures on evolution and natural history on radio. A Spanish colleague at UCR heard him lecture and offered the ultimate compliment: "You have mastered our language."[22]

At the end of the first semester, Carr prepared a report on the development of the course *Fundamentos de Biología*. Not surprisingly, Carr recommended *El Hombre y El Mundo Biológico*, the Spanish translation of *Man and the Biological World*, written by his mentors J. Speed Rogers and Theodore Hubbell. Because the price of the textbook was rather high (six dollars), Carr suggested purchasing several copies for the library so that it would be accessible to all students. Other critical components of the biology curriculum included film strips (which had to be ordered from England), films (with Spanish commentary), and slides.[23]

Carr also argued for the contribution of biology to the program in general studies:

> It is the obligation of the Program in Biology to provide a course that speaks extensively to the strengths and weaknesses of the scientific method as an instrument of knowledge, of the extent and principles of the discipline of biology and the base of information that supports it; and something of the reasoning by which those principles are derived. The students should receive a new comprehension of the dimensions of the biological world, a new appreciation for the professionals dedicated to its research, and some new concepts of their proper place and role as a living creature, member of a community of living creatures.[24]

Later Carr argued for the role of direct experience in the curriculum:

> Another problem is the necessity of teaching a natural science, constructed on an observational base, without resorting to the experience in the laboratory. Without permitting the student to have in his hands a living creature in the process of life, we have to instill in him a valid concept of the phenomenon of life. This indisputable disadvantage, inherent in the Program of General Studies, is the part of the Program most often attacked by its critics. In reality, given that the defect does not have a complete remedy, the themes will be reinforced by the use of the student as his own specimen, taking advantage of the abundance of material continually presented.[25]

Carr also recommended extending this approach to include the diversity of the region:

> When one offers the student the amazing findings of paleontology, he should also offer an elephant's tooth from the Pleistocene era of Honduras or of Florida, to sustain his confidence. The fantastic colors of the genus of little venomous frogs from the coffee plantation behind the University can be used to focus discussions of adaptation, of natural selection and of such debated theories as that of the coloration of warning. Use of such expedients like a fount of objectivity furthers those themes at a rate completely disproportionate to the rapid increase of the task of the personnel/staff and to the worthless equipment needed.[26]

From these generalizations, Carr developed a plan of study in biology for the development of secondary school teachers. In the first year, would-be biology teachers studied history, Spanish, philosophy, sociology, general chemistry, general biology, and general mathematics. By the second year, they would specialize somewhat, studying education, general botany, general zoology (invertebrates), general physics, and organic chemistry. And in the third and final year, they would specialize further with the following course load: education, genetics, plant anatomy, general zoology (vertebrate comparative anatomy), and Costa Rican biology.

As in Honduras a dozen years earlier, Majorie and the children joined Archie in Costa Rica. While only Mimi has distinct memories of their time in Honduras, all of the Carr children have memories of Costa Rica and Tortuguero. Tom Carr recalled that his passion for tropical nature was kindled at the age of five years. During the summer of 1957, Tom spent three months on the beach at Tortuguero with his family. He learned how to turn a turtle (no small feat for a five year old) and discovered Caribbean fishing. The tagging program was in its infancy at this point (the previous summer, Leonard Giovannoli had set the still-standing record for turning fifty turtles in one evening).

His work at the School of Biology at the University of Costa Rica extended Carr's reach. Under the tutelage of Rogers, Hubbell, and Byers, Carr had absorbed the principles of biology and society, which incorporated humans in the study of biology. Through his work as technical advisor at the University of Costa Rica, Carr took this philosophy and approach to Costa Rica, where it became a standardized curriculum. It would be difficult to overestimate the impact of Carr's efforts to develop the School of Biology on the study of biology in Costa Rica and Central America. One historian recently noted: "The School of Biology became one of the best of its kind in Central America and has served as a springboard for research into tropical studies for Costa Rican and other Latin American students."[27]

The Great Trunkback

Outside of Florida, Carr's research on sea turtles was largely limited to the summer months. However, during a reconnaissance trip to Tortuguero in May 1954, Carr found trunkback turtles (now known as leatherback sea turtles, *Dermochelys coriacea coriacea*) in the surf, as well as their tracks on the beach. Carr surveyed the tracks he found over a 10-mile stretch of beach, but did not discover any signs of colonial nesting or evidence of anything other than random stranding by individuals. This struck him as odd because he had read published accounts of small groups of female leatherbacks climbing onto beaches simultaneously along the coast of Ceylon.[28] Other accounts of encounters with schools of leatherbacks convinced Carr that the species probably traveled to select nesting beaches.[29] En route to Tortuguero in 1955, Carr and his research assistant conducted an informal track survey along the coast in a small airplane from Limón (about 40 miles to the south). From the mouth of the Matina River (about

10 miles along the beach and continuing for 4 or 5 miles), Carr saw numerous tracks, some of which ran to and from completed nests. Retrospectively, he realized that the tracks were probably made by leatherbacks because the green turtles were not yet nesting at Tortuguero.

On May 4, 1957, while he was on leave from UCR, Carr traveled to Tortuguero. During the flight from Limón, Carr once again spotted turtle tracks on the beach starting at the mouth of the Matina River. This time, Carr asked the pilot to circle the area while he studied the tracks with binoculars. He determined that the tracks were too broad to be those of green turtles and had to have been made by leatherbacks. The density of tracks indicated that the leatherback was migratory and not a casual wanderer as had been suspected previously.[30] Unfortunately, it was not possible to land and confirm his suspicions then. Two weeks later, Carr rode a train to Matina station and took a mule car to the mouth of the river. During the trip, he noticed that the cars traveling away from the coast carried turtle egg hunters with sacks of leatherback eggs. At the end of the line, Carr walked for about half an hour to reach a small lagoon and took a small dugout canoe from there to the point on the beach where the concentration of nests was heaviest. He immediately confirmed that the beach was a leatherback nesting colony. He found eighteen recent nests (fewer than two days old) in 1.5 miles, but egg hunters had robbed all the nests.[31]

Having established that Matina beach was a nesting colony for leatherbacks, Carr decided that someone needed to return to the beach for an extended period, so he recruited Larry Ogren, an undergraduate who could not begin work until the end of the semester at the University of Florida. Ogren finally arrived at Matina beach on June 9, 1957. He set up a camp 4 miles north of the mouth of the river, even though he and Carr knew that the height of the nesting season had passed and that competition from egg hunters would be intense. That very evening, Ogren discovered five leatherback nests, each of which was immediately emptied by egg hunters. The reason for such severe depredation was simple: few other turtles were nesting in early June, so the leatherback eggs brought a good price (seventy-five centimes per dozen) in the towns along the railroad. Remarkably, Ogren persevered and ultimately prevailed in obtaining data on the colony: "In spite of this competition, and other than the fact that the crest of aggregate emergence had passed, by outrunning, browbeating, and bribing the egg collectors Ogren was able to accumulate data and observations that extend our knowledge of leatherback reproduction on American shores."[32]

The next summer, Carr returned to the beach at Matina and made detailed notes on the nesting behavior of three leatherback turtles. His observations of the nesting behavior of leatherbacks were as meticulous as his study of nesting in the green turtle (perhaps more so given that he completed three different studies for the leatherback).

Carr's study of leatherbacks broke new ground in that he and Ogren observed the emergence of leatherback hatchlings from the nest. On July 28, 1958, Ogren alerted Carr that he had discovered a single hatchling emerging from the sand. When Carr reached the site, he found more hatchlings emerging in small numbers and congregating at the base of a makeshift wall constructed around the nest. Once the board closest to the water was removed turtles began to work their way to the shore. The majority maintained a course toward the water, but a few at the edge of the main group occasionally stopped and reversed direction. Nevertheless, the three remaining sides of the wall kept most of the turtles on track. Carr and Ogren conducted a series of simple experiments to analyze orientation in hatchling leatherbacks. Over time, the turtles exhibited a diminished response to the beach and courses to the water. At the end of two weeks, the hatchlings had lost their orientation ability altogether: "On the fifteenth day, although released at a point only 10 feet from the surf, none reacted or did anything more than raise its flippers in the air and make irresolute efforts to rotate its body. It seemed evident that this development was not due to general physical decline but to a loss of specific receptivity to orientation clues."[33] How sea turtles oriented continued to be a central topic of Carr's investigations.

Carr and Ogren took additional notes on the leatherback hatchlings in captivity. They were struck by the way that the young turtles swam constantly during daylight hours. Leatherbacks were in this respect similar to green turtles. Hawksbill turtles, in contrast, swam purposefully only when in search of food. The behavior of constant swimming suggested to Carr that leatherbacks and green turtles probably migrated a significant distance from their nest: "It seems likely that green turtles and leatherbacks spend the first hours or days following emergence from the nest in steady travel away from the place where they enter the surf. This is a point which clearly bears upon the troublesome question of what young sea turtles do, and where they go, in nature."[34] In addition to notes on the behavior of leatherback turtles, Carr and Ogren considered the systematics of the species. Based on the massive size of the adult animal and the difficulty of preserving specimens, Carr believed that taxonomic analyses would probably be based on young individuals, so he and Ogren took detailed notes on the colorations, dimensions, and scalation (scales) of young leatherbacks.

Results from Tortuguero

By 1959, the fifth season of the Tortuguero tagging program, Carr had collected a large amount of data on the nesting biology of green turtles (*Chelonia mydas*). Along with his research assistants, he had marked 1,178 turtles, and they had received 35 international tag returns. For the most part, the returns supported Carr's original assumptions that the large population of Mosquito Cays was derived from the Tortuguero breeding ground and that Tortuguero was the reproductive center for turtles from a large area. As Carr reviewed the data from the five years of tagging, he noticed that the green turtles did not return to nest during the season following one in which they had nested and that the turtles seemed to follow a three-year breeding cycle (that is, nesting at Tortuguero was followed by a two-year hiatus). In a few cases, turtles missed only one season (a two-year breeding cycle). There were, however, numerous cases (92 from 1955–1959) of females returning to the beach within the same season to renest after an interval of twelve to fourteen days.[35]

Besides the reproductive cycle, Carr could also delineate the nesting season. At Tortuguero, green turtles laid most of their eggs between July and September, with the peak of activity in August. A few green turtles nested in June and October. Leatherbacks, for the most part, nested in April and May and rarely nested on the beaches used by green turtles. Hawksbills nested between leatherbacks and greens (although Carr noted anecdotal evidence that some hawksbills nested in October and November after the greens had left their nesting grounds. Such division of nesting season suggested to Carr an adaptive response to interspecific competition for nesting space (between different species of turtle), particularly in the case of the hawksbill, which was much smaller than the green, leatherback, and the loggerhead. Although he added little additional information to the original description of mating, Carr was able to divide the nesting process into eleven separate stages.[36] Carr also offered a possible explanation of why green turtles "cried" while nesting. While acknowledging the findings that in marine reptiles such crying was associated with salt regulation, Carr believed crying was a useful adaptation in another way: "Sand is bound to get into a turtle's eye while she is digging her nest. A simply moist eye would go blind on shore from sand stuck on it. A dry eye such as that of the snakes might serve but turtles do not possess such an eye. The wetting of the adhering sand by copious tear flow causes it to cake into masses that keep falling away repeatedly and leaving the eye clean."[37]

Through a series of informal experiments, Carr and Ogren attempted to determine what cues hatchling sea turtles used when emerging from the nest and making their way to the sea. Initially the hatchling relied upon negative geotaxis (using gravity to orient) to reach the surface of the ground. Once on the beach, the newly emerged turtles seemed to use what Carr called modified phototaxis (using a source of light to orient). The rudimentary field tests performed by Carr and Ogren suggested that the turtles sought out an "openness of outlook," but Carr interpreted this search to be the sort of illumination that comes from an extensive view. Environmental conditions did not appear to play a significant role in the ability of the turtles to orient themselves: they continued to find the sea by day or night, sunlight or moonlight, in clear or overcast conditions. Carr and Ogren further isolated the orientation mechanism to a complex relationship between light intensity and area. They noted that the sky over the sea drew hatchlings away from the sun or from a full moon, when either was over land. Yet, just as they had found among leatherback turtles, a strong gasoline lantern overrode any other attraction when set by water at night. This kind of light would draw hatchlings back to the beach after they had entered the water.[38]

Leatherback turtle hatchlings faced challenges similar to those of green turtle hatchlings in making their way out of the nest and down the beach to the edge of the water. In studying these two species, Carr and Ogren suggested that leaving the nest might depend on some sort of cooperative effort on the part of hatchling sea turtles. They even went so far as to claim that it was unlikely that a single sea turtle could dig its way out of the nest without the assistance of its siblings.

By 1961, Carr and his assistants had determined that female green turtles nested every second or third year and in the process laid between 500 and 1,000 eggs in several nests of about 100 eggs each. Because a sea turtle was limited by the size of her leg and foot in digging her nest, 100 eggs filled the nest to capacity without overflowing. Carr suspected that the number of eggs correlated in some way with the behavior of the hatchling turtles in the nest. Ogren and Carr observed survivorship in several dozen artificial nests. If four or more turtles hatched, 98 percent or more of the hatchlings emerged from the nest. When only a single turtle hatched, no turtles emerged from the nest. Thus, it seemed to Carr that successful emergence from the nest depended on unconscious cooperation among all hatchlings.[39]

By installing glass sides on artificial nests, Carr and Ogren could monitor the social interactions of the hatchlings in the nest. Carr's graduate student Harold Hirth refined the experimental design of the nest study to monitor how hatchlings found their way out of the nest. Hirth observed how the first turtles to hatch did not start digging at once but lay still until others had hatched. Here is how the turtles assisted in the escape from the nest according to Carr and Hirth:

> The vertical displacement that will carry the turtles to the surface is the upward migration of this chamber, brought about by a witless collaboration that is really a loose sort of division of labour. . . . Turtles of the top layer scratch down the ceiling. Those around the sides undercut the walls. Those on the bottom have two roles, one mechanical and the other psychological: they trample and compact the sand that filters down from above, and they serve as a receptor-motor device for the hatchling super organism, stirring it out of recurrent spells of lassitude. Lying passively for a time under the weight of its fellows one of them will suddenly burst into a spasm or squirming that triggers a new pandemic of work in the mass. Thus, by fits and starts, the ceiling falls, the floor rises, and the roomful of collaborating hatchlings is carried toward the surface.[40]

Thus, it is the joint activity of mutually stimulated hatchling turtles that moves the nest chamber toward the surface. Synchronized effort resulted in more effective progress for the entire group of hatchlings, or in Carr and Hirth's vivid terminology, "the hatchling superorganism." With this study, Carr explained another mystery in the natural history of sea turtles. As aspects of the green turtle's life history emerged, Carr considered other species, such as the loggerhead turtle.

The Loggerhead

In embarking on research into the ecology and migrations of sea turtles, one of Carr's assumptions was that the adult green turtles' herbivorous food choices underlay its biology. Herbivory, Carr believed, was at the root of green turtles' long-distance migrations between breeding and feeding grounds. Such a feeding strategy made the green turtle unique among sea turtles and necessitated a highly developed ability for navigation. Yet, as his study of other sea turtles developed, Carr realized that the green turtle was not unique. Herbivory was not the decisive factor that Carr had once supposed. In fact, the Atlantic loggerhead turtle (*Caretta caretta*) consumed a carnivorous diet but seemed to migrate significant distances to and from its nesting beaches. Carr developed his ideas in a suite of papers published with his former research assistant David Caldwell, who had joined the U.S. Fish and Wildlife Service since completing his degree, and with Ogren.

As already discussed, the tagging program yielded valuable information about green turtles within a few years of its inception. Carr hoped to develop the knowledge of other sea turtles by extending the tagging program to additional species. Once again, tagging produced quick results. Caldwell and a friend tagged a large female loggerhead at Henderson's beach (on the Atlantic coast of Florida) on March 27, 1957. A shrimp trawler recaptured the turtle off the mouth of the Mississippi River (near Pass-a-Loutre) during the last week of March 1958. To travel from one point to the other without crossing the open Gulf of Mexico, the turtle must have swum at least 1,000 miles down the Atlantic coast and around the tip of Florida and up much of the Gulf coast. Given the time elapsed since tagging, Carr, Caldwell, and Ogren surmised that the loggerhead may have actually taken an even longer and more circuitous inshore route. Wouldn't a turtle traveling this route have to swim against the current of the Gulf Stream for much of the trip? Carr and his students dismissed this notion by studying the seasonal variation in Gulf Stream currents, which revealed that the current may have assisted the turtle in its migration, especially if she waited to leave the waters off the breeding grounds until August (as Carr suspected turtles did). Lest the significance of this loggerhead turtle's voyage be lost, Caldwell, Carr, and Ogren noted: "This is one of the longest journeys ever proved for any reptile, being rivaled only by trips made by female green turtles tagged in Costa Rica."[41]

As the life histories of the various turtles began to take shape, Carr concentrated his efforts on the most challenging problems remaining. One such problem was where the turtles spent the first few months of their lives after hatching and struggling to the sea. To Carr's knowledge, there were no records of hatchling turtles captured at a distance away from the nesting beach. Thus, he was excited to report on three hatchlings taken in two captures, one of a turtle at about eleven weeks of age and one of two turtles twelve to thirteen days old (these were removed from the stomach of a white-tipped shark).[42]

Like the green turtle and other species, loggerhead turtles seemed to confine their breeding to a few beaches that Carr called "rookeries," using the preferred term for nesting aggregations of colonial birds such as herons, egrets, terns, and gulls. Over the course of their research, Carr and his assistants had located three such rookeries on the Atlantic Coast: Hutchinson's Island (where the loggerhead tagging program commenced), Little Cumberland Island, near Brunswick, Georgia, and Cape Romain (propitiously located within a National Wildlife Refuge). Aerial reconnaissance on the coast of Florida north of Matanzas Inlet and of the entire coastlines of Georgia and South Carolina confirmed that although loggerheads nested in small groups elsewhere, the largest concentrations formed at the sites listed above.

Ongoing studies enabled Carr and Caldwell to emend their earlier studies of the nesting behavior of loggerheads. Carr had been able to follow a single turtle through the nesting cycle from the point of exiting the water to the point of reentry, but he and Caldwell had to construct a composite sketch of nesting for the loggerhead due to the difficulty of finding turtles as they emerged from the sea. Nevertheless, Carr felt that the portrait provided an accurate reflection of the nesting biology of Atlantic loggerhead sea turtles due to the consistencies among the different individuals observed. Throughout the study of additional turtles, Carr and Caldwell failed to induce the nesting loggerheads to use their flippers to "keep the wall from falling in," as Carr and Giovannoli had recorded in a paper two years earlier. Nor did any of the nesting loggerheads use their plastron to pack or pound the nest after laying eggs and covering them with sand.[43] Thus, repeated observations had sharpened the picture of loggerhead nesting.

Since Carr's first visit to Tortuguero, locals told him stories of a few loggerhead turtles nesting with the green turtles. Yet Carr could not verify such reports until the summer of 1957, when one of his assistants found and turned a loggerhead for later tagging. This was the southernmost record of loggerhead nesting, and Carr and Caldwell were interested in the implications for the evolution of sea turtles:

> Breeding aggregations of animals usually develop integrating bonds that place the aberrant individual with an urge toward solitary breeding at a selective disadvantage. The chances for consummation of any reproductive venture apart from the group effort, especially when a long migration to the nesting grounds is involved, would appear slight. With sea turtles we can at present only point out the phenomenon and hope that with better understanding of the organization of the breeding group an explanation will eventually emerge.[44]

Is it possible that nesting within a colony of green turtles provided some measure of collective protection like that afforded by other loggerheads? Carr and Caldwell turned to the question of multiple and group nesting in their next paper.

As was so often the case, anecdotal evidence sparked Carr to embark on a research problem. Fishermen, turtle poachers, and conservation officers, despite their disparate agendas, all agreed that the Atlantic loggerhead turtle laid more than once during a given nesting season and that groups of turtles stayed together during the nesting season. Evidence from tagging programs confirmed multiple nesting and group nesting in the loggerheads. To interpret the data, Carr and Caldwell teamed up with two Georgia biologists. Frederick H. Berry was a fishery research biologist with Caldwell at the Bureau of Commercial Fisheries Biological Laboratory, and Robert A. Ragotzkie had directed the University of Georgia Marine Institute before moving to the University of Wisconsin. Clumped tag returns suggested group movements in loggerheads similar to those of green turtles. On May 27, 1957, Caldwell tagged seven turtles on Hutchinson's Island, including no. 78, and the following night they tagged eleven more, including no. 23. An agent for the Florida State Board of Conservation discovered both of these turtles nesting nearby about a month after they were tagged. Carr and Caldwell interpreted this finding as significant: "In spite of the small size of the sample, the fact that this double return involved two out of a group of only 18 turtles tagged makes it highly improbable that the dual recovery was due to chance."[45]

The next summer Carr, Caldwell, and their associates conducted another tagging program on Jekyll and Little Cumberland islands just off the coast of Georgia. Of seventy-two loggerheads tagged between June and July 1958, the researchers recaptured twenty-six one or more times, and tourists reported seeing about ten tagged turtles nesting on the beaches over the course of the summer. With the concrete data of the recaptures and the anecdotal tourist sightings, Carr and Caldwell confirmed that the approximate interval between nestings for loggerheads was about twelve to fourteen days, similar to that for the green turtle. Further evidence for multiple nestings came from the evidence of unlaid eggs. One dead female turtle contained 120 shelled eggs of mature size, 182 eggs of large yellow yolks without shells, and 25 eggs of much smaller yolks, also without shells. The different types of eggs provided circumstantial evidence of at least two separate layings in one season. However, Carr and Caldwell recognized the limits of their data: "While these data strongly suggested that multiple nesting occurs in loggerheads, they did not furnish firm grounds for calculating the number of times an individual might nest in a season, nor do the tagging results."[46] Years would pass before researchers addressed these problems. At the same time, Carr continued to search for the breeding site of the Atlantic ridley.

Return to Ridleys

As he collected data on the breeding sites of the five known species of sea turtles, Carr and his co-workers sought information regarding additional breeding colonies. They were particularly interested in any information on the breeding of the Kemp's ridley, which as of 1958 remained obscure: "The ridley of the western Atlantic, *L. kempii*, continues to hold a place as one of the more puzzling members of the North American vertebrate fauna."[47] In the two years since he had written about the ridley, Carr and assistants had used aerial reconnaissance of the island beaches in Florida Bay in the hopes of locating the breeding colonies of ridleys there. However, there was no evidence of ridleys breeding on these beaches. Carr realized that the large population of nonbreeding ridleys must migrate to the waters of Florida from some distant nesting ground.

To sort effectively through possible sites, Carr reviewed and expanded upon the range of the Atlantic ridley. Particularly striking was a gap in the range of ridleys that encompassed all of the West Indian Antillean Islands, with the possible exception of Trinidad. After examining a specimen and discussing it with local residents, Carr decided that a lone Cuban ridley was a stray from the waters off West Africa that was carried on the westward current. He was surprised to discover that there were no records of ridleys in the Bahamas or Bermuda (despite a relatively short crossing from the coast of Florida). But the most interesting element of the range of ridleys were stories of the Trinidadian form, known locally as the *batalí*. Despite considerable effort, Carr was unable to obtain a specimen, but he was inclined to accept folk talk of the *batalí* as fact.

For Carr the most important element of ridley taxonomy was the similarity between Pacific and West African forms:

> In any event, the most striking zoogeographic aspect of the West African ridley population is not its variability, but the fact that it predominantly agrees with the Pacific stocks instead of with its West Atlantic congeners, and that West African *Caretta* does the same thing. It seems likely that these two are reflecting zoogeographic, or paleozoogeographic, influences of deep-seated significance, perhaps to be finally elucidated only through study of the distribution of fishes and other groups in which large series can be collected in all the critical localities.[48]

Thus, for both the ridley and the loggerhead, the populations of the eastern Atlantic and the Pacific shared more similarities with each other than the population of the Caribbean and the western Atlantic.

The similarity between West African and Pacific ridley forms prompted Carr to conjecture that an undiscovered colony of ridleys might exist along the West African coast. Two facts supported his hypothesis. First, Carr knew that the zoology of the West African coast was virtually unexplored. Second, drift bottle evidence demonstrated a continuous flow from Africa to the Americas.[49] Also, during one of his trips to Africa, Carr flew from Mauritania to Morocco and observed that most of the coast was uninhabited. He thought that ridleys could nest in many places on the West African coast and escape detection. Carr may have developed his hypothesis in part because he had spent many years canvassing the Caribbean for evidence of turtle nesting and had not been able to glean any information on the nesting biology of the Kemp's ridley.

During August and September 1960, Carr traveled with his family to the Pacific coast of Mexico to conduct a reconnaissance of Pacific ridley sea turtles with the hope that a better acquaintance with the Pacific or olive ridley might shed light on the behavior of the Atlantic ridley. Carr's trip took him from Kino Bay to San Blas, a distance of about 650 miles. He found only two species of turtles: the green turtle and the Pacific ridley. Surveys of numerous beaches revealed the nests of relatively few ridleys. One reason was the size and weight of the ridley, according to Carr: "An observer making an aerial survey of the beaches would get the impression that they are almost completely devoid of nesting activity. The ridley is so relatively little and light, and the sand of these shores so loose and flaky, that tracks may disappear completely during the day following the emergence of the turtle."[50] Even allowing the possibility that the evidence of turtles was ephemeral, Carr still felt that the turtle count was low. The rarity of nests prompted Carr to hypothesize that solitary nesting might be the rule for the ridley throughout its geographic range: "Part of the mystery that has surrounded the breeding habits of the Atlantic ridley may merely be a result of its disinclination to congregate in big gatherings at nesting time."[51]

With this hypothesis in mind, Carr was not surprised to learn about evidence of Atlantic ridleys nesting in Veracruz, Mexico. Based on the evidence of three nests in Veracruz, Carr focused his attention on Padre Island, Texas, a gulf island that stretches from Corpus Christi to Brownsville. In September 1960, Carr discovered five ridley shells at a gift shop on the southern end of the island. Inquiring about the shells, Carr learned that they had been collected by Mexican shrimp boats, so he visited the shrimp docks and showed the shells to the crews, who were very familiar with the turtles and called them *lora* or *cotorra*, as did the people of Veracruz. Some recalled that they had discovered eggs in shells in the turtles after butchering. When he returned to Florida, Carr was stunned when Ogren showed him a copy of a paper written in 1951 that described a ridley nesting on Padre Island. When Carr called Jesse R. Laurence, who discovered the first nest, Laurence said that he found one nest in 1948 and another in 1950, and he sent photographs of the nest to Carr. From this evidence, Carr drew several conclusions: "It now seems reasonable to offer as partial explanation of the durability of the 'Ridley mystery' these factors: (1) scattered nesting emergences, with no rookeries formed, (2) lightly cut emergence trails, easily obliterated by natural causes; (3) a nesting range comprising little visited, and in some cases almost inaccessible shores, such as Padre Island and parts of the Gulf coast of Mexico; (4) nesting by daylight."[52] It turned out that Carr was wrong about the West Coast of Africa, as he was about the absence of rookeries and their accessibility. The solution to the riddle of the ridley did not arrive until 1961, when an old film reached another biologist. Twenty years of searching for nesting ridleys did not prepare Carr for what he saw in the film (see chapter 9). In the meantime, there were other sea turtles to study.

The "Black" Turtle

During the course of his exploration of Mexico's Pacific coast, Carr also saw the black turtle, which was a dark race of the green turtle (*Chelonia mydas*). He found the arcane language of the Mexican fishermen to be confusing with respect to different forms of sea turtles, so he initially attempted to sort the various terms for sea turtles. In other parts of the Caribbean, the word *Caguama* referred to the loggerhead, while in southern Mexico and Central America, it meant the ridley, but north of Acapulco, *Caguama* was a generic term for sea turtle. When Carr realized how vague the word *Caguama* was, he would ask fishermen to describe precisely the kind of turtles they knew. Eventually, Carr constructed a list of equivalent names used by the Mexican fishermen along the Pacific Coast:

carey: the rarely observed hawksbill

galapago: the leatherback

mestiza: the exceptionally light-colored examples of the ridley or of the "black" turtle, and possibly the loggerhead

golfina: the ridley

caguama prieta or *tortuga negra*: the "black" turtle.[53]

To develop this inclusive list, Carr systematically questioned local fishermen and turtle and egg hunters. He also walked beaches, searched dumps for turtle shells and bones, visited markets, crawls, and docks where turtles were landed, and cruised 300 miles in offshore waters. Carr first encountered the "black" race of the green turtle at Kino, where he found a man splashing water on a group of sea turtles. The turtles seemed familiar yet strange to Carr: "The turtles looked familiar, but somehow a little wrong. They had the smooth, neat, unmistakable heads, long front flippers, and general air of the green turtles I know from Atlantic regions; but they were off-key in color, and the shells seemed too deep and straight-sided."[54] When he asked what kind of turtles they were, the man said, "*Prieta*," or black turtles. The coloration of the turtles provided Carr the opportunity to discuss "melanism" or aberrantly dark pigmentation. Both the carapace (top shell) and the plastron (bottom shell) were shaded almost black. Farther south at Topolobampo, Carr studied the shells of 115 black turtles that also revealed the steep-sided shells of the Kino turtles. Carr's visits to Mazatlan, Tamboritos, and Teacapan failed to yield additional black turtles.

After returning home, Carr went back to Colima and La Paz, Mexico, where there were rumors of sea turtle congregations. At Manzanillo (the main port of Colima), he heard about black turtles from fishermen who knew them well but only caught them occasionally. From Manzanillo, Carr crossed the Sea of Cortez to La Paz, where black turtles were so common that the population supported a small fishery. None of the fishermen knew where the black turtles nested. As in Kino, the distinctiveness of the black turtle struck Carr, who mused that even a novice could distinguish the black turtles of Kino and La Paz from Atlantic green turtles. He also wondered if the dark pigmentation of the black turtle was a genetic trait or somehow resulted from the turtle's diet: "If the trait is inherited, and if the turtles showing it are confined to the upper Pacific coast of Mexico, there is a good probability that someone will give the population a new scientific name."[55]

The most important information to come out of the trips to Mexico, in Carr's view, was the clear understanding of local common names: "Knowing these names clears the way for more profitable sifting of folk zoology for leads that can be used in locating and separating habitats of the Pacific turtles. With this done, we can then trace their migratory routes and stations."[56] In exploring the sea turtles of the Mexican Pacific, Carr fully exploited what had already proved to be highly successful strategy: his research began with gathering and analyzing local knowledge, and then he attempted to reconcile local knowledge and scientific knowledge to situate the data within the broader framework of his ongoing research on the ecology and migration of sea turtles.

Ascension Island Turtles

Throughout the early years of sea turtle reconnaissance, Carr concentrated his efforts on the Caribbean. Despite extensive travel by air and on foot, he discovered relatively few nesting colonies of sea turtles. Without question,

Tortuguero was the most significant colony in the western Caribbean, and it was for this reason that Carr devoted the majority of his effort to this 22-mile section of beach along Costa Rica's Caribbean coast. Nevertheless, Carr was aware of other sea turtle colonies, including one of the most inaccessible and remote islands in the Atlantic Ocean—Ascension Island. He knew that victualing ships of the British Navy captured turtles on Ascension and that the island had served as one of the original sources of green turtles for London soup chefs, but he could not visit the island until 1960. Authorization from the Royal Navy's Patrick Air Force Base and cooperation of officials of the British Cable and Wireless Company facilitated turtle research on Ascension. One of Carr's research assistants, Harold Hirth, spent about ten weeks (from February 19 to May 2, 1960) on the island. In the meantime, the tagging program at Tortuguero was in its fifth year. Still another source of data came from the research reports of Tom Harrisson and J.R. Hendrickson, both of whom studied the sea turtle colonies of Sarawak (Borneo) and Malaysia. Drawing on these sources of data, Carr and Hirth compared the features of the three colonies and the breeding habits of the turtles that visited them.

Compared to Ascension Island, the turtle colony at Tortuguero had received intensive scrutiny. Carr and Hirth described Ascension as a true oceanic island located in the middle of the South Atlantic, and they enumerated the many differences between Tortuguero and Ascension: "As a nesting habitat for *Chelonia*, Ascension is markedly different from Tortuguero. The island is a pinpoint of land hundreds of miles from other shores. The nesting ground is not a single beach but a series of sandy crescents at the heads of narrow, rock-guarded coves."[57] Ascension Island differed from Tortuguero in terms of disturbance in that no one hunted the turtles or their eggs on the island, save for the occasional baby turtle collected and sold to visitors. In addition, there were only few introduced predators, such as cats, which had no significant impact on the population. Not even ghost crabs, which were a menace to both eggs and hatchlings in the Sarawak Islands, occurred on Ascension.

In 1957, Carr visited the Atlantic coast of South America in search of information on sea turtles. Although the people who lived on the coast were very familiar with green turtles, no one could direct Carr to nesting beaches, and he found no evidence that turtles nested on the Atlantic coast of South America. Museum visits revealed no specimens of young or hatchling turtles from the shores of Brazil or Argentina. Carr viewed this absence as particularly egregious given that, in a typical museum collection, the easy-to-preserve and store small, young, and immature sea turtles made up the majority of examples. The balance of evidence suggested that the sea turtles of the South American coast arrived from someplace else. Where was the nearest center of aggregated nesting? Ascension Island. In this case nearness is relative: a sea turtle traveling from the coast of Brazil to Ascension Island must swim in excess of 1,400 miles against the current and much farther to take advantage of currents. Based on this information, Carr sent Hirth to Ascension to tag turtles.

During his time on Ascension, Hirth tagged 206 green turtles. On average, the Ascension turtles were larger than the individuals of any other population of green turtles. Over the course of the next year and a half, two of the tags were returned: both from the waters off Brazil. Considered independently, these two returns might suggest a mere coincidence of random wanderings, but Carr knew better: "Considered against the background of relevant circumstantial evidence, however, they must be regarded as important evidence of migratory contact between the resident turtles of the Brazilian coast and the Ascension breeding assemblages."[58] Just as Carr could find no evidence of sea turtle breeding on the South American coast, he could find no significant areas of turtle grass off Ascension Island. There were no shallow bays and lagoons where turtle grass could become established. In fact, at the conclusion of nesting, sea turtles promptly left the waters surrounding Ascension. Here was an intriguing parallel to Tortuguero. Tortuguero also lacked turtle grass, and over the course of five years since tagging had begun there, no sea turtle had been found near the beaches of Tortuguero outside the nesting season. After nesting, the majority of green turtles migrated to the beds of turtle grass in the Mosquito Cays about 200 miles away. In contrast, the nearest suitable foraging grounds to Ascension Island were those off the coast of Brazil. Thus, Carr concluded: "The pair of recoveries thus should be regarded not as odd facts of no quantitative significance, but as pieces in a pattern of evidence."[59]

As for other possible sites of origin for the Ascension Island colony, the West African coast struck Carr as the most likely source, but he had explored that possibility during two reconnaissance trips to West Africa. Additional information came from interviews with fisheries officers from Dakar to Liberia. Although Carr was able to confirm limited nesting by green turtles, he believed that turtles off the coast of West Africa might also nest at Ascension. Tortuguero returns suggested that one nesting beach might serve the turtles from regions more than 1,000 miles apart.

For Carr and Hirth, the circumstantial evidence strongly supported the claim that Brazilian turtles nested on Ascension Island. This deceptively simple answer raised additional questions such as, What routes did the turtles use to reach the island? As noted above, the most direct route would require mature Brazilian turtles to swim up the Equatorial Current to Ascension. Hatchling and immature turtles would experience a much easier trip traveling down the same current to the waters off Brazil. Carr noted that such travel would conform to the "classic pattern" for marine migration: upstream movement for the strong-swimming adult stages and downstream drifting for the weak and naïve young.[60] Alternatively, turtles could exploit favorable currents by swimming north to the Equatorial Current, which eventually joins the Gulf Stream and stay with it through its entire global circuit, but both the time involved and the lack of sightings of aggregations of migrating turtles along the Gulf Stream belied this possibility. After eliminating another route as implausibly convoluted, Carr and Hirth reiterated that only the direct upcurrent route seemed possible, despite the considerable challenge it posed.

Yet another question plagued Carr, and it was technically even more confusing than the mystery of the ridley. Given that Brazilian turtles nested on Ascension and that they somehow mustered the strength as adults to swim at least 1,400 miles up the Equatorial Current to reach the island, how did the turtles manage to locate the island in the vast Atlantic Ocean? During World War II, navy pilots somberly joked, "If you miss Ascension, your wife gets a pension." The problem of turtle orientation and navigation occupied much of Carr's research time during the 1960s and 1970s. Carr and Hirth refined the question of orientation into two related questions: "(1) How is the high-seas journey to the general area of the island guided? (2) What is the character of the landfall that the migrant makes?"[61] Unlike turtles searching for Tortuguero, which were headed for a point on the land-sea interface, turtles searching for Ascension Island had to find the island in mid-ocean. High-flying birds could use visual cues to correct their course to Ascension, but turtles could only look out from the water, and a low island would be out of sight even a few miles away. No one knew what cues a sea turtle might use to complete such a difficult and unlikely journey. The theory of upstream travel enabled Carr and Hirth to postulate an olfaction gradient in addition to possible cues from navigation by the sun or stars (which birds used).[62]

Zoogeography through Aerial Reconaissance

When Carr originally applied for funding for the tagging program in Tortuguero, he submitted the same proposal to the National Science Foundation (NSF) and the Office of Naval Research (ONR). When he learned that NSF would fund his research at a reduced budget, he eliminated some of the most costly elements of the proposal that included extensive travel. While revising the proposal, Carr contacted S. R. Galler, Head of the ONR's Biology Branch, in the hopes of leaving a door open for future funding: "In the form approved by NSF the project budget was cut by about six thousand dollars. In looking over my plan for spots to prune I decided, with some misgiving, to cut down on the zoogeographic aspect of the problem, since the NSF panel regarded this as somewhat peripheral, and since its exclusion would allow considerable economy of travel costs."[63] Although Galler tentatively supported Carr's revised proposal, Carr decided to focus on the central issues and postpone the extensive travel once his NSF grant was approved.

In March 1961, Carr renewed the search for additional funding to support migration and navigation research on sea turtles. In his proposal to the ONR, he reviewed the state of knowledge regarding the migrations and orientation of sea turtles. Adult green turtles offered distinct advantages for experimental study: they moved slowly, surfaced regularly to breathe, stayed near the surface while traveling, were large enough to carry tracking apparatus, and their shells could hold instruments. However, there were clear disadvantages to using adult turtles: it was difficult to assess which individuals were going to undertake the nesting migration in a given year of the three-year nesting cycle; nesting females returned to lay additional eggs four to six times over the course of the nesting season; and homing tests depended upon the particular "home" the turtle had in mind.[64] On balance, young turtles were better experimental organisms with which to study orientation. Thus, Carr's proposal included two parts: the study of zoogeographic patterns and experimental work with young turtles.

To map zoogeographic patterns in sea turtles, Carr proposed a series of reconnaissance flights to turtle basking locations in the Pacific Ocean, including the islands of the Midway archipelago, Green Island at Kure Atoll, and islets in Pearl and Hermes reefs. Reconnaissance would enable Carr to positively identify the species involved in the basking aggregations and to determine seasonality, sex, and age. Military flights were crucial to Carr's proposed research: "Such a survey could be carried out effectively, and most economically, only if MATS [Military Air Transport Service] and other military transportation could be made available. Airplane surveys in the Caribbean have allowed us to work out field characters for rapid determination of species, sex, and age group, using binoculars from low-flying aircraft."[65] Such flights would also facilitate surveys of feeding grounds. With ONR support, Carr and his students could fly virtually anywhere in the world on MATS flights for free. The two kinds of surveys would lay the groundwork for navigation studies. Specifically, Carr planned to fill out world migration routes and to locate sets of localities between which turtles made periodic journeys without reference to landmarks that would serve as the sites for future experimental studies (like the ones underway at Tortuguero and Ascension Island).

The second part of Carr's proposal involved orientation experiments with young sea turtles. As Carr had demonstrated in his earlier experiments conducted with Ogren, hatchling sea turtles had the ability to travel from a nest located far inshore, behind dunes and vegetation, to the ocean which they had never seen and which could be entirely obscured from view by various objects. This ability seemed to include some features of the navigation used by the mature sea turtles. Over the long term, Carr hoped to devise field tests to assess light perception (especially the relative importance of intensity and area in phototaxis or orienting by light), pattern discrimination (as a basis for planetarium tests), angle discrimination, sun orientation, and homing and "map" sense. But for the ONR contract, he restricted his research to light perception and angle discrimination. Carr noted that he had kept turtle hatchlings in the laboratory for up to sixteen months and that he had access to laboratory space at the University of Florida and its associated Marine Lab at Seahorse Key near Cedar Key in Florida. Moreover, Carr's close friend, James A. Oliver of the American Museum of Natural History, had offered the use of facilities at the Lerner Laboratory at Bimini, where large numbers of hatchlings could be kept through the winter months and where light angle studies could be carried out.[66] In a relatively short time, the ONR funded Carr's contract for the full amount requested, $5,564. Far more significant than the amount of the grant, however, was access to MATS flights. Carr and his students would exploit the golden age of scientific travel by air at no cost.

In developing a methodology for determining how sea turtles found their way from nesting grounds to feeding grounds and back again, Carr explored the possibilities of using telemetry, which had only recently become a viable option:"Application of telemetry to biological problems has had to await advances in miniaturization, and these have been greatly accelerated by the guidance demands of missile and space programs."[67] In March 1962, Carr presented a talk on telemetry and sea turtle navigation at the Interdisciplinary Conference on the Use of Telemetry in Animal Behavior and Physiology in Relation to Ecological Problems, a conference held at the American Museum of Natural History. After reviewing progress of the Ascension Island tagging program (four tags had been returned, and all were found in Brazil), Carr discussed the results of crude tracking tests using Styrofoam floats and helium balloons: "Results have ranged from the ambiguous to the meaningless. Gross procedural troubles cancelled most of the efforts." Everything from rough inshore waters to line snags to the curiosity of crews of shrimp boats marred the experiments, leading Carr to conclude: "It is thus clear that, when serious tests are undertaken, telemetry must be used, and, as was said, the genus *Chelonia* seems in some ways a specially good prospect for such manipulation."[68] Carr believed that the possibility of using radio telemetry to solve problems of sea turtle navigation was a sign of new times for natural history.

By 1963, eight of the tags had been returned from the 206 turtles that Hirth tagged on Ascension Island in 1960. All of the tags were recovered in Brazil. Although this represented strong circumstantial evidence, Carr could not make a definitive claim that Brazilian turtles nested on Ascension. To strengthen the

Figure 20. Tracking experiments with sea turtles using balloons and floats (ca. 1961). Courtesy of Mimi Carr.

evidence, Carr established a tag-check patrol on three of the Ascension Island beaches in 1963, three years after the original tagging. Given the typical reproductive cycle of sea turtles, 1963 should have been the first year that the turtles tagged in 1960 returned to the beaches to nest once again. Out of the original 206 turtles, 3 tagged turtles were nesting on the same beaches on which they had been tagged. Those three recoveries, plus two more in the 1964 nesting season, further bolstered the evidence that the Ascension Island sea turtles were returning from someplace a great distance away such as Brazil.

The new evidence prompted Carr to consider the evolution of the orientation system that enabled turtles to find Ascension Island (only 5 miles wide) in the middle of the Atlantic Ocean. Carr theorized that Ascension might not have been so small when ancestral turtles began making the nesting journey to the island. The island is a peak on the Atlantic Ridge and given lower sea levels may have been much larger. The gradual shrinking of the island may have given the turtle's navigation sense time to undergo a process of refinement. Carr also raised the possibility of passive transportation of the naïve hatchlings from the nesting ground to Brazil by the Equatorial Current. Such passive travel would significantly reduce the minimum requirements of instinctual navigation. Carr wrote: "If they behave as mere plankton on the westward trip, the three-knot Equatorial Current and its branches could carry them within short distances of any of the localities from which our mainland tag-returns have come."[69]

Carr also looked beyond Tortuguero and Ascension Island for additional clues about the navigation abilities of sea turtles. In September and October 1963, he was able to visit Kenya, Tanganyika, and Madagascar on the western coast of the Indian Ocean. Carr's specific aim for this trip was to determine if populations of turtles from East African coasts might nest on the remote island of Aldabra. Traveling along the coast, Carr determined that there were no significant concentrations of sea turtle nests. He did see young and subadult green turtles, but he suspected that these turtles derived from a distant rookery. From traditional turtle hunters, the Bajuni in Kenya, Carr learned that most of the turtles "came from somewhere far off."

The Bajuni used a unique technique to catch sea turtles. They attached lines to the tails of remoras (sucking fish). After a few encouraging words to the fish, they dropped one or more in the water and played out the line until the remoras found a sea turtle and attached to it. Once several fish had attached to the turtle, the Bajuni fishermen could slowly reel it in. After conducting careful surveys and interviewing local fishermen, Carr concluded: "All evidence pointed to Aldabra as the only important rookery remaining in the western part of the Indian Ocean"[70]

Gradually, Carr was assembling the biogeography of sea turtle populations and, specifically, nesting congregations and migratory patterns around the world. He realized that migratory patterns might hinge on fat content. Some turtles (such as the "yellow" green turtle occasionally found among the Pacific "black" green turtles) stored large amounts of fat, and these turtles were most likely to migrate long distances. The most promising sites for continued analysis of this issue were the Galapagos Islands, where both the Pacific "black" and the "yellow" green turtle occurred, or the coast of Brazil, where one small population migrated a short distance to Trinidad and another swam to Ascension Island some 1,400 miles away. Studies of turtle fat could be conducted on Florida's Gulf coast, where a fairly large population of nonbreeding turtles congregated each year.[71]

Conclusion

In studying Carr's research on sea turtles over the course of the first ten years, it is useful to recall the rather abbreviated and limited accounts contained in the *Handbook of Turtles*. Carr devoted his first decade of sea turtle research to initiating long-term studies, particularly of the green turtles at Tortuguero. With the assistance of undergraduate and graduate students, he was able to fill many of the gaps in the life histories of other species, including leatherbacks, loggerheads, and ridleys (though the "riddle of the ridley" lingered). Still interested in taxonomy, Carr investigated two distinct races of the green turtle: the "black" turtle and the "yellow" turtle. In addition, Carr connected with a small but growing contingent of sea turtle researchers (most notably Harrisson in Borneo and Hendrickson in Sarawak), as well scientists who studied the phenomenon of migration in other organisms, such as birds. He expanded the geographic range of his studies to include Ascension Island, the Pacific coast of Mexico, and the Galapagos Islands. Carr's zoogeographic studies were greatly facilitated by support from the Office of Naval Research and particularly by access to MATS flights. Some of his experiments such as aerial reconnaissance were hindered by lack of suitable technology.

At the beginning of this chapter, we saw the experience of working with Carr through the eyes of Larry Ogren. Carr's other students have similar memories of time in the field with their professor. David Ehrenfeld recalled how Carr would recite from the "Jabberwocky" or comic poems of P. G. Wodehouse with the night sounds of the lowland rainforest as an accompaniment.[72] In a remote town in Suriname called Alliance, Carr used a public outhouse perched over a river. He emerged quickly to show his student Peter Pritchard the coprophagic (feces-eating) fish that congregated underneath the privy.[73] It was this energy and spirit of fun and curiosity that inspired colleagues to invite Carr on their expeditions (see chapter 7). As extensive as Carr's research on sea turtles was, it is remarkable that he was able to set aside time to continue his writing for the public. Through an unlikely but fortunate series of coincidences, Carr wrote two works devoted to the nature of Africa.

CHAPTER 7

In Africa on Ulendo

Most scientists, such as physicists, chemists, molecular biologists, and geneticists, care little for where they conduct their research: a lab in New York differs little from one in Paris or Tokyo. But for a naturalist like Carr, place mattered. In his early years, his studies were local: the backyard and the southern pine and mixed hardwood forests he came to know on hunting trips with his father. College and graduate school (and an automobile) meant greater independent mobility, and he took extended trips throughout Florida. When he completed his education and settled into a full time appointment, Carr enjoyed still greater mobility: summers in Cambridge with his mentor Thomas Barbour and ultimately travels to Mexico, the Caribbean, and Honduras, where Carr would spend nearly five years with his young family exploring the tropical highlands and lowlands. Further wanderings gave Carr critical data on the status of sea turtles in the Caribbean and Costa Rica. Such travels convinced Carr that the breeding grounds of sea turtles had been severely reduced and that Tortuguero in Costa Rica was the largest colony of green turtles in the Caribbean.

During his early career, Carr's research was largely confined to Florida and the Caribbean, but two factors motivated his desire to extend his explorations to other places. First, it had become clear to Carr that sea turtles could be found in most of the oceans of the world and that a biologist could only understand these species by following them. Carr's second reason had less to do with his particular expertise than with the passions of any naturalist. To explore life, to compare life forms, to understand the relationship between biology and geography—(which is to say, life and place), naturalists travel.

Genesis of a Journey

Carr's first journey (*Ulendo* in Chinyanja) to Africa evolved out of a long-term friendship and professional relationship with his colleague Lewis Berner, who was a graduate student in biology at the University of Florida roughly contemporaneously with Carr. Like Carr, Berner devoted himself to life histories.Berner's master's thesis and doctoral dissertation were on the mayflies of Florida.[1] Like Carr, Berner remained at the University of Florida and continued to study entomology generally and mayflies specifically. He published his dissertation as *The Mayflies of Florida* in 1950.[2] Berner's entomological expertise interested a British engineering firm, Sir William Halcrow and Partners, which was conducting feasibility studies for various major engineering projects in the British colonies around the world. Entomological surveys represented a key component of such studies. In the late 1940s, Berner contracted with Halcrow to conduct entomological surveys for feasibility studies of the River Volta project.[3] Berner's mission was to assess the risks of insects to workers and settlers. Malaria (carried by the *Anopheles* mosquito) and schistosomiasis (hosted by aquatic snails) were two of the greatest health threats to people living in eastern and southern Africa.[4] Other problematic insect-borne diseases included filariasis and blindness caused by a tiny black fly. For his second contract with Halcrow, Berner invited Carr to join him in Nyasaland (now Malawi) and Mozambique. Carr leapt at the opportunity to study African wildlife first hand: "My own secret aim, not revealed in correspondence with Sir William Halcrow and Partners, was to see my dream of Africa unfold."[5]

Carr arrived in Africa in late June 1952, after several days in London for meetings with representatives of Sir William Halcrow and Partners. Carr's letters to Marjorie balance his exhilaration in studying African wildlife and culture with his considerable regret that his partner and fellow naturalist could not accompany him. Each of Carr's many letters to Margie began: "Dearest" and closed with "I love you." In between there are many details of life and nature in Africa and an equal number of references to shared experience and testimonials of devotion. Marjorie reflected on her disappointment in not being able to accompany Archie to Africa years later:

> In the summer of 1952 when Archie went on his wonderful *ulendo* there was only one fly in the ointment—I couldn't go with him. There was no possible way I could go. We had *five* young children ranging in age from nine years to six months. We lived in a barely completed house on ten acres of land on the shore of Wewa Pond. Our piece of woods was ten miles from Gainesville and two miles from the little old town of Micanopy.
>
> We had two cows that Mimi and Chuck milked twice a day, a horse, Cricket, that had a colt named Kate, two dachshunds, and a German shepherd. . . .
>
> Of course it was a lonely summer, but the young were a good pack. We went swimming each day; for a while we had a baby raccoon and the

garden produced the biggest Beefsteak tomatoes I have ever see, before or since, in Florida. Archie's letters, written on little thin airmail stationery, were anxiously watched for and read and reread. And time did pass and he came home.[6]

Even though Africa represented a completely new environment for Carr, he found many similarities to Central America and Florida. The British colonial residents of Nyasaland reminded Carr of the atmosphere at the Escuela Agricola Panamericana in Honduras. Some of the tribes used big-headed ants to suture wounds (like Indians in Nicaragua and Ecuador), Carr reported to Marjorie.

Carr's daily activity in Nyasaland involved taking mosquito censuses with Berner. The two scientists would go to a village and find the chief and walk with him to several selected huts where they hoped to sample mosquitoes. Then the chief would explain to the residents what Carr and Berner planned to do and the woman of the house would move out and cover anything that insecticide might hurt (everything from containers of food to dried fish to chickens on eggs to puppies). After covering the floor with sheets, the scientists filled the hut with aerosol spray and waited for the mosquitoes to die.[7] Villagers, who gathered to witness the spectacle of the mosquito survey, were fascinated by the exodus of insects from the building. Many farm animals (chickens, ducks, and guinea hens)

Figure 21. Archie Carr and George Hopper taking mosquito survey in Nyasaland. Photograph by Lewis Berner, courtesy of the Department of Special and Area Studies Collections, George A. Smathers Libraries, University of Florida.

Figure 22. Archie Carr collecting snails in Nyasaland. Photograph by Lewis Berner, courtesy of the Department of Special and area Studies Collections, George A. Smathers Libraries, University of Florida.

joined the fray. But the climax of the event arrived when Carr and Berner opened the door to the house and revealed scores of dead insects, including many mosquitoes. Carr fondly recalled hearing the villager's response: "They stretched their necks to see all the dead little creatures lying so still on the cloth, and the grand word *udzudzu*, which to any ear could only mean mosquito, was buzzed about by every adult and child."[8] It is clear, however, that while mosquito, fly, and snail surveys had gotten Carr to Africa, his real interest lay with the diverse wildlife and ecosystems of the continent.

The Desire to Hunt

In Africa, Carr felt his desire to hunt ebb. As a child in Georgia and Florida, he had hunted regularly with his father and continued through his early career, but in Africa, he began to question his motivation and the necessity of hunting generally. In one way, Carr mused, killing things was childish, but it represented a mature and necessary craft: "And when all the need to kill for survival was gone, the old blood stirring urges stayed with us, as intact as our useless wisdom teeth. Like wisdom teeth they have been far slower to yield to changing times than the need to use them was."[9] Carr's ambivalence regarding hunting became most pronounced during an elephant hunt in Mozambique. During the course of an entire day of tracking elephants on the move, Carr noted numerous details of the behavior and significance of elephants and the impact of a herd on the environment around them. By the time ten hours had passed searching for elephants in vain, he began to lose heart for the hunt, and his thoughts turned to past and future generations: "We are Cro-Magnon no longer; we are no longer Minute Men or Forty-niners. Whatever needs, joys, and rights our grandchildren might have, these will not include felling trees or felling bodies."[10] Carr recalled this hunt fondly when he wrote to Marjorie: "I had a wonderful time this weekend—the kind you and I seem better able to enjoy than most people, and perhaps the pleasantest unsuccessful elephant hunt anybody ever went on."[11] Here Carr suggests that satisfaction from hunting had less to do with success or failure than with the experience of tracking and studying wildlife.

Like most visitors to Africa, Carr expressed his amazement at the impressive diversity of Africa's megafauna: predators such as lions and prey species including wildebeests, gazelles, kogonis, and zebras. But herpetology was still Carr's foremost passion. Everywhere he went Carr found snakes, turtles, and crocodilians that never failed to impress him. Nevertheless, Florida and its varied herpetological fauna was never far from his mind. For example, while exploring Lake Nyasa (now Lake Malawi), Carr discovered numerous large Nile monitor lizards (*Varanus niloticus*). The monitors were particularly abundant on Boadzulu Island at the south end of the lake, and this reminded Carr of the abundant cottonmouth snakes he had found on Snake Key.[12] In one of his earliest published papers, Carr noted that the cottonmouths fed on fish and nestling birds dropped from cormorant nests. It seemed that the monitors, like the cottonmouths, fed on fish and nestlings from the cormorant nests as well. Despite considerable swimming ability, Nile monitors hunted only on land, and there was no other source of food for a large carnivore on Boadzulu Island.

Given the small size of the island and the tidy relationship between cormorants and monitors, Carr recommended that Boadzulu be designated as a preserve or monument for some form of protection. A little research revealed that there had once been a python preserve on Chidiamperi Island in Lake Shirwa. If people had once thought to preserve a space for pythons, perhaps they could be persuaded to save Boadzulu. Carr noted that wilderness preservation emphasized "spectacular spreads of earth, the big wild lands in which the public can play."[13] Boadzulu Island called for a different kind of preservation. In addition to the places that facilitated recreational pursuits such as fishing or hiking, there were from Carr's perspective "the Boadzulus to think of saving. . . . They are delicate bits of balanced landscape that can be kept only if nobody ever sets foot upon them."[14] Here Carr added his voice to ongoing wilderness debates.[15]

In comparing the story of monitor lizards and cormorants on Boadzulu to cottonmouths and cormorants on Snake Key, Carr's narrative progressed from species to interactions to place and then to conservation. Here is a critical component to his philosophy of biology. The relationship among species, places, and conservation is a fundamental aspect of conservation biology. In moving from organism to ecology and from habitat to ecosystem, Carr bolstered the case for conservation. Boadzulu was a small island in Nyasaland that harbored a species at little risk of extinction. Nevertheless, in Carr's view Boadzulu was unique. Following a nearly identical logical progression, Carr had made the case for the preservation of Tortuguero, a twenty-mile stretch of black sand beach in Costa Rica, but the physical description fails to capture the ecological and biological significance of Tortuguero: it was the largest sea turtle nesting colony in the western Caribbean. To make this argument, Carr proceeded from a species (the green turtle) to the interaction (nesting) to the place (Tortuguero) to conservation (the largest remaining colony). In this respect, Florida, the Caribbean, and Africa were essen tially identical. Each supported threatened species in endangered spaces in need of protection from a burgeoning population of human beings.

Whither *Kungu*?

Carr had gone to Africa to survey mosquitoes, snails, and other disease-bearing insects, but turtles, snakes, and lizards occasionally distracted him from his surveys (or perhaps his insect surveys distracted him from his study of things herpetological). Nevertheless, invertebrate life in Africa was no less compelling than vertebrate life, and Carr held credentials in invertebrate zoology. His master's degree at the University of Florida was in limnology (the study of lakes), and his professors J. Speed Rogers, Theodore H. Hubbell, and Charles Francis Byers were all entomologists. Rogers in particular tried to inspire a fascination with insects in Carr. Carr recalled this while witnessing the great swarms over Lake Nyasa called *kungu*. To convey the spectacular nature of *kungu*, Carr quoted from David Livingstone's *Zambezi*: "But next morning we sailed through one of the clouds on our side, and discovered that it was neither smoke nor haze, but countless millions of minute midges called 'kungo' (a cloud or fog)."[16] To say that Carr was impressed when he witnessed the clouds of midges risks understatement: "You can name me the wonders of Africa, and tell of Ruwenzori or Serengeti, of the great falls of the Zambezi, or of Cape Town Harbor—or of any of the other great things there are to see there. But I know of nothing in Africa that more wholly astounds the mind than the kungu clouds of the Great Lakes. There is nothing anywhere that so overpoweringly seems to show the mindless drive of life as these vast up-pourings of protoplasm show it."[17]

Seeing *kungu* reminded Carr of his days as a graduate student in zoology at the University of Florida when Rogers showed him the larval form of a midge of the genus *Corethra*, the same genus as the *kungu* flies (Carr noted that the name of the genus had been changed to *Chaoborus*). Rogers' demonstration intrigued Carr, but not enough to draw his interest from turtles and other reptiles. Although no one had studied the biology of the *kungu* flies in depth, Carr discovered detailed notes on *Corethra* in the writings of the Wisconsin limnologist Chancey Juday (1871–1944), who found 33,800 *Corethra* larvae per square meter in Lake Mendota, Wisconsin. When Juday multiplied the number of larvae by their average body weight, he calculated 1,070 pounds per acre. Carr noted that that amount was more meat than beef could produce on an acre of cultivated pasture![18]

Carr wondered what caused the swarms and what ends they served. What did *kungu* flies stand to gain by forming such massive aggregations? When he posed this question to Berner, his colleague gave a concise response based on extensive knowledge of Ephemeroptera (the mayfly order), which form breeding aggregations: "to mate." But the swarms of *kungu* gnats soared thousands of feet into the air over Lake Nyasa. What was the point of the swarms if not to mate? Carr received further clarification from the director of the Entomological Research center of the Florida State Board of Health: "We believe that insect migration and swarming are consummatory rather than appetitive behaviour, to use the current ethological idiom. As such they serve no special survival purpose."[19] That comment continued to bother Carr until he realized that evolutionary biology was only part of the reason for *kungu* swarms: "whatever the initial reasons for the Nyasa kungu swarms, they soon stop being a biological phenomenon and become a part of the weather. Like some buzzards . . . the flies have entered and have become a part of twisting columns of rising air. The trade wind comes in over the hot Mozambique country and the air warms, lightens, and rises; and here and there goes up in spiraling eddies."[20] Perhaps the spectacular *kungu* assemblages reflected and made apparent the unique weather patterns over Lake Nyasa. According to Carr, Edward Young, the captain of the *Ilala*, the first steamboat on the lake, described immense storms and waterspouts.[21]

Neither biological nor meteorological explanations completely satisfied Carr's curiosity, so he turned to folklore. The cook on board the *Ilala* (the vessel on which Carr traveled named in honor of the original *Ilala*), told Carr that the people living on the western shore of Nyasa, where, because the water was deep and the fishing difficult, protein was always in short supply, believed that the *kungu* was sent in with the clouds for benefit of their diet. The day after this conversation, Carr was able to sample *kungu* for himself when the cook presented him with a *kungu* omelette. Despite their best intentions, Carr and Berner failed to finish their "fly omelettes." Nor did anyone else aboard the *Ilala* that day. And yet, Carr recognized that the prejudice against terrestrial invertebrates in the diet was an issue of familiarity. Citing an article by Marston Bates in which he examined what he believed to be an irrational taboo against insects in the diet, Carr noted that for many Americans the taboo extended from terrestrial invertebrates to "animals with too many legs," such as insects, and "animals with too few legs," such as snakes. To test the taboo, Carr served *kungu* on crackers when he returned to Florida after his trip to Africa. He commented that there were few compliments but no complaints: "Later, when I divulged what the spread was, no one struck me, but they quite clearly thought they had been imposed upon. After all, the consensus was, the rules for *hors d'oeuvres* are pretty slack, but gnats.... And these were people, mind you, who were stuffing themselves with smoked oysters and anchovy paste. And anyway, they said as a last thrust, the gnats taste like they're spoiled."[22] Here was a case where a practical joke (for which Carr was renowned) underscored a cultural prejudice that Carr parodied.

In describing the natural wonder of *kungu* clouds over Lake Nyasa, Carr deftly wove several narrative strains into a seamless story that incorporated biology of animals in Africa (and Florida and Wisconsin), autobiography, evolutionary biology, meteorology, historical accounts and folklore, and African and western dietary imperatives. The result is a tour de force of science writing. The reader obtains a deeper appreciation of African wildlife and culture that resonates with an appreciation of American wildlife (particularly in Florida) and American culture.

The Diversity of Cichlids

Carr's evolutionary theory served him well when he shifted his gaze from the skies to the water of Lake Nyasa. *Kungu* flies were impressive for their phenomenal abundance; the fish of Nyasa were fascinating for their incredible diversity: no fewer than 180 species out of more than 200 belonged to a single family, Cichlidae. As in much of Carr's musings on natural history, his first encounter with the cichlids of Nyasa linked nature and culture. Carr spent an afternoon watching a group of villagers maneuvering a large net (a half mile long by Carr's estimate). Eventually, he noticed the largest crocodile he had ever seen watching the fishermen. What made the crocodile's presence a cultural event, however, was the way the the villagers responded to it. According to Carr, the villagers accepted the crocodile as a necessary part of their fishing. Unlike Europeans, the African villagers did not have large enough guns to destroy the crocodile, so they settled into an uneasy truce.

As the villagers drew the net to the shore, Carr saw that the fish haul was relatively small (only 50 pounds or so) given the size of the net and the effort of numerous villagers. In a moment of great ambition, Carr tried to write down the names of all the different fishes as the villagers transferred them from the net to baskets. Some species were familiar and deceptively easy to identify: lake mullet, lake salmon, several catfish species, and the elephantfish, but Carr had difficulty identifying the vast majority of the catch "because the main body of the catch was a tumbling galaxy of shining cichlids. Among these the diversity that I was able to make out seemed to be nowhere near as great as that of the [vernacular] names I was taking down. . . . After a while I gave up the census in disgust and tore the pages of names from my notebook as worse than useless."[23] Only later when he corresponded with an ichthyologist at the British Museum did Carr discover the root of his difficulties with cichlids:"The trouble is real. The fishes of the lake are an incredibly finely subdivided spectrum of kinds of life. The plethora of vernacular names corresponds to a real and far greater plethora of species, to an extravagant redundance of fishes that has no counterpart anywhere in the world."[24]

Incredible biological diversity drove a corresponding diversity of culture. In *Ulendo*, Carr described several kinds of nets used by the villagers, but in his book *The Land and Wildlife of Africa*, the editors included diagrammatic drawings of four different nets in the margins.[25] The largest of these was called the *khoka*; this was the net used by the villagers in Carr's first encounter with cichlid fishes. Carr considered the *khoka* to be a testimony to African ingenuity and resourcefulness. Whereas the nets had once been woven from plant fiber, when Carr saw the nets, they were tiny threads of rubber carded from old truck tire treads and laboriously knotted together: "There is no way of escaping the surety that on Lake Nyasa mile-long nets are made of threads salvaged from tire cord."[26] Other nets were smaller but no less ingenious. Nonetheless, cultural creativity could not compare with the diversity of the cichlid fishes.

The great riddle of the Lake Nyasa cichlids, as Carr put it, was "to know what sort of isolation, what kinds of barriers, built and maintain this unequaled exorbitance of species."[27] Carr imagined that the Nyasa fishes would in time join the mammals of Australia and the birds and reptiles of the Galapagos Islands as a classic example of the process of evolution at work. The Nyasa cichlids seemed to have broken Jordan's law, which Carr described as follows:

> What Jordan's Law said was that the nearest relative of a kind of animal should not be looked for in the animal's own environment, or in some far-off place, but in an adjacent place, living separated from the first animal by some kind of barrier. Obviously your feeling about the words "place," "environment," and "barrier" will determine the amount of good you see in Jordan's Law. But all the law really says is what Darwin himself said, and what every biologist believes today—that to have speciation there has got to be isolation.[28]

Given that these cichlids occupied a single freshwater lake in East Africa, it was not clear to Carr (or anyone else) what constituted the barriers that drove speciation in this case. Evolutionary biologists were stymied when it came to Nyasa fishes. Carr noted that different species of cichlids spent much of their time doing essentially the same things, occupying almost the same roles. Such overlapping behavior perplexed ecologists.

One way to explain both challenges would be to demonstrate that Lake Nyasa contained a rich complex of environments, but Carr noted that in fact Nyasa was anything but varied. The lack of dissolved oxygen below 250 meters meant that nearly half of Lake Nyasa was uninhabitable by animals in need of both oxygen and contact with the bottom of the lake (like the cichlids). The consistency of environments left Carr wondering about the diet of the fishes. No fewer than three different genera of cichlids survived on a diet of the scales of other fish. Carr thought that this extraordinary state of affairs reflected a form of mimicry. After describing several kinds of mimicry in nature and classic cases of mimicry, Carr suggested that the scale-eaters of Nyasa evolved to share a resemblance with other species so that they could blend into the school of a host species and parasitize it at will. Carr acknowledged that this was not a perfect example of mimicry.

Nor did it meet the strict definition of parasitism or other forms of symbiosis, which Carr dismissed in turn.

The biological phenomenon most clearly exhibited by the scale eaters was that of social parasitism (exhibited by cuckoos in the Old World and cowbirds in the Americas). Carr knew of only two examples of social parasitism in freshwater fish species, both of which had been documented by Marjorie while she was working on her master's degree in biology. While studying bass embryos, Marjorie noticed that eggs from a single nest sometimes hatched into larvae that grew into two different kinds of fishes (large-mouth bass, *Micropterus salmoides floridanus*, and chub suckers, *Erimyzon sucetta*). She later noted the same phenomena in eastern stumpknockers (*Lepomis punctatus punctatus*) and golden shiners (*Notemigonus chrysoleucas*).[29] Like cuckoos and cowbirds, parent chub suckers and golden shiners took advantage of the nests of other species and had their offspring raised by those species with minimal investment.

There were, of course, other possible explanations for the miraculous diversity of Nyasa cichlids. In addition to the diet explanation, Carr explored the age of the lake and the effect of periodic increases and decreases in the water level that connected (and separated) Lake Nyasa from other bodies of water and created opportunities for speciation. Also, the shoreline of Lake Nyasa was not as homogeneous as it first appeared, and given the considerable size and length of the lake, a series of micro-habitats existed within it. Predators may also have played a role in the spectacular evolution of cichlids for at least three reasons, according to Carr. First, numerous predators would keep populations (and thus competition) low. Second, predators would cull weak or ill adapted individuals. Finally, it seemed to Carr that predation along the shoreline might favor the evolutionary phenomenon known as genetic drift.[30]

A unique and highly evolved reproductive strategy was yet another possible element in the evolution of cichlids. Most of the cichlids cared for their young by mouth brooding, the advantages of which Carr compared to internal fertilization in mammals. Unlike bass, which broadcast thousands of tiny eggs into an often hostile environment, female cichlids lay a few large eggs which develop more quickly into mature fry and thus stand a greater chance of continued survival. Carr cited the research of Geoffrey Fryer, who argued that mouth brooding led to schooling by slowing the development of the fry. Because young fish would return to the mother's mouth for safety, the tendency to school would become a fundamental part of survival.[31]

Like the explanation for *kungu*, the explanation for cichlid diversity had to be multifactorial and required the integration of several theories.[32] As with *kungu*, Carr's discussion of cichlid diversity deftly combined natural and cultural history, evolution and ecology, theory and practice, personal research and that of others (including Marjorie Carr's). Carr explicitly expressed his confidence in the continued survival of the cichlids by comparing the prospects of the fish and *kungu* with the prospects for more visible elements of Africa's cultural and natural history:

> The Lake Nyasa fishes, like the *kungu* there, are out of the main rush of
> ruin that is changing the African earth. The time is not far off when the

last lion landscape will be fenced about or scraped away. The last Masai will put on shoes one of these coming decades, and the last crocodile will bake in his drained bog or go off to the zoo. But long into those times, two of the old wonders of Africa will last on in spite of us. The *kungu* will teem up out of the cold mud and climb to the clouds. And the cichlid fishes will go on living the fine-spun secrets of their lives.[33]

In light of the rapid decline of cichlid diversity in Lake Victoria following the introduction of Nile perch (*Lates niloticus*), Carr's prediction now seems somewhat optimistic.[34]

Africa and Florida

When Carr returned to Florida, Africa had become a part of his vision. As a naturalist and a biologist, Carr felt the influence of Africa. More than any place else, Paynes Prairie evoked Africa. The Carr family lived in a house (christened Wewa when Carr returned from Africa) to the south of Paynes Prairie, and Carr worked at the University of Florida to the north. This meant that Carr crossed Paynes Prairie thousands of times and became acquainted with its natural and cultural history.

Carr described Paynes Prairie as 50 square miles of level plain in north-central Florida, let down by collapse of the limestone bedrock. But lest anyone be left cold by that technical description, Carr clarified the significance of the place: "The Prairie is about the best thing to see on U.S. Route 441 from the Smoky Mountains to the Keys. . . . But everybody with any sense is crazy about the Prairie."[35] Carr attempted to capture the essence of Paynes Prairie: "The Prairie is a solid thing to hold to in a world all broken out with man. There is peace out there, and quiet to hear rails call, and the cranes bugling in the sky. It looks like Africa, out on the Prairie, looking off through the tawny plain to far bands of cattle like wildebeests in Kenya."[36]

By chance one afternoon during his daily passage from one side of the prairie to the other, Carr glimpsed an event that forever linked Florida and Africa in his mind. Carr had noticed snowy egrets (*Egretta thula*) and buff-backed or cattle egrets (*Bubulcus ibis*) out on the prairie with a small herd of cows, when one of the snowy egrets began to hunt in the wake of a large, diesel dredge. In a moment of gestalt, Carr imagined that rather than a large dredge, the egret was following a large animal like those he had seen on African savanna. And this connection raised a question in his mind: why did the snowy egret, a marsh bird, leave the wetlands to follow the cowherds? After considerable rumination on the subject, it struck Carr that perhaps the snowy egret had at one time survived by following large herds of animals such as the cattle egrets of the African savanna. Carr started on this line of thinking while traveling by boat down the Lower Shire in Africa:

> There where David Livingstone saw elephants, lions and buffaloes, there were only cattle for egrets to stand with. I watched them, and an idea

dropped into my shifting daydream that I have not been able since to reason away. The idea was this: the snowy heron is, as the buff-back is becoming, an old game heron with the game all gone. Both of these small, white herons are today walking with cattle as a compromise with the grand living of the past, as the best they can do in a world all changed around them. The buff-back is becoming a cattle heron because the savanna fauna of Africa is being wiped out—the snowy because the Pleistocene herds are gone.[37]

Carr ran through the list of large mammals that the cattle egret followed in Africa: elephants, rhinos, zebras, giraffes, wild asses, and many antelope. Such opportunities had not existed for snowy egrets in Florida for millennia, but Carr's vision stretched to the prehistoric past:

> Through millions of years Florida was spread with veld or tree savanna much like the Zambezi delta land today. Right there in the middle of Paynes Prairie itself, there used to be creatures that would stand your hair on end. Pachyderms vaster than any now alive grazed the tall brakes or pruned the thin-spread trees. There were llamas and camels of half a dozen kinds, and bison and sloths and glyptodonts, bands of ancestral horses, and grazing tortoises as big as the bulls. And all these were scaring up grasshoppers in numbers bound to make any heron drool.[38]

Carr wrote that cattle egrets had arrived in Florida only in 1941 but had undergone a rapid expansion in range and number, and in 1956 there were more than 1,000 cattle egret nests at Okeechobee alone. Wherever cattle were abundant, cattle egrets fared well. But while the survival of both the cattle and snowy egrets revealed their remarkable ability to respond to environmental changes, Carr worried what the necessity of such changes meant for wilderness in Florida and Africa:

> There is a growing emptiness around us, and we fill it with noise, and never know anything is gone. But the buff-back remembers other times, with great game thundering through all the High Masai. And back at home you come upon a raging dragline with a wisp of a snowy heron there, dodging the cast and drop of the bucket as if only mammoth tusks were swinging—and what can it be but a sign of lost days and lost hosts that the genes of the bird remember?[39]

It was also in Florida that Carr began to worry about the future of Africa's wildlife and culture. The inspiration for Carr's concern was innocent enough. He was enjoying a lazy afternoon tubing down the Ichetucknee River. Tubing is a popular pastime in the spring-fed, freshwater rivers of northern Florida, and the Ichetucknee is one of the most popular sites. The springs keep the water at a constant 72° Fahrenheit, which sounds warm enough until you consider that 72° is generally about 20° cooler than the air temperature during the summer when most people go tubing. Floridians find the springs either refreshingly cool or too cold. Carr was fortunate to find teeth from two different mammoth species at the bottom of the river as he floated along, and his discoveries started him thinking: "There was a time when the spring-run fauna of Florida was live elephants of four kinds, sloths bigger than steers, sabertooths and jaguars, and a dire wolf the size of two German Shepherd dogs together. There were camels in it, and horses in herds and glyptodonts like armor-plated Volkswagens; and to show the endless bounty of the grass, there were even tortoises in slow shoals there—herds of giant tortoises that grazed among herds of mammals."[40] In short, Pleistocene Florida was an ecological dead ringer for the African savanna, or as Carr eloquently concluded: "So the spread of beasts that today we call the African veld fauna is not from a long view indigenous to Africa at all. As an ecological organization it is really a version of the grand plains community of the Ice Age."[41]

Conservation in Africa

During the first two of his four visits to Africa, Carr was "astonished and depressed" to see relatively few large animals throughout much of the rural areas of the continent. Just as surprising to Carr was his general lack of knowledge regarding the plight of African wildlife and the even more profound lack of connection to the problem among Americans. Even enlightened Americans tended to be more concerned with the loss of wild lands in the United States: "In Florida these days, you can still hear, under the mindless, glad din over industry coming in, the voices of the old ones—or of the young ones who have listened to the old ones—grieving over the passing of the wilderness. They are no longer watching landscapes wasting away. That happened long ago. What is going on now is just a lot of little cleanup operations."[42] Carr went on to review some of the most egregious examples of the loss of wilderness and wildlife in the United States that kept most Americans worrying about American wild places rather than the plight of Africa. But Carr underscored his point that the problem of Africa's wilderness was not Africa's alone; the global community needed to join the fight to save Africa: "The world has never come to grips with a preservation problem of the stature of this one. Nobody ever set out on a conservation project that could be compared with the job of keeping intact a delicate, rowdy, perishable relic like the plains-game landscape. It is a problem that puts both the humanity of man and the skill of the ecologist to test."[43]

Overpopulation and politics contributed to the urgency to save Africa's wildlife: "So now, when population explosion and nationalism are bringing the crisis of Africa to a climax, the laborious studies on which to ground control techniques needed to save the most complex mammal community in the world are only barely getting underway."[44] But even though Carr knew that a series of questions had to be answered to discover how to best preserve Africa's wild lands, he argued that Africa's challenge involved more than science: "it is no use making plans that exclude people. Like other saved-up bits of wilderness everywhere, any African preserve that lets visitors in will, from the start and increasingly as time passes, have people to contend with."[45] The national parks in the United States were literally being loved to death, he noted in passing, but Carr believed the problem of too many park visitors to be soluble as an economic problem that demonstrated the value of wild places. In theory, docile park visitors could be herded like sheep and fleeced of their tourist dollars to the benefit of parks.

Poachers, however, were another matter. Carr had read that one band of poachers had killed 3,000 elephants in Tsavo National Park. That seemed like a large number, so he queried officials at the local game department, who confirmed that it was a careful estimate and excluded many more calves that had died after losing their mothers. Organized poaching was a significant problem, but it seemed to Carr that legal hunting, whether with bow and arrow or snare, could also be threatening. Carr's worst scare in Africa (and he survived close encounters with crocodiles and pythons) occurred when he was in pursuit of a hyrax (a small rodentlike mammal closely related to elephants). Having heard what he believed to be a hyrax, he crawled into the bush only to find that he had put his head into a snare. Having set snares as a child, Carr determined that there was no trigger on the snare (the struggles of the quarry would likely entrap them). Vast quantities of wildlife in Africa did not share Carr's instincts for snares.[46]

As in Florida and the Caribbean, Carr narrowed the problem of conservation in Africa down to people. Were Africans concerned with preserving the wildlife and wild places? Carr was skeptical:

It is not usually sensible to generalize about the races of men—especially the African race. But it has to be said that the usual African citizen, the dark-to-light man of varied phenotype, with round-to-long head, short-to-tall stature, straight-to-peppercorn hair—this hamitic, nilotic, negritic, australoid or forest-type fellow, of blood group O, A, B, or AB, who beats hell out of drums and out of white men in a lot of their own athletic games, will distinguish himself in various ways in the world to come, but not for a while for any passion for preservation of the wilderness.[47]

As a biologist, Carr knew that generalizations about "the African race" revealed more about racist preconceptions than about the people themselves. And Carr recognized the plurality of types that constituted Africans. Nevertheless, speaking as someone who had spent enough time among Africans to know, Carr found little interest in wilderness preservation among the remarkably heterogeneous group of people known collectively as "Africans."

By the time Carr wrote *Ulendo*, other biologists were concerned about conservation in Africa. Several reports appeared in both the popular and professional literature in the years around 1960. Two of the most influential of these were written by Frank Fraser Darling. As vice president of the Conservation Society, Darling participated in three expeditions to Africa. Like Carr and Berner, Darling's trips were conducted to survey insects that carry disease (specifically tsetse flies), and the government of northern Rhodesia supported the trips. Darling also recognized the tension between a growing human population and the survival of wildlife in Africa: "Conservation of wild life in Africa is essentially, and in many ways unfortunately, a human ecological problem."[48] A full appreciation of the ecology of wildlife in Africa required an understanding of cultural activities of humans: "Lastly, and of extreme importance, the habits, customs, and practices of man, and their changing character, must be studied in order to reach significant orientation in a terrain of so many variables."[49] In another account, Darling questioned the role of colonial rule in conservation in Africa: "This, ultimately, is one of the major problems of colonial rule; are we by firm administration to help the peoples to conserve their habitat for their own posterity, or are we by addlepated diffidence and *laissez-faire* to leave these peoples with ruined lands?"[50]

A year after Darling's reports, the prominent British biologist Julian Huxley published a report on a trip to Africa he and his wife took between July and September 1960 for the United Nations Educational, Scientific and Cultural Organization (UNESCO). Their travels were somewhat more extensive than Carr's. Huxley's sole focus was the wildlife and natural resources of Africa. For Carr and Darling, their chief objective was the study of insect-borne disease, which limited their mobility somewhat. Huxley limited his travels to national parks, botanical gardens, and wildlife preserves. Despite the differences in coverage, Huxley reached similar conclusions regarding the role of people in African conservation: "The ecological problem is fundamentally one of balancing resources against human needs, both in the short and in the long term. It must thus be related to a proper evaluation of human needs, and it must be based on resource conservation and resource use, including optimum land use and conservation of the habitat."[51]

Like Carr and Darling, Huxley commented on the conflict between humans and wildlife. He noted that most tribal Africans regarded wild animals either as pests to be destroyed or meat to be killed and eaten. According to Huxley, the close association between wildlife and meat was substantiated linguistically: "This latter point of view is semantically fostered by the fact that in Swahili, as in several other African languages, the same word (*nyama* in Swahili) does duty both for wild animals and meat; and it is physiologically encouraged by the shortage of animal protein in the area and the prevalent meat-hunger of its African inhabitants."[52]

Clearly there was a growing consensus regarding conservation in Africa. The ever-growing population and limited food resources conspired with a utilitarian view of nature to create a crisis of conservation in Africa. In comparing conservation concerns in America to similar worries in Africa, Carr's thoughts turned once again to the prehistoric record of extinction: "With that sort of clutching at straws going on at home, I suppose it is no wonder the plight of the Pleistocene seems far away in Africa, where an unreal aura hangs anyway, part gin, part cordite smoke, part sex on a canvas cot."[53]

Above all, Africa reminded Carr of the tenuous relationship between nature and culture. In Africa, as in Florida and Central America, Carr recognized that an expanding population of humans threatened wildlife both directly and indirectly. But if culture represented a significant part of the problem, it was also a critical aspect of the solution:

> It would be cause for world fury if the Egyptians should quarry
> the pyramids, or the French should loose urchins to throw stones in the

Louvre. It would be the same if the Americans dammed the Valley of the Colorado. A reverence for original landscape is one of the humanities. It was the first humanity. Reckoned in terms of human nerves and juices, there is no difference in the value of a work of art and a work of nature. There is this difference, though, in the kinds of things they are. Any art might somehow, some day, be replaced—the full symphony of the savanna landscape never.[54]

Carr, the lifelong scientist, exhorted his readers to preserve the African wilderness as a critical part of the humanities. Above and beyond the ecological and scientific value of African wild lands, they provided a real value to humanity. Throughout his *ulendo* in Africa, Carr enjoyed engaging local culture. Most places where he found wildlife, he also found people, who he recognized as the decisive factor in the conservation of African wildlife and wild lands. By comparing the savanna to the Egyptian pyramids and the Louvre, Carr suggested that the African grasslands were more than national or even continental treasures. Rather, the savanna was a significant element of international heritage, and its loss would be a loss to the world. Thus, if culture threatened the future of nature in Africa, perhaps culture and its preservation could be a part of the argument for conservation in Africa.

Carr refined his argument somewhat in his second book on Africa, *The Land and Wildlife of Africa*:

> But for me it is sad that the intangible aspects of wilderness are being so dangerously ignored—or not just ignored, actually deprecated. One hears on every hand, "We can't ask Africans to save game for any starry-eyed esthetic motives. One has got to be realistic, you know."
>
> And to be sure, one has to be. But one has to be foresighted, too, and foresee times when tourism will be disrupted, when new techniques of land use make game husbandry as obsolete as blacksmithing is. One must think a long way beyond the life and any material value for wilderness. Thinking that far ahead, the only worth of wild land is the wonder in it, the splendor of old Africa, the look and feel of an unspoiled bit of the original earth.[55]

This call for vision and appreciation stood on its own terms, but Carr emphasized it with a series of photographic images: A baboon sitting on the hood of a jeep looking directly into the camera through the windshield. A pride of lions languidly walking on a road with cars stopped behind them. Game wardens removing a decapitated hippopotamus. A mortally wounded elephant struck by a locomotive lying beside railroad tracks. Poachers in handcuffs, their spears and camp in the process of being burned. Carr and the editors of Time-Life Books clearly selected images for the greatest impact on Americans and others learning about Africa and its plight. The image that best captures Carr's point, however, is an artist's rendition of the African savanna ("as it was a hundred years ago" and "as it is becoming today"). The first image (a fold out covering three full pages) depicts an impressive herd of a variety of large animals: zebras, gnus, impalas, antelope, giraffes, elephants, ostriches as well as lions, rhinos, and cattle egrets. The second reveals only goats, cows, guinea fowl, and humans (Masai herders). Clouds of dust rise above the virtually grassless plain from across the remains of the savanna. It is possible to see a small village of huts in the great distance. The symphony of the savannah has been extinguished in this image.

Conclusion

When compared to a lifetime in Florida and the many years in Central America, Carr's African sojourn was relatively brief (less than six months). And yet Africa became a part of Carr's perspective as a naturalist: one of the most spectacular landscapes in the world was also one of the most threatened. Carr believed that significant parts of Florida and Central America were beyond salvation: the landscape had undergone such a profound alteration as to make recovery impossible, but in Africa Carr saw extensive areas for which there was considerable cause for optimism, and in his writings on Africa, both the wonder of the continent and the hope for its conservation shine through. In a review of *Ulendo* for the *New York Times Book Review*, Marston Bates wrote of Carr's musings on the Pleistocene: "Above all, however, he has written a plea for everyone to take an interest in saving this remnant of the Pleistocene that has survived in Africa— crocodiles and pythons as well as elephants and gorillas—so that our children as well as African children can know something of the wonders of the wilderness."[56] Carr's first trip to Africa was the only one that pulled him away from sea turtle research; he devoted the other three trips to aspects of the study of sea turtles. Between Carr's first trip to Africa and the publication of *Ulendo* in 1964, his research on sea turtles began to yield important results that would ultimately strengthen the case for conservation.

CHAPTER 8

"And for the Turtles!"

International Conservation Efforts

The Caribbean Conservation Corporation provided Carr with an activist organization dedicated to the protection of sea turtles in the Caribbean and Latin America. Joshua Powers used his publishing connections to broadcast Carr's conservation message throughout Latin America. Nevertheless, Carr knew that sea turtles faced threats to their continued survival throughout the world. To truly benefit the sea turtles of the world, a conservation campaign had to be global in focus. As a scientist, Carr hoped that solid data would strengthen the case for sea turtles around the world and that national bodies, both governmental and private, would come to see sea turtles as more than a commodity or food but also as a significant natural and cultural resource. To that end, Carr spearheaded Operation Green Turtle with the aim of introducing breeding stocks of green turtles to beaches where they had historically nested.

At times, Carr's struggles with international conservation committees provoked outrageous comments or actions. David Ehrenfeld recalled one particularly contentious meeting of the International Union for Conservation of Nature (IUCN) marine turtle group in Miami in 1983. After the meeting had ended for the day, participants were gathering for dinner, when a "ragged hag muttering and cursing," walked through the group. A few of the more astute bystanders realized that the "hag" was Archie Carr.[1] In moments of frustration over the continued slaughter of sea turtles around the globe, Carr would shock his graduate students by saying of bureaucrats: "We need to get a scrotal hold on them."[2] Carr meant that science and reason only went so far in trying to convince national governments of the need for international cooperation for conservation. After that, advocates for the conservation of sea turtles needed to exert some leverage. As in his scientific research, Carr's conservation efforts covered a range of initiatives that included his development of Operation Green Turtle to return sea turtles to abandoned beaches around the Caribbean, his role as chairman of the IUCN's Marine Turtle Group, his work to develop Tortuguero as one of Costa Rica's first national parks, and his initially positive and later negative views of attempts to farm sea turtles for profit.

Operation Green Turtle

From its incorporation in 1959, the Caribbean Conservation Corporation (CCC) provided critical support for Carr's research in the form of small grants for equipment and supplies. Such support was critical for ongoing research and conservation projects. Joshua Powers and Ben Phipps used their connections in media to publicize Carr's work. In 1961, CCC's restoration program began to take flight. During the 1959 season, Carr's assistants Harry Hirth and Larry Ogren had expanded the hatchery, collected the baby turtles when they hatched, transferred them to tanks, and fed them chopped fish until they could be transported to various locales around the Caribbean, entirely at the mercy of the capricious schedule of a small Cessna four-seater. All that changed two years later when several of Carr's contacts offered their assistance. The first was Sidney Galler of the Office of Naval Research, who believed that research in animal orientation might reveal clues to new ways of obtaining guidance information or even new navigation guideposts. As a result of Galler's continued interest, Carr was able to obtain funding for aerial reconnaissance flights through ONR (see chapter 6). Through the efforts of James A. Oliver, the Office of Naval Research contracted with the Navy to provide a plane and crew to transport the hatchling green turtles around the Caribbean.

By 1961, the turtle hatchery had grown dramatically. On July 7, Phipps joined Hirth and Archie F. ("Chuck") Carr III to open the camp at Tortuguero and initiate the hatchery. During July and August, the team placed 30,000 eggs in the hatchery and more than 18,000 young emerged, the lowest mortality ever recorded in natural or artificial nests. Of these, 5,000 were released in Costa Rica. The rest, some 13,000 baby green turtles, were transported around the Caribbean in a Navy UF 2, a large, four-engine Grumman *Albatross* amphibious plane (often used in air–sea rescues). In the opening pages of *So Excellent a Fishe*, Carr recalled the circumstances that led to Navy transport on an imposing water plane:

> So the Albatross roaring out of the hidden river was partly a sign of the open-mindedness of ONR, and partly proof that *Chelonia* is a many-sided reptile. I am going to talk a lot about green turtle virtues later on. For now, I just want to recollect how, that day, I looked back out of the bubble-port of the plane and saw Tortuguero down there sprinkled among the palms and breadfruit trees, more like one of my dreams than like any real place anywhere; and thought through the devious happenings behind Harry and me being up there with the eight-man crew of a military airplane and eight thousand baby turtles in stacks of wide, flat boxes.[3]

Figure 23. Operation Green Turtle: Archie Carr boarding a Navy Grumman Albatross seaplane to deliver hatchling green sea turtles, 1961. Courtesy of the Caribbean Conservation Corporation.

As part of the CCC restoration project, in late September 1961, Carr and his associates delivered 13,000 turtles to the beaches throughout the islands of the Caribbean: British Honduras (Belize), Columbia, Genada, Barbados, St. Lucia, Antigua, St. Kitts, Puerto Rico, and Miami (for distribution to representatives from still more localities).[4] Carr hoped that the hatchling turtles introduced to various Caribbean beaches would imprint on the particular sand or environmental factors and return to the same beach to breed. By releasing thousands of turtles year after year, perhaps some of the long-since destroyed nesting colonies might be restored. It seemed to Carr, however, that in the shorter run Operation Green Turtle could have an ameliorative effect on U.S.–Latin American relations, due once again to the Navy planes: "The direct transportation afforded by the Navy airplane not only reduced hatchling mortality to almost nothing, but had obviously favorable public-relations effect in the places visited. In almost every case the fisheries officers who received the turtles pointed out that the dramatic means of delivery enhanced local sympathy for the restocking effort and awakened interest in conservation generally."[5] Whether the turtles would in fact return to the site of their release to breed was impossible to determine from one significant distribution.

Carr explained his long-term goals for the project in a letter to Señor don Adriano Urbina G.:

> It is hoped that, as introductions are repeated and extended, new breeding colonies will be established at the places in which the little turtles are released. This hope is contingent upon the eventuality that, as salmon have been shown to do, the turtles will return to the place at which they were liberated to nest when they reach sexual maturity. The green turtle is a strongly migratory animal, and it is of course possible that the hatchlings released elsewhere will go back to the ancestral home in Costa Rica to nest. In either case, however, the net gain for the Caribbean would be the same.[6]

In addition, Carr noted an ancillary benefit of the program that directly affected the turtles at Tortuguero:

> For instance, during the 1960 season, eggs were gathered only during July, when perhaps 75% of those laid in the northernmost three miles of the beach were removed for incubation. Nesting is even heavier in August, when no eggs were taken. Our presence and work on the beach resulted in almost complete protection for August and September nests, whereas formerly (up to five years ago) the combination of commercial exploitation, in which the females were turned on their backs before they laid, and destruction of nests by predators, resulted in virtually complete blocking of productivity on this section of the beach, as on all the rest of the shore from the Tortuguero bar to Parismina. Thus, a very conservative estimate of the gain in production for Costa Rica from these three miles alone would be 50,000 hatchlings.[7]

Based on this dramatic improvement in productivity, Carr predicted that Costa Rica would see increased turtle populations within four years. Moreover, as turtle populations increased throughout the Caribbean, neighboring countries would look favorably upon the generosity of Costa Rica in this endeavor.

In 1962, CCC workers raised 28,000 hatchlings, of which 20,000 were distributed as part of the restocking program, and the remainder were released in Costa Rica. Operation Green Turtle hatched and distributed 20,000 sea turtles around the Caribbean in 1963. In his annual report of the technical director, Carr suggested that new resident populations of green turtles appeared to be developing in three of the localities where introductions had been made. The data for 1964 revealed even greater evidence of a growing population of sea turtles near the beaches where they were introduced. As usual, Carr concluded with his sense that the significance of Operation Green Turtle transcended its potential importance to science and conservation. He wrote: "For example, their collaboration in the restoration project has stimulated Mexican fisheries officers to adopt a program of green turtle studies as a major part of their official activity for the coming years."[8] Moreover, it seemed to Carr that the program improved the profile and standing of the United States among the Caribbean people: "All along the Caribbean coast people are approvingly aware that an airplane of the U.S. Navy came to their locality for no reason other than to deliver young green turtles, with the aim of improving local living standards. Such favorable reaction among our Caribbean neighbors must be reckoned an asset to our country."[9]

In November 1962, Carr visited Great Inagua Island in the Bahamas to examine the Union Creek area. In cooperation with the Bahamas National Trust, he planned a turtle-culture venture at the site. The following year, David Ehrenfeld worked with game wardens Sammy and Jimmy Nixon to deliver 3,017 hatchlings to Union Creek, a fenced-off ocean inlet, despite the imminent arrival of Hurricane Flora.[10] When Ehrenfeld visited the area in February 1965, he reported that, despite the hurricane and associated flooding in December 1964, approximately 350 turtles from the 1963 shipment were alive and in good health. There was some evidence that some of the turtles that escaped the pen during the flood had remained in the area to feed on the extensive turtle grass beds. By 1965, Carr reported that the Union Creek Experimental Turtle Culture Project was a success.[11]

As Carr suggested, most of the encounters with people outside the organization during Operation Green Turtle were positive, though some were humorous.

Figure 24. Ray Curry, pilot, Sibella (standing), Archie Carr, Harry Hirth (plaid shirt), and crew discuss Operation Green Turtle over dinner, 1961. Courtesy of the Caribbean Conservation Corporation.

Ogren, who flew on many flights, recalled being asked what they were carrying. When he replied, "sea turtles," other Americans assumed he was speaking in code about a secret mission. In 1965, the restocking program delivered more than 13,000 turtles to 22 different receivers around the Caribbean. Carr added a research component to the program by dropping 1,200 drift bottles in three localities off the beach at Tortuguero in the hopes of determining where the current might carry the hatchling turtles during their first years after release. Increasingly, he focused on drifting mats of sargassum seaweed as a refuge for turtles and their food supply.[12] More than a decade would pass before Carr found the opportunity to test this hypothesis directly.

In 1966, the number of turtles distributed by Operation Green Turtle fell to just more than 5,000 delivered to 18 receivers. Once again, Carr focused on water currents (dropping more than 800 drift bottles in the waters off Tortuguero) and the search for mats of sargassum. In his annual report, Carr relayed the grim news that after seven years of support in the form of naval planes, there was a real possibility that such assistance would not be possible as of 1967. Despite his concerns, the Navy continued to provide a Grumman Albatross for two more seasons (1967 and 1968). With the Vietnam War escalating, the Office of Naval Research discouraged Carr from requesting additional assistance from naval planes in 1969. Nevertheless, Carr transported 600 eggs to Bermuda, where hatchlings would be released on Nonsuch Island. This was a continuation of a project begun by CCC board member Clay Frick.[13] Operation Green Turtle continued in Bermuda for several more years, as Frick, his wife, and his daughter Jane tried to establish a breeding colony on Nonsuch Island. Jane Frick was particularly devoted to this effort, and she provided some of the first evidence to explain where hatchling green turtles went after they left the nest. Frick, who was an accomplished swimmer, would literally follow the hatchlings for several miles by swimming behind them.[14] Also, the Union Creek project seemed to be thriving. Colin Phipps (John H. Phipps's son) and Stephen Carr had made separate inspection trips to the site and discovered a healthy population of about 100 green turtles thriving on the turtle grass beds.[15] The released sea turtles did not survive at Union Creek. However, one of Carr's graduate students, Karen Bjorndal, continued to study the naturally occurring population.

As a final statement on the restocking program, Carr acknowledged the uncertain effects while underscoring cultural benefits: "While no strong evidence of the reestablishment of migratory breeding can be clearly demonstrated, the collateral benefits of this project, both scientific and conservational, have been undeniable and many. Their total value will be assessed only when time has shown whether some of the local resident colonies planted by the distribution flights will eventually mate and nest where they now are living."[16] Nearly twenty years would pass before Carr unraveled the complexities behind the probable failure of Operation Green Turtle to reestablish nesting colonies, but he regularly encountered people whose lives had been touched by the arrival of a U.S. Navy seaplane filled with baby sea turtles (see chapter 9). Operation Green Turtle was one of the ways in which Carr and the CCC (with the support of the Office of Naval

Research) became directly engaged in sea turtle conservation efforts. The 1960s were also a time of increased international conservation activity. and Carr joined these efforts as chairman of the Marine Turtle Group of the IUCN.

The Marine Turtle Group of the IUCN

The prospect of truly international conservation for sea turtles improved considerably late in 1963, when Sir Peter Scott of the IUCN invited Carr to join the Survival Service Commission (SSC).[17] Carr accepted immediately, noting that it would be difficult for him to attend a meeting in Morges, Switzerland (where the IUCN was based) that November. Scott, an internationally renowned ornithologist and conservationist, was chairman of the SSC, which assessed the threat of extinction to many species and races of plants and animals.[18] In his second letter to Carr, Scott explained his appointment as well as the objectives of the commission. The primary purpose of the SSC was to collect and marshal as much information as possible on any forms of animals or plants that were faced with extinction, to make that information available for the use of conservationists, and, finally, to take measures to avert extinction. As Scott explained to Carr:

> [The Survival Service Commission's] first task is to *bring to light* the relevant details with maximum scientific accuracy. If it errs it should do so on the side of caution for although there are many dangers in crying "wolf, wolf" they are less than the dangers of discovering that a species is threatened when it is too late to save it. If it is found to be commoner than was thought, no harm is done, but if it is rarer the result may be irrevocable. Its next and even more important task is *to do* something to ensure the survival of the threatened animals and plants.[19]

The SSC was established in 1949 as part of the IUCN.[20] Until 1963, SSC scientists collected significant quantities of data on just two taxa: mammals and birds, the latter as part of the research of a related organization, the International Council for Bird Preservation. For the most part, the SSC neglected other groups of animals, except for a few of the largest and most spectacular reptile species. At the 1963 eighth general assembly of the IUCN in Nairobi, the delegates elected Peter Scott chairman of the SSC. Scott saw his election as an opportunity for a general reorganization to address the gaps in the knowledge of threatened species and to cope with the increasing volume of work. Carr attended the general assembly in Nairobi, met Peter Scott, and lobbied for a resolution commending Costa Rica's efforts to protect sea turtles (see below).

Scott and the IUCN envisioned a range of specialist groups and liaison committees devoted to zoos, rhinos, orangutans, apes and monkeys, whales, seals, European bison, wild horses, marsupials, lemurs, deer, antelopes, reptiles and amphibians, freshwater fish, marine fish, plants, insects, and marine turtles. Meanwhile, the International Council for Bird Preservation would develop nine specialist groups on various bird taxa. Each of the specialist groups planned by

Figure 25. Archie Carr at the General Assembly of the IUCN in Nairobi, 1963. Courtesy of Mimi Carr.

Scott would assemble data on a group of animals for the SSC's *Red Data Book* (a catalog of threatened species around the world). An entry for a species (or sub-species) covered no more than two pages of text. Thus, species accounts could easily be updated individually as scientists revised their contributions. Meanwhile, the liaison committees (like one for zoos) provided links with the other commissions of IUCN and with outside organizations.

After Carr accepted Scott's invitation to join the SSC on November 18, 1963, Scott replied almost immediately on December 4 and invited Carr to chair the Marine Turtle Group.[21] Carr would have discretion to select the members of the group, and once in place the group would need to compile a list of rare or threatened forms according to several categories such as "giving cause for very grave anxiety," "considerable anxiety," or "some anxiety." Scott included a list of possible members of the Marine Turtle Group along with his letter. The list included Tom Harrisson of the Borneo research, John R. Hendrickson from the University of Malaya, P. E. P. Deranyagala from Ceylon, Ernest Williams, and

H. Wermuth. In his response, Carr accepted the appointment and all but two of Scott's recommendations.

The Marine Turtle Group of the SSC quickly became an important vehicle for Carr to report on threats to sea turtle populations around the world. The summer after he became chairman of the group, Carr forwarded a letter to Scott that described rising interest in the commercial exploitation of sea turtles around the world at levels that Carr knew the existing populations could not withstand for very long. No one nation could be isolated as the main culprit in the exploitation of sea turtles. Fishermen of three major nations (Germany, Japan, and the Soviet Union) actively sought turtle meat and calipee, and the Japanese also pursued hawksbill turtles for tortoiseshell. The renewed demand surprised and troubled Carr, particularly in the case of the hawksbill, because plastic imitation tortoiseshell had replaced genuine tortoiseshell in the United States. Carr believed that the situation in Costa Rica was especially controversial because the national government viewed the turtle nesting grounds as a source of revenue. But the real problem could be plotted on a supply-and-demand curve: "The awful trouble is that this renewed demand for sea turtles comes at a time when there are too many people in the world to supply with turtle soup and tortoiseshell. If they are to have these things the only possible way is to produce them on turtle farms."[22] The diminishing supply of sea turtles simply could not keep pace with the demand of a growing population of humans.

Data for the *Red Book*

By 1964, Carr had begun to work in earnest on the manuscript that would become *So Excellent a Fishe*. Before he could complete the accounts on world populations of sea turtles, Carr believed that he needed to complete the book, which would review the latest information regarding the sea turtle population levels around the world. This meant that Carr delayed providing material for the *Red Book* for several years while he sorted through the available data. Scott wrote an urgent letter to Carr on February 1, 1966, exhorting him to forward any information he had from the members of the Marine Turtle Group. Scott's sense of urgency stemmed from the research conducted by other groups of the SSC: "An immense amount of research that has taken place on certain bird and mammal groups has shown that the plight of many more species than is readily appreciated can only be described as desperate, and that no time can be spared before setting the Second Stage in motion."[23]

Two weeks later, Carr replied and apologized for his delay in sending the data for the *Red Book*, but he planned to write "Sea Turtles for the Future" as the final chapter in *So Excellent a Fishe*, so he continued to postpone the *Red Book* entries until he could complete the longer and more detailed analysis. His letter alluded to the plight of sea turtles and the challenges of identifying the turtles in need of protection:

> Except for the clearcut case of the Atlantic ridley, which seems obviously doomed, it is not going to be easy to make fair comparative appraisals of

the predicaments of the various kinds of sea turtles. In the first place, the taxonomy of the group is poorly known, and it is not always clear whether it is a whole species that is in danger, or just one of many populations of a species. In fact, I believe this is one of the important difficulties of any survival program: to decide whether the survival to be fought for is at the species level, or whether subspecies are to be saved too, or even demes— or possibly even every separate nesting colony.[24]

Carr cited two problems in this statement: the difficulty of drawing comparisons between the plights of various sea turtle species and deciding whether to protect subspecies and nesting populations as well as full species.

Carr clarified the problems with defining the status of sea turtles around the world when he published *So Excellent a Fishe*. It seemed obvious to Carr that all sea turtles were endangered in the long run, but the purpose of the *Red Book* was to identify the animal species that needed protection most. To make distinctions of this kind, Carr knew that he needed to have a fairly complete knowledge of the biology of sea turtles. Knowledge of natural history sometimes provided the key to saving a particular species from extinction. Carr wrote: "Some localized troubles can be remedied by laws and regulations. Some species can be husbanded in enclosures, bred there, and in a sense be saved for posterity. But to plan a strategic campaign to save for the future any species in its natural habitat, you have to have a working knowledge of the life cycle of the species."[25] Before Carr could address these problems, he needed to have a clearer sense of the taxonomy and distribution of sea turtles. Although sea turtle systematics had been determined with considerable precision to the level of the genus, the distinctions between species were less clear. Recall that Carr's first publication on the taxonomy of the genera of sea turtles appeared in 1942.[26] No one had completely mapped the ranges of the five sea turtle genera, much less the individual species. Carr cited a record of the hawksbill turtle for Scotland. It was clear to Carr that this individual turtle had to be a stray far from its normal range. However, records of leatherbacks in Nova Scotia seemed to define the northern limits of the range of this tropical breeder.

Problems of speciation were linked to mapping difficulties in complicated ways. Most sea turtle research depended on museum specimens, and Carr knew from considerable experience that such collections were inadequate on the one hand and incomplete on the other. The state of sea turtle collections resulted in misidentifications. As of the 1940s, most zoologists did not recognize the Atlantic ridley as a separate species (nor could they distinguish it from the loggerhead), and Carr listed a number of published mistakes.[27] Yet another complication in establishing the natural ranges of sea turtles was the gaps created by humans. Carr noted that many of the mass nesting sites had disappeared, leaving vast stretches of turtle-free beaches around the world. The definitions of complex concepts such as extinction and survival were just as problematic as speciation and distribution. Carr recognized that for some people the survival of the five genera of sea turtles would circumvent the extinction of sea turtles, but most survival action focused more narrowly on the level of species. The emphasis on the species level brought Carr full circle to the original problem of sea turtle classification and the subtle distinctions between species, subspecies, and populations.

Beyond these theoretical issues, Carr feared that all sea turtles (even all of the genera) faced endangerment over the next ten to twenty years. Carr doubted that sea turtles could maintain their populations without protection from the development and exploitation of the seacoasts. For this reason, he believed that the IUCN should strive for protection for all sea turtle nesting grounds. Arguments for sustainable harvest of turtle eggs failed to convince Carr, who viewed the ever-rising population of humans to be too much of a threat to sea turtle colonies. Carr listed several of the most threatened turtle nesting grounds:

> The Karachi green turtle rookery and that of Maruata Bay in Michoacan, Mexico, badly need attention. The trunkback colonies of Trengganu, Tongaland, and Matina (Costa Rica), also though not the only breeding colonies of *Dermochelys*, nevertheless should be wholly protected at once. Above all, the nesting assemblage of the Atlantic ridley (*Lepidochelys kempii*) on the Gulf Coast of Tamaulipas, Mexico, one of the most vulnerable colonies of any animal I know that is not already reduced to low population levels, must get strict surveillance at the earliest possible date to prevent its sudden disappearance.[28]

Carr's review of sea turtle nesting beaches suggested that most of the known locations needed protection as soon as possible if there was to be any hope for continued survival.

New and Renewed Threats to Sea Turtles

Part of Carr's difficulty in responding to Scott's repeated requests for species data for the *Red Book* stemmed from his conviction that the threats to sea turtles transcended biological distinctions such as species or subspecies and that all sea turtles needed concerted protection from commercial interests. Carr continued in his letter to Scott:

> At the moment, however, the main problem in sea turtle survival is not just deciding which "species" are in peril, but rather fighting to provide every scrap of protection possible for all populations of all the five genera. The egg markets of the Sarawak Islands and those of Eastern Malaysia ought to be stopped. So should the calipee trade; the expanding commerce in turtle skins; the Japanese exports of stuffed green turtles to be used by morbid Californians as household furnishings; and the worldwide traffic in young hawksbills, polished and mounted for hanging on the wall. The aimless harpooning and netting of sea turtles by commercial and sport fishermen should be illegal. And so on.[29]

In short, Carr's extensive research indicated not only the threatened forms of sea turtles but also the range of threats affecting them. Controlling such threats depended on international activism and coordination, which Carr clearly believed that Scott and the IUCN could provide.

From March 27 to April 2, 1968, IUCN organized a conference in San Carlos de Bariloche, Argentina, with the sponsorship of the United Nations Educational, Scientific and Cultural Organization (UNESCO) and the Food and Agriculture Organization (FAO).[30] The conference was called "The Latin American Conference on the Conservation of Renewable Natural Resources." Carr presented a version of his recommendations to the IUCN under the title: "Sea turtles, a vanishing asset." At the outset of his talk, Carr confessed that the problem was less how to manage an asset than how to save the species from extinction. After a brief review of the limited knowledge of taxonomy and geographic distribution, he turned to the threats to sea turtles, including local exploitation of eggs and meat and the belief in parts of Latin America that the eggs enhanced sexual vigor. But one of the most disturbing trends was the development of a market for calipee, which Carr described as the "cartilaginous material cut from among the bones of the belly-shell." Calipee was the essence of turtle soup: "the one irreplaceable part of the turtle that goes into the making of clear green turtle soup, giving it the gummy consistency considered essential by epicures."[31] Traditionally, turtle hunters had to transport the entire turtle, weighing more than 300 pounds. In contrast, calipee was light, durable, and very valuable. With the bare minimum of experience, a turtle hunter could remove a turtle's belly shell, extract and cure the cartilage, and carry the resulting 3–4 pounds of calipee to market. The calipee trade posed a threat to all sea turtles, not just to green turtles, as the most skilled chefs could not distinguish the calipee of one species from another.

Sea turtles faced new and renewed threats in addition to the calipee trade. Markets for two other products developed at about the same time as the worldwide market for calipee: turtle leather and turtle oil. Leather could be cured from the turtle's neck and upper forequarters, and it was used in expensive shoes and handbags. Olive ridleys were a popular species for turtle leather, and Carr reported that his son had seen hundreds of dead ridleys that had been killed for leather on the Pacific coast of Mexico. A marketing campaign by a minor celebrity sparked the demand for turtle oil. In parts of Central America, people believed that turtle oil made women beautiful, and Carr had traced this superstition to Alexandria, Egypt, where women visited turtle butchers on the dockside to drink fresh turtle blood in hopes of restoring lost youth. Even before Columbus made his world-changing voyages, lepers traveled to Cape Verde Island to cure their disease by consuming turtle meat.

Markets were reemerging for species that had been spared from exploitation. During the 1940s and 1950s, the development of plastics destroyed the market for genuine tortoiseshell (traditionally fashioned from the shell of hawksbills). However, a backlash against plastics reinvigorated demand for tortoiseshell. By the late 1960s, when Carr was evaluating the risks to sea turtles, demand for hawksbills was again on the rise. When combined with calipee and leather (both of which could also be recovered from hawksbills), a fisherman could earn fifteen dollars or more for a mature hawksbill. What could the IUCN do against all of these new and renewed threats?

The Role of IUCN

Despite Carr's grim prognosis for all the species of sea turtles around the world, he remained convinced that IUCN could support a range of programs to help sea turtle populations recover: "It seems clear, thus, that the marine turtles of the world demand attention for two reasons: because their existence as natural species is endangered; and because if intelligently treated they could bring great benefit to man."[32] Carr's ambitious program of action included five elements: range reconnaissance, exploitation surveys, migration and population studies, international regulation and protection, and pilot culture projects.

Late in 1967, Carr finally completed the sea turtle accounts for the *Red Book*, which drew on his Bariloche paper. In the meantime, one of the other members of the Marine Turtle Group, Tom Harrisson, had prepared a short paper on problems of sea turtles. Scott felt that Harrisson's recommendations could be translated into more immediate, small-scale action, while Carr's advice would become part of the long-range plans for sea turtles. With considerable tact, Scott suggested that Carr might strengthen the Marine Turtle Group by appointing Harrisson vice-chairman of the group.

Carr had greater ambitions for the IUCN and indicated his disappointment that the union could not offer more definitive support for his proposed action plan. Carr's efforts to purchase the lease for a 10-mile stretch of the beach at Tortuguero, Costa Rica, were progressing, and international support from a group would be invaluable for such efforts: "I believe we are going to accomplish this, and can only pray that IUCN will put some effort into this kind of intervention, as well as into the other sorts of moves that I proposed in my recent communication."[33] More than support for specific international projects, IUCN should, in Carr's view, provide a vision for conservation strategy: "It seems to me that when an organization of the stature of IUCN decides to give special attention to the plight of a group, it should go far beyond tactical stop-gap measures, and put into effect a major strategy of intervention. If IUCN can't mount a cooperative international campaign to protect and restore international resources then nobody can. So I strongly hope that the plan that I outlined for Herbert Mills may not be set completely aside."[34]

Having suggested the extent of his expectations for IUCN, Carr acknowledged his own inability to contribute to the organization's work and encouraged Scott to ask Harrisson to accept vice-chairmanship of the Marine Turtle Group. "In fact, it would probably be better for the group if he [Harrisson] were chairman. My time is so divided among research, conservation, teaching, and writing that my chairmanship duties have been badly neglected."[35] Carr's statement captured his frustration regarding the Marine Turtle Group. On the one hand, he had lofty expectations for the role of IUCN in international conservation efforts, but on the other hand, his standing commitments did not permit him to participate as fully as he would have liked.

Carr met with Harrisson and Herbert Mills in March 1968 to iron out the differences between their proposals for sea turtle conservation. The main objective of the meeting was to develop a tentative project plan that might qualify for support from the World Wildlife Fund (WWF), which provided a funding body to complement the scientific research generated by the IUCN.[36] In a letter to Colin Holloway (secretary to the SSC), Carr reported on the productive meeting with Mills and Harrison. The three scientists quickly clarified their differences:

> I think Herbert and Tom now understand that any adverse reaction to their original paper I may have seemed to show was not based on dis-agreement with their proposals but merely on my feeling that in the aggregate they seemed to fall short of the wide-ranging, long-term, "action treatment" sort of campaign that I thought we had been asked to organize. And they made it clear to me that their dissatisfaction with my schemes was simply that they were too broad to be worked up as a practical proposal to the WWF.[37]

Later in the same letter, Carr underscored his claim that he, Mills, and Harrisson generally agreed on possible action plans. The lack of full support from IUCN must have frustrated Carr, given his hopes for the organization's potential role in international conservation programs. This certainly seems to be the case when he returned to the subject later in the same letter. After exploring several possible avenues for research such as basic research, turtle farming, population censuses, carrying-capacity studies, or exploitation surveys, Carr again raised the issue of general recommendation: "But when I am asked, as I was, what the SSC should do to bring sure alleviation for the world turtle situation, I only send back my Bariloche recommendations."[38] Carr even wanted to extend these to include an effort by the IUCN to reduce international interest in sea turtle products, such as cosmetic oil and leather.

Carr, Harrisson, and Mills developed a general plan to focus conservation efforts on four island nesting sites (probably Ascension Island, either Aldabra or Astove Islands, and an undetermined island in the western Pacific). By focusing on just four islands, they believed they would improve their chances of obtaining support from WWF and achieving the kind of quick results that would provide a basis for continued support. However, Carr still harbored the hope that IUCN would spearhead a massive international attack on the sea turtle problem. For this to happen, the FAO would have to be involved, and Carr exhorted the leadership of IUCN to engage that of FAO to work to develop a vast campaign. Carr wished to incorporate the FAO into international efforts to save sea turtles, and he argued for a prompt ban on killing female turtles at the nest as well as all taking of eggs. Given the precarious state of sea turtle populations, egg hunting would soon come to an end due to the demise of the turtles.

Carr thought that, to be effective, the IUCN needed to balance its research aspirations with legislative action:

> Thus, although I too am sure that the most needed commodity in our sea turtles work is information derived from research, it seems to me that IUCN cannot afford to hold aloof from problems of legislation. In many

of the places in which turtles are in bad shape, unless prestigious outside voices urge the adoption of legislation there will not be any, or at least there will be no enforcement, which is the same thing.[39]

During its early years, the Marine Turtle Group functioned as a loose organization of scientists who represented different parts of the world. In his annual report for 1968, Carr delineated the limitations of such an arrangement. He noted that during 1968, despite advances in research and legislation, the population of green turtles in the Caribbean had declined significantly. Without listing the signs and cause of this demise, Carr explained that the Caribbean situation called for quick tactical intervention rather than group action. With similar situations in areas represented by other members of the group, the Marine Turtle Group's efforts tended to be compartmentalized rather than coordinated and truly international. Carr's solution to such disjointedness was a dedicated staff member: "The Marine Turtle Group badly needs a full-time traveling member who combines a thorough understanding of the zoogeographic, ecologic, economic, and political problems involved; and who for a period of, say, three years, could devote all his efforts to critical field investigations, or to negotiations for legislation, enforcement and international planning."[40] As professional zoologists, none of the members of the MTG could afford the time needed for international coordination with the other members of the group and their projects, and for this reason, it seemed to Carr that the group would remain a collection of individuals devoted to separate problems in separate areas.

Challenges of International Conservation

The IUCN's Marine Turtle Group gathered officially for the first time in March 1969 in Morges, Switzerland. The purpose of the meeting was to review existing information on marine turtles, determine national and international research and conservation priorities, and examine the scope for future cooperation in the field. The Marine Turtle Group also hoped to develop a coordinated plan for sea turtle conservation and discuss the implementation and funding for such a program. The meeting was divided into four parts: national (regional) situation reports, national requirements, international requirements and scope for cooperation, and priorities for action and finance. Carr presented a status report dealing with the sea turtles of the Caribbean or, more accurately, Costa Rica. But in Carr's view the situation in Costa Rica was as an exemplar of the crisis facing other regions. Over the course of fifteen years, the tagging program had demonstrated that Tortuguero was the nesting site for most of the green turtles of the western Caribbean. Unlike most places in the world, Costa Rican law prohibited egg collection and any disturbance of turtles on shore. However, Carr noted that harpooning and netting was permitted and that the commercial take off the Tortuguero nesting ground totaled 4,000–5,000 turtles, while estimates of mature females nesting annually was only about 6,000. Clearly the reproductive

yield was not keeping pace with the commercial extraction. Thus, while supportive of the Costa Rican government's legislative efforts, Carr recommended three
emergency actions: discontinuing the export of green turtle products, banning the
operation of turtle boats off the northern 12 miles of Tortuguero Beach, and
prohibiting the sale or storage of calipee in Costa Rica.[41]

Carr recognized that protection for turtles at Tortuguero only addressed the
problem in part. While Costa Rica was taking significant steps to protect sea turtles
by restricting the harvest of green turtles, Nicaragua was building a large turtle
processing plant with the support of the FAO. When Costa Rican officials discovered Nicaragua's plans to expand the exploitation of sea turtles (in effect a
resource the two nations shared), they abandoned legislation that would have provided complete protection to turtles nesting in Tortuguero in 1969.[42] It had
become clear to Carr that the success of protection for sea turtles depended upon
coordinated and concerted international effort rather than legislative actions
taken by isolated nations.[43]

Burgeoning demand had altered what had been a moderately sustainable sea
turtle market. Carr described the exploitation of the green turtles in the
Caribbean as it had existed through the 1950s:

> Until a decade ago the flow of events in the exploitation of West-
> Caribbean green turtles was this: Costa Rica produced the turtles;
> Nicaragua fed them, boats from the Cayman Islands caught them; and
> the English and Americans ate them. Now the Nicaraguans, under-
> standably anxious to harvest their own marine assets, have phased out the
> Cayman captains. Meantime, the small American markets of other days
> have grown to tremendous size, and much of Western Europe is eating
> turtles too.[44]

Sea turtle populations simply could not keep pace with modern industrial
production and with turtle consumption by North Americans and Western
Europeans.

The Promise of International Cooperation?

By the time the Marine Turtle Group met in Morges, Carr's disappointment
regarding the IUCN seems to have faded, and he began to explore ways to
work within the limits of the organization. For example, after the conference, he
wrote to Scott exhorting him to write to the Minister of Agriculture and
Ranching in Costa Rica. Carr included a copy of a letter he had sent the same
day to serve as a model for Scott:

> Word has been received here that a proposal to provide protection for
> the important breeding colony of the Atlantic green turtle at Tortuguero
> on your Caribbean Coast was abandoned when word reached San Jose
> that a plant to freeze turtle meat was being erected in Nicaragua with

the financial support of an outside sponsor. I can appreciate your concern over this development and trust that Nicaragua can be persuaded to reconsider this unfortunate decision. Meanwhile, however, I suggest that renewed slaughter at the Costa Rican breeding ground will be a far more harmful factor than the Nicaraguan freezing plant. The West-Caribbean turtle colony has several feeding pastures, but nesting goes on only at Tortuguero. May I therefore take the liberty of urging that your government revive its plan to provide the Tortuguero colony with the complete legal protection it must have if it is to survive. I sincerely hope also that Costa Rica will initiate whatever negotiations with the Nicaraguan government may be necessary to reach an international accord that will relieve the drain on this endangered species, and eventually restore it as an important food resource for Caribbean people.[45]

In his 1969 annual report to the SSC, Carr described the outcome of the letter. He successfully initiated a meeting among the governments of Costa Rica, Panama, and Nicaragua at the end of September 1969. Representatives from all three countries agreed to suspend all turtle fishing and egg collecting for a period of three years, after which the status of sea turtles could be reassessed and international controls could be established. Carr wrote, "As far as I know, this is the first truly international agreement for sea turtle conservation and management that has ever materialized anywhere in the world."[46] In addition, Carr reported the establishment of Tortuguero National Park, which would give protection to green turtles, leatherbacks, and hawksbills along more than half of the 22-mile shoreline (see below). Just as important, Tortuguero's designation protected turtles out to sea for a distance of three miles from the beach and inland far enough to protect many endangered terrestrial and freshwater animals including manatees, crocodiles, and several species of cats. Scott warmly conveyed his respect for Carr's efforts in a letter: "As you say in your report, this is the first truly international agreement for turtle conservation which is bound to stir other countries to make comparable moves and must have aided the establishment of the Tortuguero National Park in the Caribbean region. . . . Thank you so much, Archie, for all that you do and continue to do for the Marine Turtle Group.—*and for the turtles* [handwritten addition]."[47]

In addition to the very promising outlook for the protection of sea turtles in the Caribbean, Carr announced a significant rise in the number of turtles nesting at Tortuguero. Carr attributed the increase to three factors. First, based on the analysis of remigration records, there was a very high rate of first-time nesters during 1969, and Carr speculated that this rate reflected increased hatchling production during the seasons following the prohibition of turtle turning at Tortuguero. Second, analysis of accumulated records of remigration returns indicated a shift in the breeding periodicity. More turtles had returned to breed after only a two-year cycle rather than the previously more common three-year cycle. Although causation was difficult to pinpoint, Carr assumed that the shift probably resulted from ecological changes at the main feeding grounds for the

Tortuguero colony. Finally, the extension of the legal harpooning ban from 1 mile offshore to 3 miles offshore made intercepting turtles en route to nest more difficult and probably reduced general harassment of the arriving migrants.[48] For the first time since Carr became aware of the precarious status of sea turtles in the Caribbean, the future looked hopeful. Unfortunately, such hope proved short lived.

In 1971, the Marine Turtle Group convened its second working meeting again in Morges at the IUCN headquarters. The meeting again incorporated national regional reports from around the world. In his report, "Research and Conservation Problems in Costa Rica," Carr provided a disheartening update on the status of the international ban on turtle hunting developed in Costa Rica, Panama, and Nicaragua. Although the ban received prompt ratification in Costa Rica and Panama, Nicaragua failed to follow suit. Worse, Nicaraguan turtle fishermen intensified their efforts to catch sea turtles. Dispirited, Carr concluded: "So this promising first international migratory turtle agreement has now dissolved, and efforts to arrange for renewed negotiations ought to be made immediately."[49] In addition, Mariculture, Ltd., a turtle farm in Grand Cayman, gave a report on its progress in breeding green turtles in captivity for market. Over the next few years, Mariculture would ignite heated debate throughout the turtle community (see below).

The Marine Turtle Group also evaluated grant proposals relating to sea turtles that were submitted to the WWF. Peter Scott's biographer referred to relations

Figure 26. Meeting of the IUCN Marine Turtle Group, Tortuguero, July 25, 1983. Front row (*kneeling, left to right*): Jacques Fretey (France), Anne Meylan (USA), Mario Hurtado (Ecuador), René Marquez (Mexico), Peter Pritchard (USA). Back row (*left to right*): Harold Hirth (USA), Larry Ogren (USA), George Hughes (South Africa), George Balazs (Hawaii, USA), Perran Ross (USA), David Ehrenfeld (USA), Njoman Sumertha Nuitja (Indonesia), Nicholas Mrosovsky (Canada), Colin Limpus (Australia), Leo Brongersma (The Netherlands), Karen Bjorndal (USA), Archie Carr (USA). Courtesy of Larry Ogren.

between the the IUCN and WWF as "both symbiotic and complex": "The IUCN, advised by scientists from all over the world, was supposed to pass on to WWF news of endangered species and places: which ones were threatened, how severely, and what might be done about it. WWF would then allocate funds."[50] While this arrangement sounded perfectly amenable to all parties, a considerable gulf existed between the research arm and the granting arm of the IUCN.

Carr continued as chairman of the Marine Turtle Group until 1984. Throughout the 1970s and 1980s, Carr would occasionally alert the IUCN to various threats to the sea turtles: turtle farms, poaching, shrimp nets, and so on. Invariably, at Carr's instigation, either Scott or the secretary of the IUCN would write a letter of support. During this period, Carr concentrated his efforts on Tortuguero for the most part through his ongoing scientific research on the green turtle. Nevertheless, Carr also advocated for the establishment of Tortuguero National Park.

Creating Tortuguero National Park

At the end of March 1963, Carr received a brief letter from William J. Hart, who was park planning specialist for the International Commission on National Parks for the IUCN. Hart solicited information regarding Costa Rica and the feasibility of providing international planning for a park system.[51] With typical consideration, Carr responded to Hart at length. Carr's main concern was initiating efforts in a way that incorporated the views of Costa Ricans: "It is also good to learn that your office plans to give special attention to Costa Rica in this regard. I have for some time been wondering if there might be a really effective way for some agency in this country to foment such action without drawing the ill-feeling for interference that often attends such overtures in Latin America."[52] From the very beginning, Carr recommended a course of action that enabled Costa Rica to maintain its sovereignty in natural resource management.

Carr also explained his sense that Tortuguero might not be suitable as a park in the American sense.[53] Nevertheless, he argued strenuously for the protection of the entire stretch of beach from the settlement of Parismina to the mouth of the Tortuguero River (near Carr's research camp). Though the Costa Rican government leased a 3-mile stretch of the beach to the Caribbean Conservation Corporation for Carr's tagging project, the lease had to be renewed on an annual basis (all beachfront property is government-owned in Costa Rica). Regulations limiting turtle fishing and egg collection along the entire beach were likewise renewed on a year-to-year basis. Carr argued, "This whole shore, including the beach itself and the woods or cocal between the beach and the lagoon, should quite plainly be made a permanent inviolate preserve." Such protection seemed possible given the ownership of the land: "Since this whole zone falls within the government-owned Milla Maritima there would be no difficult real estate negotiations involved in acquiring it."[54]

Despite Carr's optimism regarding the designation of Tortuguero land as a national park, he stressed the need for complete protection, which would necessitate the complete exclusion of the public, and he worried that this kind of preserve would not fit within Hart's notion of a park.[55] Since Hart had requested recommendations regarding additional areas in need of protection in Costa Rica, Carr provided information on several other areas with an emphasis on areas that would provide refuge for breeding resplendent quetzals (*Pharomachrus mocinno costaricensis*), one of the other celebrated species facing extinction in Costa Rica.[56]

In May 1963, the president of Costa Rica, Francisco J. Orlich, and the minister of agriculture and livestock, Elias Soley Carrasco, issued a series of decrees regarding the capture and slaughter of sea turtles in Costa Rica. The official decree opened with four desiderata: sea turtles represent a valuable natural resource, the green turtle is highly desirable in foreign and national markets, fishing activity on the Atlantic coast is a valuable source of jobs, and it is necessary to take advantage of resources to strengthen fisheries. In light of these considerations, the decree comprised seven articles that for the most part declared the Ministry of Agriculture and Livestock as the governing body in charge of turtle fishing. Article three, however, had significant implications for Tortuguero: "Article 3. The capture of turtles for commercial ends remains prohibited on the beach, and from there up to one kilometer into the open sea, measured from the high tide line."[57] It is not clear if Carr played a direct role in the passage of this decree, but he regularly corresponded with the minister of agriculture regarding Operation Green Turtle and other activities at Tortuguero. After the decree, Carr attended the eighth general assembly of the IUCN in Nairobi and successfully lobbied for a resolution that expressed IUCN's appreciation of the Costa Rican government's timely action and expressed the hope that such steps would be taken wherever green turtles lived.

Through the CCC, Carr lobbied for the protection of Tortuguero. In 1964, Guillermo "Billy" Cruz became CCC's legal representative in Costa Rica. Cruz, one of Carr's former students at the Escuela Agricola Panamericana, had become a good friend. As an executive for British Tobacco Company, he had many contacts in the Costa Rican government, which made him CCC's ideal representative. At the CCC's annual meeting in 1965, Carr reported on Cruz's achievements: "Cruz engineered transactions for the Green Turtle Station property and is now bringing to an apparently successful close negotiation for a ninety-nine year lease for the government-owned land on which the buildings are located, and for exemption from import duty on equipment taken into Costa Rica."[58] Cruz's efforts had secured for the CCC a permanent research station at Tortuguero.

Carr also enjoyed considerable success with Costa Rica's executive branch. During his time at Tortuguero in 1964, Carr hosted several current and former Costa Rican government officials: "During August the former president of Costa Rica, Sr. José Figueres visited the Tortuguero project. During his three-day stay Sr. Figueres walked the beach with the tagging crew, saw at first hand dead turtles that had just been killed by poachers for their calipee, and carried back to the capital a strong appeal for government interest in the preservation of the green turtle, and especially for control of the insidious calipee trade."[59] Ehrenfeld, who was at the time one of Carr's doctoral students, vividly recalled Figueres's (also known as Don Pepe) visit to Tortuguero:

> It was Don Pepe's first visit to the legendary Tortuguero—we had been watching a green turtle nest, also a first for him. El Presidente, a short Napoleonic man with boundless energy, was enjoying himself enormously. Both he and Archie were truly charismatic people, and they liked and respected each other. Each was fluent in the other's language. The rest of us went along quietly, enjoying the show. As we walked up the beach toward the boca, where the Rio Tortuguero meets the sea, Don Pepe questioned Dr. Carr about the green turtles and their need for conservation. How important was it to make Tortuguero a sanctuary?[60]

When the group encountered a female green turtle that moments before had been stripped of her calipee (the cartilage that holds the shell to the carapace), Carr said nothing. But the image of the mutilated sea turtle still struggling to reach the sea, uselessly dropping eggs on the hard-packed sand, captured the essence of the tragedy of sea turtle poaching. Like Carr, Ehrenfeld looked to their guest for his reaction:

> It was a moment of revelation. Don Pepe was very, very angry, trembling with rage. This was his country, his place. He had risked his life for it fighting in the *Cerro de la Muerte*. The turtles were part of this place, even part of its name, Tortuguero; they had been coming here long before people existed in Central America. He understood that, just as he understood the profound significance of the useless, round, white eggs swept by the retreating wavelets down the packed sand into the surf beyond. No green turtle born at Tortuguero will ever lay her eggs anywhere else. She was home, laying her eggs for the last time.[61]

Carr recollected, "Don José was out there on the beach with us that night and we were all anxious for him to get a good impression of it, or some kind of impression of it, so he would go away sure that sea turtles were in bad shape and ought to be saved. With him on the side of turtles, their outlook was bound to be better."[62] In addition to the former president, Cornelio Orlich, brother of the president of Costa Rica at that time, and Teodoro Quiroz, the former minister of agriculture and a long-time supporter of the Tortuguero program, visited Tortuguero. As Carr was mounting the campaign for the preservation of Tortuguero, he knew he would need strong support from Costa Rican government officials. By providing politicians with firsthand exposure to the plight of turtles, Carr hoped to secure political support for his efforts.

In October 1965, Carr and Billy Cruz met with the management committee of JAPDEVA (*Junta de Administracion Portuaria y de Desarrollo Economico de la Vertiente Atlantica*; the Administrative Board of Ports and the Economic Development of the Atlantic Slope), which was the quasi-independent government body responsible for the economic development of the Atlantic slope of Costa Rica. JAPDEVA officials were interested in how they might contribute to

Figure 27. Archie Carr studying a sea turtle recently slaughtered by poachers. Courtesy of the Caribbean Conservation Corporation.

the efforts of the CCC. To that end, Carr suggested a second tagging camp be set up 8–12 miles south of the station at Tortuguero. Carr believed that a new station would automatically reduce poaching on neighboring sections of the beach.[63] Distinct from various government organizations in San Jose, JAPDEVA's regional support was extremely valuable in developing the national park.

The Third Roundtable on Renewable Resource Conservation was held in San Jose, Costa Rica, from February 26 to March 4, 1968. In anticipation of the meeting, Cruz intensified his efforts on behalf of CCC to secure a lease for Tortuguero in the hopes of providing long-term protection for the area and its wildlife. In the official petition for the lease submitted to JAPDEVA, Cruz identified the main objective of the lease: "The object of this lease is to conserve the flora and fauna found in it, which is a sample of that which exists in the area, to set it apart as a sanctuary or park dedicated to natural resources."[64] As further justification for the lease, Cruz noted that the land was not privately owned: "There are no owners and the land is completely mountainous."[65] Having filed the petition, Cruz revealed his ambition for its success to the head of the Lands and Forests Section of JAPDEVA, Fernando López-Calleja U.[66]

With the groundwork laid, Cruz notified Carr about the petition and his hope that an announcement might be forthcoming during the Third Roundtable on Conservation of Nature. He also shared his hope that the president of Costa Rica would announce the lease of Cerro Boque to the CCC as an example of the country's natural resources.[67] All went as planned. Cruz, Carr, and Joshua Powers attended the Third Roundtable on Renewable Resource Conservation in San Jose, and Carr presented a paper, "Motivation and Compromise in Wilderness Preservation." Cruz organized a fieldtrip to Tortuguero (via railroad, mule car, and dugout). On February 26, 1968, Jose Joaquin Trejos, the president of Costa Rica, announced the establishment of Cerro Boque ("Mount Tortuguero") and Tortuguero as a wildlife sanctuary. Among others, James A. Oliver, director of the American Museum of Natural History, praised the Costa Rican president. Oliver wrote: "I have been most happy to learn of your farsighted action in setting aside the Cerro Boque and surrounding area as a wildlife sanctuary. It is most gratifying to know that the wilderness of this region will be preserved so that scientific studies and observations can be made in an undisturbed condition."[68] In his annual report to the CCC, Carr exulted over this achievement, attributing it to CCC:

> Since the beginning of the conservation program of the Caribbean Conservation Corporation the Technical Director has believed that its most important contribution to the survival outlook of the marine turtles may prove to have been the stimulation of concern among Caribbean peoples for the sea turtles and other wild resources. . . . This cooperative interest by Costa Rica has culminated in the designation of Cerro Tortuguero (Cerro Bogue)—The 500 foot "mountain" standing at the northern end of the nesting beach across the pass from our station and a band of surrounding territory as a wilderness reserve under the supervision of the CCC. This is one of the few areas of relatively unspoiled woodland remaining anywhere near the station. It is the home of a troop of howler monkeys and other attractive wildlife, and its summit is an important station for observation of the movements and behavior of the green turtle colony during the breeding and nesting season. This unprecedented conservation move has opened the door for additional negotiations for the leasing, by the CCC, of six to twelve miles of the highly productive middle section of the nesting beach, where a second tagging and management station will eventually be established.[69]

Over the next few months, Cruz continued his efforts on behalf of the CCC to establish a park at Tortuguero. This work accelerated late in 1968 when Mario Boza took up the cause. Boza completed his bachelor's degree during the mid-1960s. In 1967, he went to the Inter-American Institute of Agricultural Sciences in Turrialba to study forestry with the intention to study teak, a popular commercial cultivar from Asia. In 1968, he took a course on wildlands from Kenton Miller, who took the entire class to the United States on a fieldtrip. The trip included the Great Smoky Mountain National Park in Tennessee. During the course of that visit, it occurred to Boza that he might develop a plan for similar parks in Costa Rica. When they returned to Costa Rica, Miller encouraged Boza to write a master plan for Poás Volcano, which was designated as a national park in 1939 by a law that established a 2-kilometer preserve around Poás and another volcano, Irazú. These parks and similar ones created in 1945 along the Pan American Highway became known as "paper parks" because they were preserved in name only. Boza completed the master plan for Poás Volcano, and the agriculture ministry hired him to work in its forestry directorate.[70]

Boza attended the Third Roundtable on Renewable Resource Conservation in San Jose, and in December 1968, he published an article in *La Nacion*: "The Only Solution to the Problem of the Management of the Green Turtle: Establishment and Development of Tortuguero National Park." After describing Tortuguero and the status of the green turtles nesting there, Boza cited new threats to their survival, particularly the calipee trade, and underscored his claims with Carr's data from the 1968 season, which was one of the worst nesting seasons on record. In light of this information, Boza called for cooperation between several related organizations: the Ministry of Agriculture and Livestock, JAPDEVA, CCC, the Inter-American Institute for Cooperation on Agriculture, and FAO. Boza also suggested that other groups, such as the IUCN, the Conservation Foundation, the U.S. National Park Service, WWF, and UNESCO be consulted for technical or financial assistance. Adequate protection of the green turtle across its range would require the cooperation of other nations, most notably Nicaragua, which controlled the Mosquito Cays (to support this claim, Boza cited the Third International Roundtable).[71] Three months later, Boza wrote to Carr suggesting the necessity of getting JAPDEVA to cede the necessary land and obtain a decree establishing Tortuguero National Park. The area that Boza hoped to set aside for the park was large, and he anticipated resistance on the part of JAPDEVA.[72] To motivate JAPDEVA, Boza asked Carr to send a letter of encouragement to JAPDEVA directly. Boza also noted that a certain level of funding was required to buy lands under private ownership, to maintain park rangers, and to construct necessary surveillance installations.

In the spring of 1969, Boza joined a trip to Tortuguero with José Figueres (Don Pepe), who was running for a third presidential term. Also on the trip were Figueres's wife Karen, who was passionate about the park system, Kenton Miller, Gerardo Budowski, and a biology student at the University of Costa Rica, Alvaro Ugalde. Far from direct, the trip necessitated travel by bus, train, mule, boat, and finally by foot. This left ample time for Boza and Ugalde to discuss conservation with Doña Karen. Ugalde returned to Tortuguero the following summer to tag turtles along with Figueres's son, José Maria Figueres (who, like his father, served as president of Costa Rica). A month after that, Boza called Ugalde and suggested he attend a national parks training seminar in the United States. Initially hesitant to lose a semester of study, Ugalde tentatively agreed when Boza promised to explore the possibility of a scholarship.[73]

Regarding his park development ideas, Boza had solicited advice from FAO (through his former mentor Kenton R. Miller) and from UNESCO (through another former professor Gerardo Budowski), as well as from the U.S. National Park Service. For financial support, however, Boza queried Carr about a contribution from CCC. Though Carr made no promises regarding ongoing support from CCC, he did obtain a small scholarship from the Phipps Florida Foundation that enabled Ugalde to attend the seminar on the administration of national parks and equivalent reserves, held in Great Falls, Montana, in the fall of 1969 (after which Ugalde joined Boza in the Costa Rican Parks and Conservation Office). Together, Boza and Ugalde would emerge as the two most important architects of the Costa Rican national park system, one of the finest in the world.

By August 1969, Boza and Claudio Escoto, legal advisor to the Office of Planning and Coordination in the Ministry of Agriculture and Livestock, had drafted a bill of executive decree, which upon approval would create Tortuguero National Park. The draft bill inspired Carr to send a special memorandum to the principal officers of the CCC (Cruz, Frick, Oliver, Phipps, and Powers). In the memo, Carr praised Boza's considerable efforts to secure the support of the ministry of agriculture and President Trejos's office.[74] In mid-September, Carr joined Boza at a meeting of JAPDEVA that was dedicated to discussing the park proposal. The meeting concluded with JAPDEVA approving the proposal. Carr clarified the significance of this development:

> The preserve visualized for the Tortuguero area will not be a mere paper park. The venture will go carefully ahead step by step, as the means for acquisition of the land, for disposition of included inhabitants and interests, and for surveillance of the sanctuary are found. There now seems a real likelihood that these things will extend far back inland and take in a variety of landscapes and wildlife, and the existence of the unique green turtle nesting colony on the coast has been a major point in the case for the park project.[75]

There was another highly significant legislative development during 1969. Trejos, the Costa Rican president, had moved the legal limit for harpooning sea turtles from 1 kilometer off the beach to 5 kilometers offshore. Moreover, the new law would be enforced with frequent Coast Guard patrols, which would be supported financially by the CCC. Carr believed that the new legislation would reduce the impact of fishing on the nesting turtles by reducing interference with mating pairs and disrupting collaborative efforts between poachers on the beach and harpoon boats just beyond the breaker line.[76] Trejos's law dramatically restricted sea turtle harpooning off Tortuguero, which was a profound step in preserving the nesting colony there.

In December 1969, Boza wrote again to Carr about the expected decree.[77] So confident was Boza of imminent executive approval that he devoted the rest of the letter to a careful plan describing the financial requirements of the proposed park over the course of three years (although the text of the letter was in Spanish, Boza translated the financial proposal into English). He asked Carr to present the preliminary proposal to the CCC. For the three years beginning in 1970, Boza expected to fund the park staff (one biologist/park superintendent and three park rangers) as well as various facilities and equipment (a house, horses, a boat, radios, guns, and signs) for $15,000 per year. Boza also solicited Carr's opinion for the proposal and requested the date that Carr might present it to the CCC board.

Carr emended the proposal in minor ways for clarity. It seemed that Tortuguero would soon be a national park.

Despite Boza's expectations, the wheels of bureaucracy ground slowly. Another ten months passed before the office of the president issued the formal decree. In the meantime, Boza had managed to establish sea turtle wardens at Tortugeuro with the material assistance of CCC. When Carr heard about the formal decree, he congratulated Boza on the effort: "Thank you for your letter telling of the formal decree establishing Poas and Tortuguero as National Parks and Cahuita as a National Monument. I want to congratulate you for this happy materialization of your hopes and labors."[78] Carr went on to underscore the significance of the designation to the landscapes of Tortuguero: "In the case of Tortuguero the decree has come none too soon. With the penetration of the region by the long-shore canal, the entire ecological organization of that zone will be threatened. The creation of the park makes it possible to save some magnificent samples of the natural landscapes of the Caribbean lowlands of your beautiful country, and to avert the loss of species which in other parts of their ranges are declining alarmingly."[79] Carr thus asserted his belief that preserving Tortuguero transcended the conservation of sea turtles.

Nevertheless, having noted the significance of the designation in general terms, Carr returned to the conservation of threatened species, particularly sea turtles. Boza's program had proved to Carr that even fairly limited protection in the form of guards on the beaches could significantly reduce poaching for calipee. It appears that CCC contributed supplies and other assistance but not financial backing to the protection program. Several weeks later, Carr further clarified how he envisioned CCC's continuing role in the management of Tortuguero National Park in a letter to one of Boza's colleagues in the Department of National Parks. He noted that Tortuguero would be the only national park dedicated to the protection of marine turtles and that he hoped Tortuguero received the support it deserved. This brought Carr to the potential role of CCC, which amounted to a question of sovereignty of the Costa Rican National Park System in Carr's view. He suggested that it would be presumptuous and shortsighted for CCC to accept any official role in the management of the park. Carr's statement recalls the seven principles for U.S. aid in Central America that Carr developed at the end of his time in Honduras (see chapter 4). Having advocated passionately for the development of a national park at Tortuguero, Carr refused to commit to any involvement that might suggest foreign control of a Costa Rican entity. He underscored his idea that the primary concern of CCC's tagging program was basic research out of which management technology and conservation practices might develop.[80] Thus, Carr drew a distinction between the activities of the CCC (research in support of conservation) and those of the national parks system (land protection and surveillance).

Despite the continuing efforts of Boza, Ugalde, and others in the national parks system and elsewhere, final official designation of Tortuguero as a national park languished for nearly five years. The initial steps to protect the beach and sea turtles from further encroachment and poaching proved to be inadequate. During the intervening years, Tortuguero received additional assistance when Walter

Auffenberg, Jr., son of a long-time colleague of Carr's at the University of Florida, served as a Peace Corps volunteer at the site.[81] After the requisite six months of training, Auffenberg was sent to Tortuguero, where he became concerned about the threats to the turtles offshore. For two seasons (since Boza's plan was implemented), a government PT boat had successfully patrolled the area within 5 kilometers of the beach. Since that boat had become unavailable, Auffenberg solicited funds from the Fauna Preservation Society and Wayne King at the New York Zoological Society for a basic launch with which to continue offshore patrolling.[82] Along with Steven E. Cornelius, another Peace Corps volunteer, Auffenberg drafted an additional sea turtle protection law and lobbied the U.S. Department of the Interior for endangered species status for the green turtle.

In the 1974–75 annual report to the CCC, Carr reported on the status of legislation on behalf of sea turtles and Tortuguero in Costa Rica. Carr reminded the CCC that it was illegal to kill turtles on the beaches of Costa Rica but that harpooning turtles just 5 kilometers offshore remained legal. Without a resident population of sea turtles at Tortuguero, the harvest was concentrated on breeding turtles that converged on the area from throughout the western Caribbean. The only real solution to this perpetual problem in Carr's view was the total prohibition of turtle hunting in Costa Rica from June through October. Such protection seemed imminent: "After heroic lobbying and persuasion by Guillermo Cruz, Alvaro Ugalde and representatives of the University of Costa Rica, Congress, by an overwhelming majority approved a bill that would have achieved this. This legislation would about surely have made it possible to save the turtle colony, and at the same time would have halted the mounting destruction of other wildlife and natural landscapes that completion of the new canal has generated."[83] To the dismay of Carr and conservationists everywhere, President Daniel Oduber vetoed the bill in May 1975. All hopes hung on the possibility that Congress would override the presidential veto with thirty-eight votes.

Once again, months passed. Along with two members of the legislature, José Miguel Corrales and Fernando Altman, Cruz lobbied ardently on behalf of the vetoed bill. When he heard that the Tortuguero National Park bill had once again been introduced, Carr wrote to the president of Costa Rica directly and argued passionately for support:

> Creation of a National Park at Tortuguero is a sound and creative move, certain to contribute toward a better future for Costa Rica and the world. It will save a magnificent sample of wild landscapes for Costa Ricans and for visitors of the future to enjoy. It will stabilize the depleted Tortuguero nesting colony of *Chelonia mydas*, representing a resource that is nearly exhausted and a species now in danger of extinction. The scientific information that has been gathered through twenty years of investigation of the ecology of marine turtles at Tortuguero is now being applied in management and conservation work throughout the world. Advances made by Costa Rica in protecting this international asset are a model for widespread emulation.[84]

Equally significant, Carr instigated a letter campaign beginning with his friend Peter Scott at the IUCN. In a memorandum to Scott, Gerardo Budowsky, Richard Fitter, Tom Harrisson, Wayne King, and Nathaniel Reed, Carr wrote:

> The Tortuguero National park bill has, unexpectedly, been vigorously reintroduced in Congress in San José. There are indications that if important outside endorsement can be shown, passage is likely. I am writing to ask you to consider sending a cable to Sr. Presidente, Asamblea Legislativa, San José, Costa Rica. I hope you will solicit the help of other influential people, including Prince Bernhard who is well known in Costa Rica. The cable should express satisfaction at the news of the revival of the Park project and should strongly endorse it on the following grounds:
>
> It will insure preservation of a major colony of *Chelonia mydas*, a threatened species and a vanishing resource. The scientific information gathered at Tortuguero during a 21-year study of that colony is of unique scientific significance. Information from that investigation is growing in kind and volume with each successive year, and now serves as a source of procedural guidance for management and conservation projects in other parts of the world. As now bounded in the proposal, the park will preserve, for the enjoyment and enlightenment of Costa Ricans and visitors of the future, a broad spectrum of lowland tropical faunal and forest types that are rapidly disappearing everywhere else in Caribbean Central America.
>
> It ought to be emphasized that the deliberations of the Assembly are being hopefully observed by conservationists and scientists in many parts of the world.[85]

Finally, on October 28, 1975, the bill passed with unanimous congressional approval for Tortuguero National Park. The next day, *La Nacion* announced the passage of the bill: "Yesterday the Legislative Assembly unanimously resealed the bill that created Tortuguero National Park that was vetoed by President Oduber on May 8, 1975."[86]

Cruz wrote a joyful letter to Carr:

> Still overflowing with happiness for what took place night before last in the Congress, just as I communicated to you, this has been the goal for a great while and we now have a bill that supports Tortuguero National Park. . . . Thanks to the help of all of you, to a decided undertaking and collaboration not only personally, but of various Costa Ricans who firmly believe in the conservation of our natural resources, and even against the will of a Ministry of Agriculture and of the Señor President of the Republic himself, the unanimous resealing of a bill that creates such necessary protection of the region of Tortuguero was achieved.[87]

Once Carr received the good news, he alerted the members of the Marine Turtle Group of the IUCN, the Coral Gables Task Force, IUCN, and "others concerned" (presumably the last category included a considerable number of personal contacts). On October 31, 1975, Carr sent the following announcement:

> Word has come from Costa Rica that the bill giving congressional approval to a Tortuguero National Park was passed on October 28. This was accomplished through the untiring efforts of CCC Vice President Billy Cruz and two members of the legislature, Lic. José Miguel Corrales and Lic. Fernando Altman. As a result of this legislation there is now reason to believe that the Tortuguero breeding beach, which besides the massive nesting assemblage of green turtles there is visited by hawksbill and leatherback turtles as well, can be protected not only against exploitation, but also against a mounting influx of squatters, small farms and cattle ranches. Since the opening of the canal these have been a growing threat to the future of the Tortuguero turtle colony. Creation of the Park also reinforces the efforts to negotiate with Nicaragua for protection of green turtles on their important feeding grounds in the waters of that country. The passing of the park bill over the presidential veto was a spectacular accomplishment, which deserves the congratulations of everybody interested in the survival outlook of sea turtles.[88]

In his annual report to CCC, Carr included a slightly modified version of the same text with one significant addition to the final sentence: "The passing of the park bill over the presidential veto was a spectacular accomplishment, which deserves the congratulations of everybody interested in the survival outlook of sea turtles *and the future of landscapes on the Caribbean Coast of the Republic.*"[89] Carr's addition broadened the importance of Tortuguero National Park beyond nesting sea turtles.

More than a dozen years had passed since Carr first seriously contemplated the possibility of a national park at Tortuguero. In that time, Cruz had developed as an adept advocate for CCC's concerns in Costa Rica. Barely out of graduate school, Boza proved to be a skilled manager and lobbyist for national parks in Costa Rica. He solicited technical and financial assistance from an array of international conservation organizations. Moreover, Boza appeared to have a gift for determining the requirements of national parks in terms of physical plan and personnel. Equally adept was Alvaro Ugalde. In 1983, President Ronald Reagan presented the J. Paul Getty Wildlife Conservation Prize, known as the "Nobel Prize" for conservation, to Boza and Ugalde. Boza, Ugalde, and Cruz lobbied tirelessly for the passage of a law protecting Tortuguero and other national parks. Carr encouraged the efforts of Cruz, Boza, Ugalde, and others by contributing scientific data and technical assistance. Moreover, Carr passionately promoted Tortuguero's cause through numerous international conservation organizations. Still, Carr deliberately limited CCC's involvement at Tortuguero to science and education (including training) in deference to the sovereignty of the Costa Rican National Parks system. Such restraint has enabled CCC to cooperate with Costa Rican officials and to operate an international research station at Tortuguero, which still functions today. While sea turtles seemed to be moving toward greater protection in Costa Rica, a new corporate interest was developing in Grand Cayman.

Farming Sea Turtles?

In his earliest writings on sea turtles, Carr recognized their value as a food source. Unselfconsciously, Carr wrote about eating sea turtle calipee and eggs in *The Windward Road* (1956). In an article for *National Geographic* in 1967, he described the considerable commercial value of sea turtles: "The green turtle is probably the most valuable reptile in the world and offers an expandable food resource for the future.... There is a ready market for frozen turtle meat, a growing demand for clear green turtle soup, and a rising commerce in turtle hides for leather."[90] Turtle farming was one approach to supply turtles for what could be considerable demand. Carr devoted several pages in his book, *So Excellent a Fishe*, to the promise and problems of turtle farming. Turtle farming could ideally feed people and save turtle populations from extinction.

Carr's stance on turtle farming was initially positive, but his position soon turned negative. He identified numerous obstacles that would have to be overcome before turtle farming could hope to be successful. First, green turtles grazed on large beds of underwater turtle-grass, and a farm would have to find a method to provide this food source for a captive stock. Second, there would need to be a large space designated for the hatchlings, which could not at first be released into the open feeding areas. Food for the hatchlings was another problem yet to be solved. In 1967, no one knew what food baby turtles required, but preliminary breeding experiments had indicated that young turtles would thrive on almost anything from vegetable scraps to dog food. Maintaining a constant supply of food presented yet another problem. Water temperature and cleanliness were additional difficulties to be overcome. For reasons presented below, it is significant that Carr singled out Bob and Jean Schroeder for their efforts at keeping captive turtles in Florida.[91] Nevertheless, he became increasingly skeptical of attempts to farm sea turtles.

In 1967, at the time Carr's writings on the value of turtles and turtle farming appeared, Irvin Naylor was scouting for projects to develop. Naylor was a young entrepreneur from New York who had successfully started several businesses, including a cigar box factory and a ski resort. While on vacation, Naylor encountered Robert and Jean Schroeder, who were trying to raise sea turtles in captivity. Robert (Bob) Schroeder received his doctorate in zoology from the University of Miami in 1964 and published his first *National Geographic* article the same year. It was Bob Schroeder who shot the photographs for Carr's *National Geographic* article, "Imperiled Gift of the Sea." In 1967, Schroeder had been developing the idea of a green turtle farm with turtles supplied by Carr. When he happened on the Schroeder's turtle pens, Naylor believed he had found an ideal project for venture capital.[92]

Less than three months after Naylor and Schroeder met, Mariculture, Ltd. had been incorporated on Grand Cayman island in the British West Indies, and 15,000 eggs from Costa Rica were incubating in styrofoam containers under the supervision of the Schroeders. As Carr had predicted, farming turtles had presented many technical obstacles. For example, Schroeder knew that when the turtles hatched in the sand, liquid waste would percolate through the sand, but in styrofoam containers, the waste would collect, posing a risk for the young turtles. To solve this problem, Schroeder drilled holes in the bottoms of the containers. In addition, he and other Mariculture workers rinsed off each emergent turtle and transferred them to trays where they could absorb their yolk sacs. Of the original 15,000 in the first batch of eggs, 7,500 turtles hatched successfully.[93]

In an attempt to replicate what the baby turtles would experience in the wild, Mariculture officials transported all of the hatchlings to a remote beach on the south shore of Grand Cayman. Next, the turtles were released to make their run to the sea (as if from their burrows on the beach). By 1967, Carr's Operation Green Turtle had been underway for five years. The objective of the project was to demonstrate that green turtle hatchlings imprint on the beaches from whence they emerge and make their run to the shore. To prove this, Carr and his assistants collected several thousand hatchlings from beaches throughout the Caribbean. Over the course of five years, they had released more than 100,000 turtles. Not one had returned to a single beach. For Carr, this fact was not discouraging; turtles may have required more than five years to reach breeding age.[94] Schroeder and Mariculture replicated the release on the chance that imprinting of some form took place. As the turtles reached the water, the Mariculture team collected them in nets.

Carr had also predicted that feeding captive turtles would be problematic. In Florida, Schroeder had tried dog food, chopped fish, and chopped squid, but the hatchlings exhibited signs of bloating. After a period of experimentation with the Florida turtles, Schroeder discovered that Ralston Purina Floating Trout Chow was preferred by turtles and had no negative health effects. For this reason, the Mariculture turtles received a diet of floating trout chow, squid, and fish.[95] Also, as Carr had forecast, Mariculture faced another problem: the holding containers for the hatchlings. The "live cars" that Schroeder developed in Florida proved to be inadequate for holding larger numbers of turtles. Fecal waste and leftover food collected in the tanks and threatened the health of the turtles. To circumvent this problem, the Mariculture technicians devised a large funnel (18 feet in diameter) through which water could be pumped, thus evacuating the tank.[96]

Despite the challenges, within a year of the first collection, it seemed that green turtles could thrive in captivity. Several more batches of eggs were brought to the Mariculture facility on Grand Cayman. Still, the objective of Mariculture, Ltd., to produce a captive breeding population, required successful breeding. Shortly after the company obtained twenty full-grown female turtles, turtles hatched, indicating that one or more of the adult turtles had been gravid upon arrival. In 1973, the breeding turtles at Mariculture had laid 11,512 eggs, of which 4,900 had hatched. Mariculture maintained a stock of 250 turtles (9 males from the wild, 70 females from the wild, and 180 farm-reared turtles of unknown sex). On the face of it, Mariculture, Ltd. had created a captive breeding stock, which at least indicated the potential to become self-sufficient. Although Carr was impressed by Mariculture's success in developing a breeding population, he remained concerned about the impact of new markets on wild populations.

Toward the end of 1973, an emergency bill reached the California state legislature, which would prohibit the importation of green turtle products. Naylor wrote directly to Carr imploring him to reverse his stance against Mariculture, Ltd. and to advise the California legislature to withdraw the proposed legislation. In his reply, Carr reinforced his support for turtle farming in principal. In practice, however, he felt that culture projects should be experimental, nonprofit exercises until the feasibility of a commercial enterprise had been demonstrated.[97] Turtle farming needed to prove its viability before turtle farmers could begin to consider profits. Carr went on to explain why he believed this to be true:

> While this technology was being worked out a realistic marketing prospectus would determine whether the profit necessary to keep the industry alive could be made without either expanding existing markets that could not be flooded, or creating new ones. It has seemed to me from the outset that the most crucial proof of the effectiveness of a farm in promoting green turtle survival would be its motivation and ability to lower the prices paid to poachers and legal hunters for turtles taken in the wild.[98]

For Carr, the problem with Mariculture, Ltd. transcended issues of technical legitimacy. Carr believed that the sale of turtle products would stimulate demand. Greater demand would in turn serve as an incentive to turtle fishermen to slaughter more turtles (as it was clear that Mariculture alone could not meet increased demand). As an isolated voice in opposition to Mariculture, Carr's refusal to support turtle farming may not have mattered. As the leading sea turtle scientist in the world, Carr's view influenced other scientists, conservation groups, and legislators.

Ehrenfeld also wrote against the farming of sea turtles but did not specifically attack Mariculture, Ltd. Ehrenfeld's criticism of turtle farming detailed many problems: ecological, reproductive, and economic. The possibility of genetic mixing of distinct populations and the potential spread of disease had troubling implications for wild populations.[99] The listing of the Atlantic green turtle by the Convention on International Trade in Endangered Species (CITES) and the U. S. Endangered Species Act of 1973, as well as internal strife and considerable financial problems, drove Mariculture, Ltd. into receivership in May 1975.

Almost a year passed before a buyer for the turtle farm came forth. In March 1976, Heinz and Judith Mittag (inventor of the o.b. tampon) purchased what had been Mariculture, Ltd. with its 60,000 turtles and renamed the company Cayman Turtle Farm, Ltd. Judith Mittag committed the farm to reaching self-sufficiency (requiring no additional turtles from the wild) within five years. According to Peter Pritchard, Carr derided Judith Mittag's efforts: "Archie . . . was sometimes heard to dismiss her as a 'suppository millionairess,' a rather harsh, and anatomically inaccurate, characterization."[100] Through the efforts of James Wood, who was a biologist employed at the farm, the Cayman Turtle Farm was able to contribute to the knowledge of the reproductive biology of sea turtles.[101] Nevertheless, Cayman Turtle Farm had ceased to function as a profitable organization and now relied on the financial support of tourists to the farm in addition to the sale of turtle products. Carr continued to worry that the stimulation of demand for such items would inevitably lead to increased pressure on wild populations from poaching.

Cayman Turtle Farm struggled with other problems. During the early 1980s, fibropapillomatosis, a viral tumor, was widespread at the farm. Though this fact may be unrelated to the current terrible epizootic of the disease in wild populations, captive populations should be segregated from wild ones. Moreover, since the farm drew on several populations for its group of breeding turtles, the offspring may have been of mixed genetic origin. Releasing such turtles into the wild, which the farm continues to do, may dilute the genetic strains of previously distinct populations.[102]

Have captive breeding endeavors ever benefited the long-term survival of a species while reaping profits? During the late sixties, Carr wrote an article for *National Geographic* on the plight of the American alligator, which was believed to be near extinction.[103] Protection dramatically improved the outlook for alligators. Alligator farms were profoundly successful at providing marketable alligator products without taking a toll on wild populations. Today there are more that six million alligators in Florida alone. But the biology of nonmigratory and rapidly maturing alligators differs from that of turtles. Existing entirely within the United States, the alligator population of concern required no international legislation for its continued survival. More than anything, the alligator case underscores the limitations of comparisons between different animals. What worked for alligators might not work for sea turtles. In the end, the only way to understand the role of Mariculture in the history of conservation is to recognize historical contingency. Could turtles be saved by commercial turtle farming? For Carr and Ehrenfeld, for farming sea turtles to be profitable, demand had to be created for sea turtle products. In the absence of significant supply (severely limited by the breeding biology of sea turtles), natural populations would inevitably be tapped for greater supply. It was this cycle that ultimately spelled doom for turtle farming.

Conclusion

By the 1960s, Carr's activities for international sea turtle conservation developed along several pathways. The Caribbean Conservation Corporation and support from the U.S. Navy facilitated Operation Green Turtle which aimed to reestablish nesting colonies of green turtles all across the Caribbean. This project evolved directly out of Carr's research program at Tortuguero. During the years the operation was active, Carr hoped that the image of the U.S. Navy delivering hatchling sea turtles to remote Caribbean beaches would convey good will to local residents. As chairman of the Marine Turtle Group of the IUCN, Carr envisioned an ambitious program of international activism on behalf of sea turtles worldwide, but IUCN followed a more conservative path and focused on smaller, regional initiatives. Though this approach fell short of Carr's hopes, he continued to encourage IUCN to use its international stature to encourage nations to cooperate for sea turtle conservation. The long and arduous process of developing Tortuguero as a national park revealed the challenges of sea turtle conservation even within a nation. While Carr strongly supported the establishment of the park, Billy Cruz represented CCC in Costa Rica. In addition, Mario Boza and Alvaro Ugalde became passionate advocates for the national park system in Costa Rica. Even when Tortuguero National Park became a reality, Carr restricted CCC's activities to basic research to avoid undermining Costa Rica's sovereignty. Finally, despite his early support for sea turtle farming as a possible conservation strategy, Carr criticized efforts to develop the commercial turtle farming on the grounds that in the short run it would stimulate international demand for sea turtle products without significantly increasing supply to circumvent poaching of wild stocks. Carr played an important role in each of these conservation efforts, but he was not a confrontational person, so he left direct conflicts to others. His primary focus remained on his efforts to understand the biology of sea turtles throughout the world.

CHAPTER 9

Further Results of Sea Turtle Research and Conservation Biology

The film was short. It was shaky in places, faded with time and rainy with scratches. But it was the cinema of the year all the same, the picture of the decade. For me really, it was the movie of all time. For me personally, as a searcher after ridleys for twenty years, as the chronicler of the oddness of ridleys, the film outdid everything from Birth of a Nation *to* Zorba the Greek.
—Archie Carr, *The Sea Turtle: So Excellent a Fishe*

The Ridley Riddle Resolved

Carr experienced one of the most exciting moments of his life when he discovered the amazing solution to the riddle of the ridley. At the annual meeting of the American Society of Ichthyologists and Herpetologists in Austin, Texas, in 1961, Henry Hildebrand, a professor of biology at the University of Texas, Corpus Christi, invited Carr to preview a film that Hildebrand had obtained in Mexico. Carr agreed with some misgivings: "Henry called me to arrange a preview, to make sure it was really ridleys the film showed and not just wishful thinking. I figured that his carefully understated account of what I would see must be aggrandized by his enthusiasm over finding it. It just had to be. Still, a film showing even a single ridley nesting was worth going to Austin to see, and I went."[1] Carr's skepticism turned to utter amazement when the lights went down and the film appeared on the screen. It had been made by a wealthy Mexican engineer, Andrés Herrera, who had stumbled upon the natural spectacle during a trip to Rancho Nuevo, Mexico, south of Brownsville, Texas. As Carr recalled the viewing, it was the film of the year:

> [The film] made Andrés Herrera in my mind suddenly a cinematographer far finer than Fellini, Alfred Hitchcock, or Walt Disney could ever aspire to be.... To any zoologist, however, especially to a turtle zoologist and most specifically to me, the film was simply shattering. It still is hard for me to understand the apathy of a world in which such a movie can be so little celebrated.

I am not exaggerating a bit. Not only was the film itself astounding to see, but in my case the timing of it was utterly eerie—the way it was teasingly led up to, after decades of search and blank mystery.[2]

Among turtle biologists, there is a legend that Carr promptly leapt from his seat and repeatedly shouted, "That's it! That's it!"[3] What could possibly draw such a response from the world's authority on sea turtles? At first a long, straight beach appeared, then two planes landed and men climbed out and began to dig up turtle eggs. Other men joined in the activity, and they soon amassed a gigantic pile of eggs. Carr had never seen so many turtle eggs in one place, but they were small eggs, smaller than the eggs of a green turtle or a loggerhead. In the next scene, Carr managed to see the turtle itself laying eggs, and sure enough, it was a Kemp's ridley sea turtle, *Lepidochelys kempii*. Another revelation was that the turtle was laying eggs in the bright sunlight of the Mexican morning, whereas all other sea turtles nested at night with the protection of darkness.

The next scene showed another Kemp's ridley digging its nest, and then showed two side-by-side and another covering its finished nest. Unfortunately, Carr's reverie was interrupted by a shot of one of the men standing on a turtle to ride it. Another shot showed a man catching eggs as a turtle laid them. Neither of these activities surprised Carr; in fact he had spent enough time in sea turtle nesting colonies to know that people inevitably tried to ride the turtles and caught their eggs as they dropped into the nest. "I wasn't surprised when those men did it, but I was pretty impatient for them to get it over with. Everything those turtles did was to my eyes a marvel; every slight mannerism was the material of dreams. The playful attitude of the Mexicans seemed irresponsible."[4]

After what seemed to be an interminable time, the cameraman grew tired of filming the antics of the egg hunters and redirected the camera down the beach. Of all the revelations of the film, this shot had to be the one that most profoundly affected Carr: "And there it was, the *arribada* as the Mexicans call it—the arrival—the incredible crowning culmination of the ridley mystery. Out there, suddenly in clear view, was a solid mile of ridleys."[5] Carr did not know how many turtles the film showed, but Hildebrand, who showed Carr the film, carefully estimated the numbers of turtles and determined that there were approximately 10,000 turtles on shore. By counting the number of turtles on the beach and estimating the average time required for a female Kemp's ridley to complete her nest and factoring in the total time that turtles were on the beach that day, Hildebrand calculated that the whole *arribada* consisted of 40,000 ridleys. Having studied the film, Carr saw no reason to doubt Hildebrand's estimate. In fact, Carr substantiated the estimate by citing folk wisdom: "The customary metaphor to use in telling of great abundance of beasts is to say that one might have walked across a lake (or stream or plain) on their backs, or could have walked a mile without touching the earth. In the film you could have done this, literally, with no metaphoric license at all. You could have run a whole mile down the beach on the backs of turtles and never have set foot on the sand."[6] As far as Carr was concerned, 40,000 nesting Kemp's ridleys was a conservative estimate.

Figure 28. Still photograph extracted from film of Kemp's ridley sea turtle *arribada*, Rancho Nuevo, Mexico, 1947. Film by Andrés Herrera.

As with so many of the other pieces of data regarding the ecology and migrations of sea turtles, the solution to the riddle of the ridley raised additional questions that could only be answered by witnessing the *arribada* first hand. But the film had been shot in 1947, and Carr had not seen it until 1961. Nevertheless, Carr made three separate trips and flew along the Mexican coast near Rancho Nuevo where the ridleys had emerged in the film. None of his trips was success-ful. Hildebrand made many more attempts to no avail. Carr did find four other people who had seen the arrival of the Kemp's ridley, and each of them substan-tiated what Carr had seen in the film. None had seen the *arribada* recently. But Carr and Hildebrand gleaned a few additional facts about Kemp's ridley breeding biology from these witnesses. For example, everyone agreed that the turtles emerged from the surf with the assistance of a strong wind from the northeast (*norte*). This observation suggested that the wind produced waves that assisted turtles in their progress up the beach. But Carr preferred Hildebrand's explanation that the custom of emerging with a strong wind evolved as a means of ensuring significant cover of nesting trails by the blown sand, high waves, wind, and tides.

The few facts of biology offered slight consolation because by the time that Carr answered the riddle of the ridley, the Mexican *arribada* was a mere shadow of its former magnitude. Certainly, Carr never saw it. Nor did Hildebrand. To say that the *arribada* had disappeared obfuscates the more specific, harsher reality that turtle hunters and egg collectors destroyed the congregations along Rancho Nuevo. As Carr described the demise of the turtles: "Some years ago people began lying in wait for it all along the coast, and once in a while they chanced to be there when the *cotorras* swarmed ashore. They killed turtles, distributed the meat in the interior, dried calipee for sale, and mined the eggs in masses."[7] Carr remembered that by 1964, he had not heard a definite report of an *arribada* since some time in the latter 1950s. A careful review of every piece of available evidence yielded no additional reports, which meant that no *arribada* had occurred for seven years. To Carr the conclusion was ponderously inevitable: "Now, however, there is no escaping the snowballed evidence that the great arrivals have failed. *Cotorras* [Kemp's ridleys] still straggle ashore along the Tamaulipas coast, but they have gone the way of a thousand other sea turtle colonies before them."[8] The demise of the *arribada* must have been a bitter irony for Carr, who had been pondering the breeding habits of the ridley for two decades. One of the largest aggregations of sea turtles and one of the greatest natural phenomena in the world had disappeared, leaving only a short, scratchy film as a record. Modern conservation efforts have seen an increase in the numbers of Kemp's ridleys (see chapter 11). Meanwhile, there were other species to study and to save, beginning with the little known hawksbill.

The Hawksbill Turtle

Carr concentrated most of his research efforts on the green turtle, with occasional forays into the study of other genera such as the loggerhead, leatherback, and the ridley. In fact, Carr had published papers on all of the species but one by the mid-1960s. Carr's sea turtle reconnaissance trips during the 1950s revealed that all the species were threatened except the hawksbill (although he later realized the threats to this species), so Carr felt less pressure to study this species. However, hawksbills nested in small numbers at Tortuguero, and occasionally fishermen would harpoon one just offshore. Between 1956 and 1964, Carr and his assistants tagged seventy hawksbills and examined another twenty-five that had been killed by turtle fishermen. In 1966, Carr and his former students Harold Hirth (who had become a professor at the University of Utah) and Larry Ogren (by then a biologist at the Bureau of Commercial Fisheries in Panama City, Florida) reviewed their findings regarding hawksbills in the Caribbean as the sixth installment of "The Ecology and Migrations of Sea Turtles" published in the *American Museum Novitates*.[9] The article also served as a comparative study of the breeding biology of sea turtle species. For example, Carr, Hirth, and Ogren noted that the nesting of hawksbills was more spread out in time and space than the nesting of the green turtle. Based on tagging records and on the anecdotal reports of local residents at Tortuguero, hawksbills nested between May and November. Carr and his colleagues believed that most hawksbills nested at Tortuguero in May and June, which reduced interspecific competition (that is, competition with other species) for nest sites. For hawksbills, avoiding direct competition with green turtles and other species was a key to nesting success. Occasionally, nesting green turtles would inadvertently disturb other turtles' nests in the quest for suitable sites. Although the May–June nesting period minimized overlap with green turtles, the hawksbills ended up sharing the beaches with leatherback sea turtles as well as with crocodiles and iguanas, but these solitary nesters rarely conflicted with the hawksbill.

The Tortuguero data on hawksbills enabled Carr to correct the belief that they were the smallest sea turtles (they were considerably larger than adult Kemp's ridleys). The measurement data also forced Carr to revise his earlier estimate that hawksbills might reach sexual maturity at 30 pounds or less when 80 pounds seemed to be a more accurate estimate.[10] When compared to green turtles, hawksbills laid more eggs, dug shallower nests, lived more sedentary lives near coral reefs or rock formations, and often supported numerous barnacles (a factor possibly associated with their slow-moving lifestyle). Nesting female hawksbills employed a strikingly different gait from greens; hawksbills tended to move with diagonal limbs moving together (like most terrestrial reptiles), but female greens laboriously dragged themselves along the beach moving all their legs simultaneously.[11]

Carr's concern for the conservation status of the hawksbill turtle was sparked by a renewed threat to the species. Historically, *careyeros* (tortoiseshell hunters) destroyed most of the hawksbills at their nesting beaches when the Caribbean was first colonized. By 1954, when Carr first visited Tortuguero, plastic imitation tortoiseshell had significantly dampened the market for genuine tortoiseshell. Thus, for more than a decade mature hawksbills carried no particular commercial value except as a food item for a small part of the local Caribbean population. That was changing with resurging demand for genuine tortoiseshell, calipee for the soup trade, and skins for leather. One hawksbill could yield products worth fourteen dollars or more.[12] Fourteen dollars amounted to a week's wages for many of the people living near the shores of the Caribbean, and Carr knew that any species worth that much to an international market could not survive for very long. So Carr and his colleagues scrambled to sketch out the basics of hawksbill biology to lay the foundation for a case for conservation. This approach conformed to Carr's research pattern in which science served conservation. In the case of the hawksbill, it was possible to apply the same basic approach to study that was used to explore the natural history of green turtles and loggerheads. As questions became more complex, it became necessary for Carr to collaborate with scientists who had particular expertise.

Vision and Green Turtle Migration

While Carr and his associates were compiling the first summary of the biology of hawksbill turtles, their analysis of the biology of green turtles was becoming ever more sophisticated. The problem of navigation continued to occupy much of Carr's attention, and a number of his students and collaborators were able to determine that vision played a significant role in turtle migration. David Ehrenfeld, who had received his medical degree at Harvard, took on the problem of sea turtle vision for his doctoral studies at the University of Florida with Carr. The challenge of distinguishing one animal behavior from another came down to the ability of a researcher to isolate the different senses. To identify vision as the primary sense sea turtles used in finding the sea, Ehrenfeld constructed lightweight, aluminum goggles for adult female sea turtles with exchangeable lenses. Using the goggles, Ehrenfeld could "blindfold" female turtles or modify their vision by using colored and polarizing filters. Opaque lenses prevented turtles from finding the sea except occasionally by random wandering. Deep red filters had the same effect, which indicated that turtles could not see red light. UV filters (dark to the human eye) were transparent to sea turtles.[13] He also tested vision in hatchling green turtles by constructing an "orientation arena" on the beach at Tortuguero less than 10 feet from the high-water mark. After several additional experiments, Ehrenfeld and Carr concluded: "A comparison of our results with those of previous studies suggests that fresh water and marine turtles have solved the problem of water finding in the same way."[14] Ehrenfeld also used a rheostat-controlled light bulb built into the goggles to manipulate the apparent brightness at the horizon and determine the light level that attracted sea turtles. From these experiments, he and Carr concluded that sea finding involved inspection of the beach horizon with movement toward the brightest horizon.

In October 1965, Carr conducted research with Nicholas Mrosovsky, an experimental psychologist at the University of Toronto, who had worked previously with freshwater turtles. Unlike Ehrenfeld, who mostly used filters and goggles to alter how turtles perceived natural light, Mrosovsky introduced the turtles to artificial light and controlled brightness and color with filters at the light source. His results complemented Ehrenfeld's. The findings demonstrated the importance of field experiments. Mrosovsky and Carr concluded: "Nevertheless we should like to stress that it is necessary to test animals in the field, whatever limitations that may impose, because only when this is done can one judge if laboratory tests are dealing with the same behaviour."[15] As a lifelong field naturalist, it must have been gratifying for Carr to have convinced Mrosovsky, an experimental psychologist, of the value of field work. Carr also continued to interpret his findings for popular audiences through his books.

So Excellent a Fishe

In 1967, Carr published *So Excellent A Fishe*, his only book devoted entirely to sea turtles. Like his other books, this one offset natural history and science with cultural study. Tortuguero provided the setting for much of the book, and Carr provided an intimate portrait of life and research on the island. Each of the seven chapters of *So Excellent a Fishe* explores a facet of sea turtle research. The chapter "Arribada" revealed the solution to the riddle of the ridley; "Tagging Turtles" introduced Operation Green Turtle; "Señor Reward Premio" reprinted and interpreted many of the picturesque letters from turtle fishermen throughout the Caribbean requesting the $5 reward for returned tags; "A Hundred Turtle Eggs" explored the ecological and evolutionary significance of the average number of eggs laid by sea turtles; and "The Way to Isla Meta" described Carr's research on the remarkable migrations of green turtles from Ascension Island to the coast of Brazil and back. In the last chapter, "Sea Turtles and the Future," Carr drew on his extensive knowledge of sea turtles around the world to divine the future of the various species of sea turtles and their relative states of endangerment. Unlike the technical articles in which much of this material originally appeared, *So Excellent a Fishe* displayed the full range of Carr's literary talents, ranging from descriptive narratives of exotic places to explanations of turtle senses to observations of human culture to persuasive writing about conservation.

Like Carr's travels in search of turtles that led him to write about culture in *The Windward Road*, the results of the tagging program yielded data on the migrations of sea turtles as well as glimpses of Caribbean culture. By the time *So Excellent a Fishe* was published, Carr knew that the tagging program was his best hope for "zoological immortality":

> If my name goes down in the canons of zoology it will be as the instigator of the five-dollar turtle-tag reward. The tag I started using twelve years ago—and the one still in use at Tortuguero and wherever we mark sea turtles—is inscribed with a number and with the offer of a reward to whomever finds the turtle outside the tagging locality and returns it to the Department of Zoology of the University of Florida. The Tortuguero tag says these things in both English and Spanish. The tag doesn't say so, but the reward is five dollars, and this is paid promptly and without any haggling. I imagine that the National Science Foundation had misgivings over the reward item in the budget of my research project plan; but they sent the money, and I never spent so little to learn so much.[16]

The misgivings Carr imagined were not reflected in his ability to renew National Science Foundation (NSF) grants. In fact, he was continuously funded by NSF for more than thirty years, a considerable scientific achievement. Carr reflected on possible explanations for the success of the program: "In those days the reward was almost everywhere more than the cash value of the turtle. In some places it was three times more. In some places, in fact, the turtle had no cash value at all, and was only caught to be eaten locally. Almost everywhere about the Caribbean, five dollars coming so easily out of the sea was a substantial blessing, and any tag found was likely to get back to Gainesville eventually."[17] Knowledge of the written word varied across the Caribbean, so reward seekers addressed their tag returns in a variety of creative corruptions including "Senor Premio" or "Mr. Reward." Nonetheless, Carr received the returns: "Some, however, address the letter to a person named Premio, Remite, Send, or Reward; or to various combinations of the four. This used to confuse the postal officials at the University, but all of them now know that Sr. Premio Remite and Mr. Reward Send are aliases of mine. They don't know why I use the extra names, but they send the letters all the same."[18]

Many of Carr's colleagues recall with fondness how much Carr enjoyed receiving tags from around the world. He reveled in stories that people would include along with the tags. Of equal interest was the language employed in their letters. Carr recalled:

> When a person picks up a drift bottle on the beach or catches an animal bearing a tag it is usually not so much the sudden surge of feeling less

out of touch, of being mystically chosen to receive the tag or the bottle. All finders of tags or bottles don't feel that way, but a lot of them do, and it often moves them to outdo themselves in telling about the event. Some are able to write only haltingly, some have to get a scribe or a missionary or the rare passing visitor from outside to compose a letter for them. Some write their own letters with extraordinary fluency, zest, or attention to detail.[19]

The tags and accompanying letters arrived from many of the places Carr had visited when searching for sea turtles throughout the Caribbean: Bluefields, Nicaragua, Bocas del Toro, Panama, Puerto Limón, Costa Rica, and Key West, Florida. One of the most prolific of the Key West correspondents was Allie O. Ebanks, who was among the most successful of the Cayman turtle captains (and the subject of the 1943 *National Geographic* article). Here are two examples (out of dozens) of letters requesting payment of the reward:

> Dear Prize Giver,
> I am very glad to be the one that found this tyrtle drifting down to the Nicaraguan's water it was found in the Pear Lagoun Bar on Thursday 18 of August at 6:00 o'clock in the morning, I caught it beside a shrimp boat in that place.
> I wish that I have explained every thing the right way.
> Number 3853.
> Mr. Wilmore Hodgson
> El Bluff, Nicaragua[20]

When translating letters, Carr tried to capture the author's formal tone, as in this case:

> Ciudano:
> Director Send Dept Biol U-F.
> Attentively I direct myself to you to send you a plaque that I found encrusted on a turtle that I caught on the coast of Cojoro Venezuela. This demands that it be sent to that Institue in order to receive a reward.
> Receive a cordial embrace.
> Roberto Faneite.[21]

In *So Excellent a Fishe*, having reviewed his data from the tagging projects in Tortuguero and Ascension Island, revisiting the riddle of the ridley, and contemplating the biological significance of one hundred turtle eggs, Carr examined what the future held for the sea turtles of the world. Just as tagging returns opened a window on the culture of the Caribbean, international conservation of sea turtles necessitated an appreciation of the various cultures that depended on turtles as a resource. One problem was that sea turtles were utilized differently by various cultures: "In different parts of the world, they are put to different human uses, and local conservation agencies disagree as to both the survival status of sea turtles and the steps needed to protect them. In one place only eggs are harvested; in another only the mature turtles, away from the nesting beach, are hunted; in still other places both eggs and nesting females are ruthlessly taken."[22] Once again, Carr confronted the problem of a growing population of humans with a seemingly insatiable desire to consume.

In the meantime, Carr had just begun to work out the life history of sea turtle species:"It is now clear that people are so abundant, and the life cycle of a sea turtle is so complicated, that nobody really knows what he is doing to a population when he kills a turtle or takes the eggs from a nest. The capacity of people to consume and their ability to destroy are growing beyond the tolerance of the small populations in which sea turtles live."[23] The only realistic solution, international agreements, seemed impossibly complex and subject to criticisms from all who depended on sea turtles for subsistence or income. Carr imagined the reactions of various nationalities:

> The Japanese are not likely to agree to laws that cut off supplies of tortoise-shell. German soup makers will oppose any restriction of their sources of calipee. To stop turtle-hunting in the Miskito Cays of Nicaragua would change the lives of people in New York, Key West, and London, and in the Caribbean it would bring real suffering to some. To interfere with the turtle hunting customs of Truk, Yap, Ponape, Palau, and the Marshall Islands would seem to folk there uncalled-for meddling, and would complicate the administration of the Trust Territory affairs.[24]

At the time, the only practical conservation measure, to Carr's mind, was protection of nesting sites. In 1967, biologists suspected that sea turtles matured quickly in as little as four to seven years. If that were the case, protection at the nesting beaches would be sufficient. But as it became clear that green turtles, at least, lived for as much as forty years or more before breeding, it became clear to Carr and others that it was equally important to protect adults away from the beaches.

As Carr did in his other books, in *So Excellent a Fishe* he wove together strands of sea turtle biology with the cultures that depended on them for subsistence or income and with the conservation status of the turtles. The responses to the book were very positive with one notable exception: Carr's editor at Alfred A. Knopf expressed surprise that the book had not been published by Knopf since the publisher had issued both *The Windward Road* and *Ulendo*. Even Alfred Knopf himself wrote to Carr and wondered what had happened. After several additional exchanges, it seemed that Carr felt his best work was *Ulendo* and that it had not sold well, whether as a result of lack of promotion or for some other reason. He claimed lack of experience regarding publisher's options, but he promised to submit *The Goshen Line*, a book about life in the Caribbean, to Knopf. As it happened, Carr published just one book after *So Excellent a Fishe*, *The Everglades* in the Time-Life series, but he continued to publish both scientific and popular articles. Nevertheless, *So Excellent a Fishe* would become one of Carr's most enduring works. At the same time, Carr's scientific writing benefited from a productive collaboration with his wife Marjorie.

Tagging Returns with Marjorie's Assistance

When Archie and Marjorie Carr lived in Honduras in the years following World War II, they spent long hours and even days in the field together. Their notes and letters reflect their shared passion for science. The demands of their growing family occupied much of Marjorie's time after the Carrs returned to Florida in 1949. Each of the Carr's five children recalls that it was their mother's job to keep them "out of Daddy's hair" while he wrote. By the mid-1960s, however, the youngest child, David, had reached high school, and Marjorie could once again resume her environmental activities. With the Alachua Audubon Society, she was active in leading the fight to save the Ocklawaha River by stopping the construction of the Cross-Florida Barge Canal (see chapter 10). She also resumed collaborative research with her husband. In 1968, Marjorie spent five weeks at Tortuguero collecting and tabulating from the tagging and site records. Marjorie's efforts in this area led to several papers on related subjects that Archie and Marjorie co-wrote and published during the early 1970s.

The first of the articles co-written by the Carrs appeared in the prestigious journal *Ecology*, and it dealt with modulated reproductive periodicity or the tendency of female green turtles to return to Tortuguero to nest every second or third year. With Marjorie's meticulous compilation of the tagging returns, the Carrs determined the frequency of two-, three-, and four-year returns. Modulated periodicity in sea turtles enabled the Carrs to draw an important distinction between sea turtles and migratory birds. The Carrs noted that condors and some albatrosses bred every other year, whereas the king penguin bred twice every three years. The fact that, in king penguins, the amount of time between migrations could be related to feeding ecology on the residence grounds and the pattern of ocean currents in the travel area prompted the Carrs to suggest that "the physiological problems of migratory reproduction would seem fundamentally the same in the green turtle and in migratory birds."[25] There were, however, critical differences between the two. Most notably, female green turtles renested at intervals of twelve to fourteen days, five or more times during a single season of nesting. The Carrs alluded to the Ascension Island breeding colony in noting that the breeding cycle for green turtles might involve 4,000 kilometers of round-trip travel, result in the laying of 500 eggs or more (the equivalent of approximately one-fifth of the turtle's body weight) and include a stay of several weeks at the nesting locality. Incredibly, it appeared that the Ascension Island green turtles completed the entire breeding cycle without access to a significant source of food. Thus, they ruled out the possibility of renesting in less than two years.[26]

To test this hypothesis, the Carrs reviewed the records of renesting turtles at Tortuguero. Given the relative abundance of shifts between the three-year and two-year period, it seemed likely that the shift was related to an ecological factor. Although periods of two and three years were most common, the analysis of renesting data revealed that a four-year period between nesting was also relatively common. They asked, "The interchangeability of the 2-yr and 3-yr cycles in a given female suggests that the dichotomy is ecologically regulated. If it is true that 3 yr is simply the usual time needed for a turtle to ready herself for the physiological feat of reproductive migration, and if it is true also that exceptionally favorable conditions may simply shorten the period to 2 yr, might not unfavorable conditions lengthen it to 4?"[27]

By 1970, after fifteen years of turtle tagging, Carr finally had collected enough data to begin to address fundamental questions of basic biology regarding the migrations and reproduction of sea turtles. Moreover, the fact that he published these findings in *Ecology*, the most prestigious journal of ecology in the United States, indicated that his questions regarding sea turtles could be applied to more generalized problems of migration and reproduction that interested a range of biologists and ecologists, not just those who studied sea turtles. Throughout the rest of his life, Carr worked to frame all his research in terms of ecology or conservation.

Each turtle that was tagged by Carr's team was carefully measured. By 1969, Carr and his colleague at the University of Florida, Donald Goodman, were able to review the data on size change for 447 remigration returns (as tabulated by Marjorie). Commercial turtle fishermen claimed that green turtles were getting smaller over time. With the data from recaptured turtles, Carr could evaluate this claim scientifically. The data from remigrations in fact showed no significant decline in the size of green turtles. Carr found this fact puzzling: "Because the growth of all reptiles, including sea turtles, supposedly continues indefinitely, and because continuous depredations by man must steadily lower both the maximum and the average ages of individuals in the populations involved, the failure of the data to show a downward trend in body size during the last 15 years is noteworthy."[28]

Until he analyzed the data, Carr believed that the largest turtles at Tortuguero (known locally as "wind turtles") represented the oldest individuals. It seemed that the smallest turtles (those with a carapace length of 29–30 inches) were young individuals that were nesting for the first time. The data from the remigrations returns revealed a more complicated story of growth in *Chelonia*. As Carr and Goodman noted: "It now appears that some green turtles mature at small, and others at large sizes; and that once they are mature—that is, once they have made their first trip to the nesting beach—their growth becomes negligible, as compared with individual variation in maturity-size."[29] Female green turtles did not follow a standard age–growth relationship after they began to breed. But was growth a factor of genetics or ecology? Carr and Goodman could not answer this question because it was not possible to determine the residence range. Instead, they asked whether size had adaptive value. To do this, they compared the Tortuguero green turtles with the Ascension Island green turtles and concluded: "It therefore seems reasonable to suggest that in Ascension turtles, as well as in those of the Tortuguero colony, adult body size is influenced more by maturation size and by factors involved with migratory movement than by growth after maturity, and that the Ascension turtles are large because they have to travel farther without feeding."[30] In this respect sea turtles differed from other groups of reptiles in which size was directly related to age and food consumption.

The Carrs also used Marjorie's tabulation of the renesting data to shed light on the issue of site fixity—the tendency of turtles to return to the general nesting area repeatedly. Stories regarding site fixity represented two extremes. On a stretch of fairly homogeneous beach, sea turtles might nest at random sites scattered over the entire space (in the case of Tortuguero, nesting took place across the entire 22-mile stretch of beach). At the other extreme were the views of professional turtle fishermen, who thought that the goal of any given turtle was a specific spot on the beach and that a person could return to that spot twelve nights after nesting to find the same turtle climbing up the beach to the same spot. Neither view accurately captured turtle reproductive biology, as the Carrs' data demonstrated. It seemed that the fundamental factor in site fixity was the imprinting of hatchlings, and the Carrs found support for this idea in the area of beach in front of the village at mile three of their study site. Although the shore resembled nearby areas of the beach (where there were many nests), only one-third of the turtles nested in front of the village. Human activity provided an obvious explanation for this gap in nests on the beach. According to the Carrs, residents of the village might have limited nesting in one of two ways. Lights would have scared turtles away from the beach, but bright lights had been added to a few residences in the village only two years prior to the study. Alternatively, villagers could have reduced the population directly by slaughtering turtles on the beach in front of the village.

Figure 29. Archie and Marjorie Carr in a *cayuca* with an outboard motor with Cerro Tortuguero in the background, October 1961. Larry Ogren is running the *cayuca*. Photograph by Jo Conner, courtesy of the Department of Special and Area Studies Collections, George A. Smathers Libraries, University of Florida.

Over time, fewer and fewer hatchlings would imprint on those beaches and thus fewer would return to nest upon reaching maturity.[31]

In their third collaboration drawing on the tabulation of the first fifteen years of turtle tagging returns, the Carrs synthesized the data presented in other papers to explain a striking shift in the population dynamics of green turtles nesting at Tortuguero. In 1968, the number of nesting green turtles dropped significantly in comparison with previous years. Yet, despite that decline, numbers rebounded dramatically in 1969, which produced the greatest number of nests on record. The Carrs discovered that in 1969 a disproportionately high number of turtles did not have tags. While it was impossible to determine whether the untagged turtles were nesting for the first time or had been overlooked by the tagging crew during previous searches, part of the resurgence of nests could certainly be explained by a shift from a three-year breeding cycle to a two-year breeding cycle as already documented. The tabulation of the sizes of the turtles over the entire period gave clues as to the loose relationship between size and maturity to breed. As Carr had previously determined, the growth rate of green turtles slowed considerably once nesting began.[32] Of his many collaborations, one of Carr's most satisfying must have been with his wife of more than thirty years.

Ascension Island Insights

The Carrs' research convincingly argued that site fixity did in fact occur, but their findings shed little light on how turtles could locate relatively small stretches of beach after traveling hundreds and even thousands of miles in open water. As we have seen, Ascension Island provided an extreme case requiring green turtles to swim 1,400 miles or more against the current. To develop a theory that explained this remarkable feat, Carr teamed up with Ehrenfeld and molecular biologist Arthur Koch. Koch, Carr, and Ehrenfeld suggested that sea turtles used a bicoordinate navigation system; essentially, head east as long as you can smell Ascension Island. Hammerhead sharks find bleeding fish at great distances, and research with salmon had shown that they used their sense of smell to navigate. Koch, a molecular biologist, wondered if data existed on the currents around Ascension Island. During World War II, Nazi submarine captains had collected exactly the data Koch sought. He devised models and showed that given thermoclines (temperature zones) and currents, the dilution of smells from Ascension Island would not be very great.[33]

Earlier, Koch and Ehrenfeld demonstrated that out of water, green turtles were profoundly nearsighted and therefore could not see stars.[34] This limited the elements of the bicoordinate system to a sun compass, olfaction, or magnetic fields, which had been demonstrated in birds. In the 1970s, Ehrenfeld's student Marion Manton conditioned sea turtles and found they had an acute sense of smell underwater.[35]

As questions regarding sea turtle biology continued to become more complex, Carr continued to find collaborators. In 1974, one such collaboration resulted in one of his most ambitious theoretical claims regarding the migrations of sea turtles. In an attempt to place into an evolutionary framework the migration of green turtles from the coast of Brazil to Ascension Island in the middle of the Atlantic Ocean, Carr joined with geophysicist Patrick J. Coleman. Given that marine turtles of the genus *Chelonia* had inhabited the seas between the continent that would become North America and Gondwanaland for 100 million years, Carr and Coleman theorized that over the course of thousands of generations, sea turtles had been forced to swim ever greater distances to reach Ascension Island. The reason for this was the gradual separation of South America and Africa as the seafloor spread. In the course of seafloor spreading, volcanic islands emerged along the rift. Over millions of years, the islands farthest from the ridge began to recede under the oceanic waters. One hundred million years ago, turtles traveled between their residence pastures and breeding grounds along the northern coast of South America, but by ninety million years ago, a corridor had begun to form between Brazil and West Africa. With the completion of the corridor eighty million years ago, there was considerably more open coastline, and a series of offshore islands had formed. To exploit the oceanic channel, turtles merely continued to follow their established travel paths (literally extending the vector created by the shore of the northeast coastline of South America).

Carr and Coleman offered an evolutionary explanation: "It seems reasonable to assume that this pattern of travel, constant over a long period of time would become an established, heritable part of the turtle's behavior."[36] Further, they argued that by seventy million years ago ancestors of the Ascension Island colony made seaward breeding migrations of up to 300 kilometers, which was consistent with a spreading rate of 2 centimeters per year. Thus, Carr and Coleman provided an evolutionary explanation for a complex biological event.[37] Certainly, the seafloor-spreading hypothesis represents one of Carr's boldest theoretical claims, and it was thus vulnerable to criticism.

And criticized it was. An English geologist named Martin D. Brasier attacked specific elements of the seafloor-spreading hypothesis as it related to paleoecology and evolutionary theory. Brasier presented an alternative theory that drifting juveniles completed the original migrations by floating passively westward across the Atlantic and then could return to oceanic islands like Ascension by means of a homing instinct.[38] A few years later the Harvard paleontologist Stephen Jay Gould analyzed Carr's argument in his monthly column for *Natural History*. According to Gould, the problem with Carr and Coleman's theory was not its empirical difficulties. Rather, Gould found a theoretical difficulty to be far more troubling. The Ascension Island turtles represented one breeding group out of a worldwide population of green turtles. For Carr's theory to be supported, the ancestral species must have been divided into several breeding populations and one of these went to the proto-Ascension Island to nest, but *Chelonia mydas* did not appear in the fossil record until fifteen million years before present. Gould preferred Carr's earlier argument that the ancestors of the Ascension turtles had accidentally drifted on the Equatorial Current from the coast of western Africa to Ascension (this route was similar to that followed by the West African Pacific ridley). Ultimately, for Gould, the shift in Carr's thinking was from a more radical theory of abrupt change (one or more sea turtles finding Ascension at random) to a more conservative one of gradual change (sea turtles traveling farther and farther to reach Ascension over many generations).[39]

Gould's critique of the seafloor-spreading hypothesis rested on a theoretical base, but the theory did not undergo empirical scrutiny for another decade. In 1989, one of Carr's graduate students, Anne B. Meylan, worked with two geneticists from the University of Georgia to analyze the mitochondrial DNA of several populations of the green turtle (*Chelonia mydas*). In comparing the DNA of turtles from Ascension Island, Tortuguero, and Venezuela, it became clear that the three populations were closely related, but the DNA from Hawaiian turtles was distinctly different. Presumably, that population separated from the Atlantic populations much earlier than the Atlantic populations had separated from each other. To support the seafloor-spreading hypothesis, the Ascension population should have been noticeably distinct from the two other Atlantic populations. Thus, Meylan and her collaborators suggested that Carr's earlier natal homing hypothesis was a more likely explanation for the development of the Ascension Island population of green turtles.[40] Though less theoretically ambitious, Carr's original thoughts regarding the founding of a turtle colony on Ascension stood the test of time even

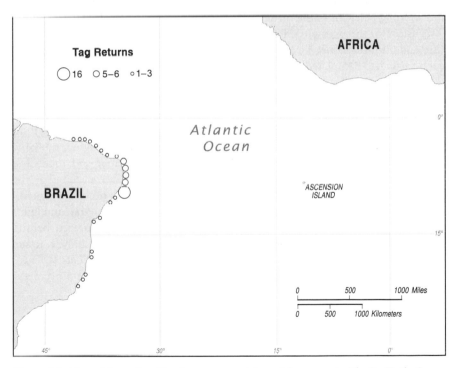

Figure 30. Map of Ascension Island tag returns. Adapted from map in *The Sea Turtle. So Excellent a Fishe* (Austin: University of Texas Press, 1984).

as his attempt to integrate contemporary island biogeography failed. While the seafloor-spreading hypothesis rested on theoretical grounds, technology was necessary to delve into other aspects of the biology of sea turtles.

Tracking Internesting Movement with Technology

By the mid-1970s, Carr, along with his colleagues and students, had filled many of the gaps in the natural history of sea turtles. Finding answers to remaining questions required increasingly sophisticated equipment and techniques of testing and analysis. Such techniques came into play when Carr turned his attention to the behavior of female sea turtles and where they spent their time between separate emergences to nest. Most turtles traveled between 200 and 2,000 kilometers from their feeding grounds to nest. Courtship and copulation also took place near the nesting grounds. Females generally emerged to nest five separate times during the nesting season, with roughly two weeks between each nest. Carr knew that the turtles could not return to their feeding grounds between emergences. So where did the turtles go? To answer this question, Carr attached a styrofoam float with a small light on it to the turtles with a 24-meter line. (Tom, his brother, was a physicist at the University of Florida and designed the prototype of the float.)

A pilot program in 1969 enabled Carr and his brother to collect data on the internesting behavior of the turtles at Tortuguero. Carr also developed one plot at Ascension for the pilot program. Carr's son Stephen and Perran Ross conducted three additional trials at Ascension in 1973. Stephen Carr and Ross waited until each turtle had completed nesting and then attached the float with the line. Then they tracked the float to plot the course of the turtle. One turtle made her way northward on virtually the same course as the turtle Archie Carr had followed in 1969. The correspondence between the two turtles suggested that they were following the same sensory cues on an established underwater pathway. A third turtle, for which only partial data could be obtained, seemed to shadow the same course. However, the fourth turtle took a completely different and much less direct course that included several periods of inactivity. Carr noted that the behavior of this turtle compared favorably to that of the turtles off of Tortuguero. Unfortunately, all of the tests were interrupted by the failure of the tracking lights on the floats and had to be terminated when the turtles disappeared behind inaccessible promontories, but Carr was satisfied with the results and expressed considerable optimism regarding further experiments with the floats.[41] Carr's doctoral student Anne Meylan would take up the study of internesting movements shortly thereafter.

Hawksbill Ecology and Survival

Carr collected data on all the species of sea turtles that spent some portion of their life cycle in the seas of the Caribbean. Carr's interest in any given kind of sea turtle was directly related to the degree of endangerment faced by the species. Early in his career, hawksbill turtles seemed to face little in the way of direct threat, but all that changed quickly when demand for genuine tortoise shell rose dramatically in Japan during the 1960s. As of the mid-1970s, there were still many gaps in the natural history and life cycle of hawksbills. While Carr and his colleagues devoted their efforts to studies of other species, hawksbills had declined precipitously. In 1975, Carr and Stephen Stancyk reviewed the state of knowledge regarding the ecology and survival of the hawksbill turtle. Up until this time, no one knew the composition of the hawksbill's diet. From the reports of turtle fishermen, Carr knew that hawksbills fed on one of two patches of rocky bottom off the northern end of Tortuguero. One of the patches was called Tortuguero Bank. This was a site of courtship and mating for the hawksbill. Both activities rendered the turtles vulnerable to harpoons, and the area was popular with turtle fishermen. Divers avoided the area, however, because visibility was poor due to its position off the mouth of the river and because of the regular presence of bull sharks.

To determine the dietary preferences of hawksbills, Carr and Stancyk examined the stomach contents of twenty-nine turtles, of which twenty actually contained food while the others were empty. Carr hoped that the stomach analysis might yield information regarding the ecology of Tortuguero Bank and the feeding habits of the turtles. The results revealed a high diversity of invertebrate species in the area and showed that encrusting sponges predominated in the diet of hawksbills. Carr and Stancyk used statistical analysis to interpret the stomach content data. Using a chi-square test, they established that the difference between the most common genus of encrusting sponge (*Geodia*) and the next most common sponge (*Styela*) was not statistically significant. They also found that there were no statistically significant differences between the diets of male and female hawksbills. However, Carr was still skeptical about overreliance on biostatistics: "Nevertheless, the trend shown in the Table suggests that a larger sample would reveal that males feed more actively at this season, and also that they are more inclined to select the sponge *Geodia* in their foraging."[42]

Carr and Stancyk tried to bring the knowledge of the life history of hawksbills up to the level of knowledge of green turtles. They examined renesting, tag recoveries, site fixity, and remigration. Although hawksbill nesting was diffuse at Tortuguero (and for that matter throughout the Caribbean), Carr's associates had been tagging nesting hawksbills since 1956, shortly after the National Science Foundation tagging program began. Between 1956 and 1974, there had been six long-distance tag returns from hawksbills. The data from these returns suggested that hawksbills never nested in consecutive years, and they probably nested only every third year like the green turtles. The correspondence in the breeding biology between the two species surprised Carr and Stancyk in light of the basic differences in ecology and genetics: "The evidence of a possible three-year nesting cycle is especially interesting because of the lack of any obvious physiological or ecological cause of either its existence or its predominance in both the Tortuguero and Ascension colonies of *Chelonia*. The agreement of two such divergent genera as *Chelonia* and *Eretmochelys* in such a peculiar and fundamental attribute ought to provide clues in the search for the cause of a non-annual breeding periodicity in sea turtles."[43]

Along with the critical additions to the understanding of the life cycle of the hawksbill, Carr and Stancyk provided an overview of the species' survival outlook. Most troubling in this analysis was a table that compared hawksbill tagging at Tortuguero during two four-year periods. During the first period (1956–1959), the number of hawksbills tagged per patrol-hour, per mile averaged 3.7, but during the more recent period (1970–1973) the numbers dropped to less than one turtle per patrol-hour, per mile. This precipitous drop occurred despite increases in the number of miles patrolled nightly from two to five and the approximate period of patrolling per night from five hours to seven. Carr and Stancyk noted that the contrast would have been even greater if earlier beach patrols had been aware of the hawksbill's ability to right itself after being turned (a virtual impossibility for the much larger green turtle).[44]

In terms of direct threats to the population of hawksbill turtles, the scientists reviewed the resurgence in demand from Japanese markets for genuine tortoise-shell, which was compounded slightly by demand from markets in other Asian countries where tortoiseshell was prized. A more recent threat had developed in the form of spearfishing, which rose proportionally with the popularity of snorkeling and scuba diving on the reefs of the Caribbean. As with the green turtle, Carr believed that each hawksbill colony represented a distinct breeding population. Given that hawksbills mated near their nesting beaches (like the greens), gene flow between colonies might be severely limited, and thus in Carr's view each breeding colony might represent an incipient subspecies or species: "The philopatry implied by the incipient speciation of island colonies makes each a separate reproductive enterprise whose fate is in no way linked with that of any other colony. If this concept is valid it has important bearing on any effort to establish a rational balance between taxonomic lumping and splitting as they bear on intervention procedures to prevent the loss of natural species."[45]

The West Caribbean Green Turtle Colony

Carr had continued to focus on the ecology and migrations of sea turtles, but twelve years had passed since he had written for the American Museum of Natural History series, "The Ecology and Migrations of Sea Turtles," although one could argue that *So Excellent a Fishe*, published by the Museum, represented a summary contribution to the series. In 1978, the Carrs and Anne Meylan published the seventh paper in the series subtitled "The West Caribbean green turtle colony." In this article, they reviewed data from twenty-two years of tagging at Tortuguero. In that time, nearly 12,000 female green turtles had been tagged. Of 2,522 resightings of this group, 1,412 appeared on Tortuguero beach as return migrants, and 1,110 had been captured on feeding grounds or in migration. Carr and his collaborators repeated a claim that had grown more impressive after two

Figure 31. Archie Carr and Albert Taylor (left) tagging a green sea turtle at Tortuguero. The turtle was "turned" in the night so that the taggers could return in the morning. This practice was discontinued in the 1980s when it was determined to cause undue stress to the turtles. Courtesy of the Caribbean Conservation Corporation.

decades: "No turtle tagged at Tortuguero has ever been reported from any other nesting shore."[46]

Along with his co-authors, Carr sought to fill some of the remaining gaps in the knowledge of the biology of green turtles. Tag returns had revealed that the majority (957 out of 1,110) of turtles nesting at Tortuguero migrated to the Mosquito Cays and adjacent parts of Mosquito Bank off the Nicaraguan coast. When coupled with careful analysis of the monthly recovery frequencies in Nicaragua, this finding prompted Carr to conclude that Mosquito Cays provided turtles with feeding grounds as opposed to a migratory stop off. Based on the 1,412 records of turtles that returned to Tortuguero after previous appearances, the Carrs and Meylan ascertained the relative interval frequency of remigration (returning to nest) as follows: 49 percent returned after three years, 21 percent returned after two years, and 18 percent returned after four years. In twenty-two years of turtle patrols, only six turtles returned to nest after a single year. The data showed that certain individual turtles shifted between two-, three-, and four-year cycles, whereas others maintained a more regular interval of three years. Rather than offering hypotheses regarding cycle shifts, Carr posed a series of questions that might be answered once the data from remigration records could be correlated with that from terminal, distant tag recoveries.[47]

Carr acknowledged local and regional folklore in establishing the sea turtle research program at Tortuguero:

> These ideas were embodied in the local folklore when the present program of research began in 1955. In fact, most of the ideas on which the Tortuguero research was based were derived from interviews with turtle hunters. The Cayman Island turtle captains, who used to sail regularly from Grand Cayman to Miskito Bank to net turtles on the extensive grass flats there, were fully aware that the green turtle is migratory. They knew also that Tortuguero Beach in Costa Rica was the breeding ground of the Miskito turtles. Similarly, at Tortuguero, the veladores, who captured nesting females on the beach for commercial sale, knew that the turtles that assembled in July came from far away, and often nested more than once during their migratory season, and that their successive nesting returns were more than randomly close together on the beach.[48]

Twenty-two years into a scientific research program that had substantiated many of the claims of indigenous peoples, Carr continued to acknowledge the value of their observations.

One of the most ambitious elements of the 1978 data analysis was an attempt to estimate the size of the West Caribbean population of green turtles. Carr's brother Tom developed a simple, algebraic formula that allowed for the remigration intervals (two-, three-, and four-years). When the Carrs and Meylan applied the numbers of turtles tagged in 1976 to the formula and multiplied the resulting figure by two (acknowledging the probable fallacy that the ratio of male to female turtles was one to one), they came up with a figure of 126,818 mature green turtles in the western Caribbean. The year 1976 had the highest turtle counts of any year on record (higher even than 1969). By averaging the counts of female green turtles tagged from 1971 to 1976, the team reached a more accurate estimate of 62,532 green turtles in the western Caribbean. Such an estimate struck Carr as reasonable, but the scientists still did not know when female green turtles reached sexual maturity, nor did they know the reproductive longevity of the population. At the Cayman Turtle Farm, officials estimated that green turtles might take ten to twelve years to reach maturity rather than the generally accepted estimate of four to six years. This estimate may have been influenced by the diet of the turtles, which seemed to mature faster in captivity. Also, it failed to account for duration of reproductive life.

At Tortuguero, Carr had encountered one individual green turtle over a period of nineteen years and two others over the course of seventeen years. The database of turtle records contained data on numerous turtles over a span of a decade or more. None of this information gave any hint of the total life span of a turtle. How long a female green turtle lived beyond her first nesting, whether ten years or one hundred, was still open to further debate. The uncertainty regarding turtle age at maturity and longevity had serious implications for estimating the number of submature adult green turtles. Subadult turtles served the vital role of reserves for the breeding population. Turtle researchers and hunters knew that turtles had grown to at least one hundred pounds by the time they migrated to the turtle-grass beds at Mosquito Bank. Without basic information on longevity, there was no way to determine the age of these turtles or to assess how many adolescent and juvenile turtles were required to maintain the adult and subadult colony at Mosquito Bank.[49]

By the time a turtle had reached a weight of one hundred pounds or more, it had outgrown most of its natural predators. Yet one hundred pounds was the approximate weight at which green turtles became attractive to human turtle hunters. While the commercial turtle-packing industry delayed the release of precise figures, they did reveal a take of about 10,000 male and female green turtles. Other reports indicated the annual take in Costa Rica to be roughly 4,000 green turtles before the hunting season was closed permanently in January 1976. Subsistence hunters in Latin America (especially in Mexico and Colombia) slaughtered an additional 1,000 or more turtles. Carr placed these figures regarding the annual take of turtles in context: "Thus, until 1976, 15,000 mature and subadult turtles were probably being lost annually to human exploitation. This drain was going on during the period when our data show the sexually mature breeding population of the area involved to be 62,532."[50] The obvious implication of these numbers was that nearly a quarter of the mature breeding green turtles were being slaughtered for international commerce and subsistence up until 1976. The reserve of juvenile and immature turtles would have to be immense to provide sufficient replacement of such losses on an annual basis. Data like these demanded consideration of the relationship between science and conservation.

Distinguishing between Direct and Indirect Catch

Carr regularly addressed the issue of conservation in his technical papers as well as in his popular books. With the passage of the Convention on International Trade in Endangered Species in 1973, it became possible to control the import and export of sea turtles. Carr believed that the decline in international markets would improve the survival status of sea turtles in certain regions. He also reported that as of 1978, five turtle packing plants had closed (two in Costa Rica and three in Nicaragua). Moreover, the slaughter of turtles on nesting beaches was prohibited nearly everywhere. Each of these factors could and should have a salutary effect on the turtle colony at Tortuguero. Yet there were continuing threats to the world's sea turtles, particularly in the western Caribbean. Since his initial studies of sea turtles, Carr had been concerned about the general lack of information regarding the natural history of the five genera. In the absence of data regarding life histories, it was difficult to develop programs of control, management, and protection for sea turtles. Despite the various protection measures in place, sea turtles still faced exploitation by humans even near Tortuguero. Though advised to stay several kilometers offshore, harpoon boats waited just off the nesting beach to intercept breeding turtles, especially those preoccupied by mating.

Despite considerable advances in the attempt to control turtle hunting, Carr reported a new and inadvertent threat to sea turtles: destruction by shrimp trawlers: "Sea turtles have always been caught accidentally in pound nets, traps, and trawls operated to take other species. With the recent drastic rise in the price of shrimp, however, trawlers have moved into new ground; the trawls now used are much larger than they once were, and the usual haul-time nowadays is long enough to drown many of the turtles caught."[51] Although the impact of incidental catch had once been comparatively minimal, the expansion of the shrimp industry, combined with the decline in worldwide populations of sea turtles raised new concerns: "Now, however, with the drastic decline of world populations of all the species, trawlers are a dangerously adverse factor that may before long deliver the *coup de grace* to some of the species."[52] Carr recognized that the shrimp industry was a powerful one and would not appreciate restriction. Nevertheless, he called for increased efforts to develop devices for trawlers that would reduce the impact on turtles without significantly curtailing the shrimp catch. Incidental catch of sea turtles by shrimp trawlers soon developed into a heated political controversy and to a campaign in the 1980s for Turtle Excluder Devices (TEDs) to enable adult turtles to escape from shrimp nets.

Western Atlantic Turtle Surveys

As the contracts expired, the National Science Foundation (NSF) and the Office of Naval Research renewed Carr's grants, but he also explored additional funding opportunities. In 1978, Carr and his growing team of researchers embarked on yet another ambitious research program with profound implications for the status and conservation of sea turtles. Carr developed a grant proposal for the National Marine Fisheries Service (NMFS) titled, "Survey and Preliminary Census of Marine Turtle Populations in the Western Atlantic." The goal of the project was to be the first attempt to census marine turtles over a major world area. Carr hoped to census entire populations of sea turtles rather than just some portion of populations.

From the outset, such an ambitious project was fraught with problems and Carr acknowledged these in a letter submitted with the proposal. Lack of statistical design and sampling heterogeneity were just two of the most significant problems. In fact, the issue of statistics had appeared in recent reviews of his NSF grant. After one particularly critical review, Carr wrote to his NSF program director: "I was particularly impressed with the detailed comments of the author of the four pages of notes—the one who complained of my lack of statistical sophistication. This is a valid observation. Some of his comments were incorrect but mostly they were very helpful. If you know him and chance to see him some time please convey my thanks."[53] The letter proceeded to explore the possibility that NSF might provide funding for a biostatistician to develop more useful statistical analysis.

One of Carr's former graduate students reflected on Carr's skepticism regarding the growing hegemony of statistics and numbers in biology: "He had little patience for the numbers—they were nonreplicable and would be different next year anyway. Rather he sought insight into the phenomena at work: what was the evidence of change in an ecosystem, of the arrival of exotic species, and the key members of the animal and plant communities functioning as they should or showing signs of distress?"[54]

Rather than ignoring problems of statistics in his proposal for NMFS funding, Carr acknowledged that no standard statistical design had been developed. Nevertheless, he noted that quantitative assessment would depend upon locality, habitat, species, and age group. Moreover, a primary objective of preliminary surveys was to develop appropriate forms and protocol for counts and interviews in different parts of the survey area. Twelve tagging programs run in cooperative arrangement with Carr's research could provide useful data regarding breeding density at rookeries. Also, Carr hoped that the goodwill generated during the course of Operation Green Turtle would facilitate the NMFS research.[55]

With approval at the full requested amount of $50,000, Carr deployed students and his children to conduct on-site interviews around the Caribbean and to conduct aerial surveys. Results from the surveys and interviews were mixed. Carr reflected: "It soon became evident that quantitatively useful data on sea turtles in non-nesting habitats are almost impossible to obtain. The occurrence of the various developmental and reproductive stages, and the habitats they are seen in, can be anecdotally recorded, but to translate the observation into population estimates now appears almost impossible."[56] Still, Carr believed that the surveys were contributing to knowledge of the ecology of sea turtles and that they should continue for the duration of the contract. Nesting surveys proved to be the most useful sources of data for quantitative assessment. Whether made from the air or on the ground, track counts could be converted to numbers of turtles nesting and even converted to the number of sexually mature females in a site-fixed population for a given season. Carr's collaborators (including graduate students Karen Bjorndal, Jeanne Mortimer, and Meylan, and his sons Chuck, Tom, and David) conducted aerial surveys wherever appropriate. They also devoted considerable effort to conduct standardized and exhaustive interviews. Fishermen proved to be better versed in the habits of sea turtles than fisheries officers, for the most part. Between March 1980 and August 1981, Carr's team canvassed the Caribbean, including Costa Rica, Colombia, Panama (*Bocas del Toro*), Nicaragua, and Honduras (*Miskitia*), as well as many smaller island nations.

Western Atlantic Turtle Symposium

A separate but complementary initiative developed in 1979: the Western Atlantic Turtle Symposium (WATS). In December 1978 officials from the Southeast Fisheries Center in Miami, Florida, met to plan a meeting regarding sea turtles in the Western Atlantic region. The following February, officials from the

Intergovernmental Oceanographic Commission Association for the Caribbean and Adjacent Regions (IOCARIBE) and the Food and Agriculture Organization/ United Nations Development Fund Western Central Atlantic Fisheries Project met with SEFC to develop the organization of the planned symposium. IOCARIBE would be the WATS sponsor, while all three groups supported WATS with funds. WATS would draw on sea turtle research in progress, particularly Carr's Western Atlantic Turtle Survey.

After more than two years of additional planning, WATS was scheduled for July 17–22, 1983. Carr's contribution to the planning was relatively limited. Nevertheless, one of the most significant developments of WATS was the preparation of the *Sea Turtle Manual of Research and Conservation Techniques*. One of Carr's former doctoral students, Peter Pritchard of the Florida Audubon Society, was the lead author of twelve contributions. Other authors included Carr and another of his former students, Larry Ogren of the Southeast Fisheries Center (National Marine Fisheries Service). The *Sea Turtle Manual* provided a very detailed and well-illustrated field manual to identification and surveying techniques as well as an overview of conservation policies. Drawing on Carr's extensive experience and the collective experience of those who conducted surveys for the Western Atlantic Turtle Survey, the manual recommended aerial surveys and included a key to the silhouettes of the various species from above as well as a key to the turtles on the beach, with detailed line drawings. In addition, the *Sea Turtle Manual* included descriptions and illustrations of sea turtle crawl tracks, including false crawls ("half moons"). By noting the shape, size, and path of the track, observers could be fairly confident of species and of whether or not it had actually laid eggs. Standardized forms guided observers in their conduct of aerial, ground truth, and vessel surveys. Beach surveys were designed to corroborate aerial surveys. The manual also included measurement and tagging techniques and a brief summary of conservation status of the various sea turtles. It is difficult to imagine a clearer or more concise guide to sea turtle surveys.[57]

The Lost Year

As the 1970s progressed, Carr became increasingly focused on an anomaly in the biology of sea turtles. After hatchling loggerhead turtles left their nesting beaches, they disappeared for at least one year. Anecdotal evidence suggested that hatchling loggerheads swam and drifted to convergence lines, especially where sargassum collected. Carr, however, hoped to collect data of a more substantial nature. In 1978, he and two of his graduate students took advantage of a research opportunity of a lifetime. University of Indiana scientists had proposed a Central American expedition to study various problems related to physiology onboard the research vessel *Alpha Helix*, which was operated as a national oceanographic facility at Scripps Institution of Oceanography in La Jolla, California. Carr, Bjorndal, and Meylan submitted proposals to join the expedition to study specific aspects of sea turtle biology.

Between September 16 and October 20, 1978, the *Alpha Helix* was stationed in the Caribbean and spent most of the time off Tortuguero. Along with Meylan, Carr cruised to Mosquito Bank off the Nicaraguan coast, which was the main territory for the green turtle colony at Tortuguero. As Carr later reported, it was enlightening to see the turtles on their feeding grounds, where males and females as well as mature and subadult turtles were abundant. Contact with Miskito Indians gave Carr hope that tag returns from the region might increase. Most of the tag returns originated in that region, after all. Meylan had been studying the behavior of female green turtles during the period between nestings at Tortuguero (the internesting period) through visual and radio tracking techniques. On the *Alpha Helix*, she was able to test the orientation behavior by displacing three female green turtles 25–70 miles out to sea after interrupting their nesting emergence. A combination of radio and visual tracking techniques made it possible to follow each of the turtles for twenty-four to thirty hours. The results were dramatic: "The headings taken by the turtles were clearly oriented; the travel of all was non random by day as well as by night   Though preliminary in nature, and in part merely a test of equipment, this was by far the most successful open-sea tracking of sea turtles that has ever been done."[58] In another experiment, Carr and Meylan took twenty female turtles from the nesting beach before they had nested and carried them 67 miles offshore. At least one of the turtles returned to Tortuguero within five days only a mile from the site of where she had initially attempted to nest.

On September 11, 1978, Carr and Meylan stopped in the sargassum weedline some 23 miles off the Panama coast to search for hatchling green turtles, and, just as Carr had predicted, they found three in little more than ten minutes. Evidence was accumulating that green turtles spent their first year in drifting sargassum. This evidence correlated with the behavior of turtle hatchlings as they swam from the beach at Tortuguero and with preliminary experiments in which Carr and his associates released hundreds of drift bottles from the beach. Increasingly, it seemed to Carr that the movement of hatchling sea turtles was largely a factor of prevailing currents in the Caribbean Sea.[59]

Carr elaborated on his suspicion that sea turtles spent their first year traveling with sea currents in a paper presented at a symposium dedicated to him in honor of his seventieth birthday, "Behavioral and Reproductive Biology of Sea Turtles," presented at the Annual Meeting of the American Society of Zoologists in Tampa, Florida, December 27–30, 1979. The topic of Carr's paper was "Some problems of sea turtle ecology." Noting the proximity of the Southwest Caribbean Gyre to the Tortuguero nesting shore and the discovery of hatchling green turtles there, Carr suggested that juvenile green turtles might spend the entirety of the "lost year" circulating with the eddy rather than being carried much greater distances by the North Equatorial Current and the Gulf Stream system. Sargassum rafts that contained invertebrate food were readily available. Moreover, Carr noted, the northwestern arc of the eddy approached the edge of the Mosquito Bank in Nicaragua, which was the main foraging ground for most of the turtles that nested at Tortuguero. The eddy's southern edge came close to the coasts of Panama and

Figure 32. Thomas Carr, Archie Carr, and Anne Meylan, aboard the *Alpha Helix* in 1978. Courtesy of Thomas D. Carr.

Colombia, where some of the Tortuguero turtles went to feed outside the nesting season. Large beds of turtle grass were present in both locations, as well as extensive appropriate developmental habitats in which young green turtles regularly occurred. From this information, Carr concluded: "It is therefore possible that both the northern and the southern contingents of the Tortuguero breeding population pass their entire life cycles within the Southwest Caribbean Gyre and the region around its perimeter."[60]

Carr's former student Ogren had conducted surveys of the Gulf of Mexico and speculated that especially advantageous currents might explain the occurrence of the Kemp's ridley sea turtle *arribadas* at Rancho Nuevo.[61] Nevertheless, Carr noted that the complexity of the currents in the Gulf of Mexico made it impossible to determine what advantages might be involved. Before the discovery of the Rancho Nuevo Ridley *arribada* in 1962, most ridleys had been collected on both coasts of Florida and along the Atlantic coast as far north as New England. Museum specimens revealed a "down-stream size-gradient." By this, Carr meant that the farther along the current turtles were collected, the smaller they were. A small ridley-tagging program had yielded additional information about the life cycle of the turtle. There were two foraging grounds where the mature turtles went after nesting at Rancho Nuevo. One was in Campeche, Mexico, and the other was off western Louisiana. The location of these sites prompted Henry Hildebrand to suggest that perhaps the entire life cycle of the Kemp's ridley was spent in the Gulf of Mexico.[62] Though Carr acknowledged this as a definite possibility, he commented that the entire population could not be so contained in light of the steady stream of records of ridleys from the nearshore waters of New England. On average, these turtles were smaller than those known from any other section of the coasts of Mexico or the United States. Carr attributed the appearance of ridleys in New England to the warm waters of the Gulf Stream. For evidence, he drew on oceanographic research that charted the currents of the Gulf of Mexico, the Gulf Stream, and the Atlantic Ocean.

By assembling all the available data, Carr was able to sketch a general picture of the life history of the Kemp's ridley: "Thus, the picture that fuzzily emerges includes a single nesting place in Tamaulipas, adult feeding grounds in Louisiana and in the Campeche-Tabasco area; a seasonal way-station for young and subadults in Florida; somewhat smaller size groups distributed northward into New England waters; and occasionally, even smaller individuals in East Atlantic waters."[63] Having suggested a plausible mechanism for the travel of Kemp's ridleys up the Atlantic coast of the United States and possibly on to the coast of Europe, Carr wondered how the turtles found their way back to the waters of Florida and the Gulf of Mexico, since there were no readily apparent guide posts or currents for them to follow. It seemed that the turtles could only return to Florida and the Gulf of Mexico by swimming against the current. Alternatively, there was the possibility that the Kemp's ridleys collected in New England waters represented a population of waifs that never returned to Florida at all. Further evidence for this alternative hypothesis derived from the small size and precarious state of the ridleys that appeared occasionally in European waters: "Most European specimens range in shell length from 10 to 25 cm, and they are usually cold-stunned or dead when found."[64] Carr concluded that these findings merely deepened the mystery of Kemp's ridleys: "It is hard for me to think of the Atlantic Coast ridleys as demo-graphically dead, but evidence that the Gulf is the entire range of a part of the population is growing. The parallel with the postulated dual cycle of the Tortuguero green turtle is striking—with some hatchlings maturing locally, and others passively emigrating. For me this adds mystery to the ridley cycle."[65]

On August 12, 1985, Carr expounded upon the first year of turtle life in his plenary lecture to the American Institute of Biological Sciences, which held its annual meeting at the University of Florida. Thirty-one years earlier, in 1954, Carr had addressed the same group, also in Gainesville. For the original talk, which was after the two small American Philosophical Society grants but before *The Windward*

Road and the NSF grant, Carr discussed "the passing of the fleet." He tempered four centuries of sea turtle destruction at the hands of humans with his excitement over a research program that might reveal long-distance migrations and navigation ability. By 1985, Carr and the hundreds of other sea turtle researchers around the world had answered many of the questions of basic biology. Nevertheless, there were several fundamental questions as yet unanswered to Carr's satisfaction. For example, did female sea turtles always return to their natal beach to nest? At Tortuguero, many of the 28,000 tagged females had returned to the beach to nest again, but none of them had ever been recorded nesting anywhere else in the world. Turtle no. 3438 returned to the beach twenty-six times over the course of seventeen years, starting in 1965. Across that long span of time, this turtle managed to lay her eggs within a few hundred meters of her original nesting site. Such site fidelity led to another question: Was the female turtle hatched on the same beach fifty or a hundred years ago? Mounting evidence suggested that green turtles might not reach sexual maturity until forty years or more.[66]

The third and final question (and the subject of Carr's American Institute of Biologial Sciences address) was, where do baby sea turtles go, and how do they live after they leave the nesting beach? Even in 1954, it was apparent to Carr that hatchling sea turtles could travel straight to the water from the nest (even if their view was impeded). Once hatchlings left the beach and passed through the surf, they would not be seen again until they had reached a size somewhere between a saucer and dinner plate. Tank-reared hatchlings would swim furiously without rest for two or three days. Carr called this the "initial swim frenzy" and it was fueled by a reservoir of yolk that swelled the turtle's belly. The logical conclusion from these two pieces of evidence was that hatchling sea turtles made their way to open ocean when they left the nest. But Carr wondered what the hatchlings could find to eat out in the oceanic waters. Green turtles at Tortuguero and loggerhead turtles on the Atlantic Coast of Florida exhibited the same behavior of swimming out to open water. Carr decided that both kinds of hatchlings might be swimming out to sargassum rafts, and that such rafts served as the first-year refuge.

Carr's working hypothesis thus became: Hatchling sea turtles (green and loggerhead) swim out from their nest until they reach sizable concentrations of sargassum where they remain until they have grown. Through anecdotal reports and many discussions with swordfishermen, evidence mounted to support this case. Carr wrote: "So the sargassum raft hypothesis held promise. And when I started talking about it with people at the Florida marinas and with swordfishermen on the Salerno longline boats, it turned out that many of them knew that little turtles lived in the 'berry grass driftlines.' Virtually every habitual weedline fisherman between Cape Canaveral and Cape Florida had seen hatchlings in sargassum or in the stomachs of fish caught near sargassum."[67] Collectively, the stories of the fishermen confirmed Carr's impressions. With the help of the *Alpha Helix*, Carr was able to confirm that at least some hatchling green turtles found their way to the sargassum rafts off Panama (some one hundred miles south of Tortuguero). The sargassum hypothesis seemed airtight, but two problems arose. First, sagassum mats would disintegrate in bad weather. Where did that leave hatchlings? Second, some sea turtles nested on beaches far out of range of any sargassum rafts. How did the hatchlings from those colonies survive the pelagic phase?

It was these two questions that drove Carr to the study of physical oceanography, at the age of 76. Ever unassuming, Carr admitted that he was taxed by the effort of taking on a new subject:

> The effort to reconcile these incongruities cost me a lot of struggling with the arcane prose of physical oceanography—a form of communication that is fiendishly hard for an outsider to read. In the end, however, I came to realize that the fundamental factor in the pelagic stage of [turtle] development is not a sargassum raft, but the gathering of resources that takes place at a front, a convergence where different bodies of water come together. Apparently, horizontal friction or collision there generates sinking, or downwelling, and this mobilizes and aligns anything buoyant in the vicinity. This may not be intuitive, but it happens, and knowing it does make other observations easier to understand.[68]

Not only did downwellings make other observations comprehensible, but the same phenomenon addressed the problems with the sargassum hypothesis: "Wherever fronts occur, the debris and food resources of the surface waters will be assembled, trophic levels will multiply, overt colonization will take place, and life in the open sea will be feasible for an epipelagic, planktonic, air-breathing little animal that forages on smaller animals."[69]

After making this assessment, Carr turned to an overview of the importance of rips and driftlines as feeding habitat for billfish and seabirds. Of all the oceanic convergences, the ones that most intrigued Carr were the Langmuir bands, which were produced when wind blew steadily at seven knots or more. This phenomenon set up a series of evenly spaced vortices. The axes ran in the direction the wind blew. Where the associated eddies collided, the water sank, and anything floating was drawn in (much like the larger convergences). Carr attributed the reassembly of the sargassum rafts to Langmuir circulation, and their parallel formation would make them easier to locate for the dislodged turtle. Parenthetically, Carr noted that Langmuir bands were easier to locate as a result of pollution: "It is not a cheering thought that these days Dr. Langmuir's bands are made more conspicuous in the seas of the world by the abundance of styrofoam scraps and plastic bags and other human garbage that they hold."[70]

Other naturalists had noticed the extraordinary diversity and numbers of marine organisms within rips. One of the most eloquent descriptions of the phenomenon appeared in William Beebe's book, *The Arcturus Adventure* (1926). Carr quoted from the ship's log (April 1, 1925), which described the spectacular abundance of life: seabirds, marine mammals, turtles, and invertebrate life. Despite Beebe's elegant prose, Carr had been dissatisfied with Beebe's explanation for the convergence, even as a youngster. But there was no doubt in Carr's mind of the significance of rips and convergences: "It would be hard to overestimate the ecological importance of convergences. If there were none there would probably be no sea turtles—certainly none of the kind we know, with the racial custom of sending their young away on developmental migrations as pelagic plankton."[71]

Finally, Carr turned to a specific case study: loggerhead turtles and the ways in which their life history was defined by oceanic currents, particularly during the "lost year." The loggerhead was particularly well suited to such study because four age classes occurred with regularity along the Florida coast: sexually mature breeders and foragers, subadult populations in Canaveral Channel and Indian River lagoons, lost-year juveniles swept ashore by storm waves, and hatchlings. When he graphed the size range for each of these groups, Carr discovered a significant gap between the largest American pelagic juveniles and the smallest of the subadults in the lagoons. To the best of his knowledge, the missing class size occurred nowhere in U.S. waters. However, word of an itinerant seasonal colony reached Carr. Helen Martins, an oceanographer at the University of the Azores and a collaborator in the tagging program, sent Carr measurements of forty small loggerheads. Carr immediately wondered whether lost-year turtles were maturing on the other side of the Atlantic, possibly carried by the Gulf Stream. These data fit Carr's graph perfectly. Carr concluded that the "lost year" was actually several years: "In any case, the lost year of the Atlantic loggerhead is more like four or five years—and this bears directly on the even longer absence of the young of sea turtles from nests out on the Pacific Coast."[72]

With the life history of loggerhead turtles mapped, Carr again raised the somewhat more perplexing problem of sea turtles in the Pacific Ocean. Pacific ridleys and leatherback hatchlings swam out from the beaches to the oceanic currents (much like loggerheads). Once caught up in convergences, sea turtles of the Pacific did not find sargassum as a refuge. Nevertheless, flotsam did collect along the convergences, and this provided shelter for a host of animals, just as the sargassum did in the Atlantic. Drifting timber ranked among the most common form of flotsam, and the NMFS lab in Honolulu had replicated drift logs and anchored them around the Hawaiian Islands in water up to 1250 fathoms deep. These fish aggregating devices (FADS) attracted a range of pelagic fishes, which in turn attracted sport fishermen and commercial tuna boats. FADS seemed to create a bright spot for the future of all pelagic-dwelling species, as Carr concluded: "So the FADS are here to stay—for as long as the big ocean fishes can stay out of the Star-Kist cans and sushi bars."[73]

Carr's optimism regarding FADS was tempered by his realism about the impact of an endless accumulation of waste in the oceanic systems of the world. To those who suggested that oceans provided an ultimate sink absorbing all waste, he responded:

> I doubt the sense of that, but right or wrong, it ignores the growing threat to the fronts, where life is arranged in lanes. In a way, driftlines are like English hedgerows or like the zones along which terrestrial habitats meet. The comparisons are superficial, though, because the rips draw in not just organisms and their food but everything else that floats as well.

And of all the driftline inhabitants, little sea turtles seem the most vulnerable to the pollution the fronts gather. Along the coasts bordered by the Gulf Stream, young turtles choked with tar or with impacted digestive tracts frequently wash ashore. . . .

So the satisfaction of finding viable American loggerheads on the other side of the Atlantic is made a little less by the thought of the additional years they face getting home in waste-burdened driftlines.[74]

In his last paper, Carr explored the implications of oceanic convergences for the conservation and management of sea turtles. Recall that Operation Green Turtle operated on the belief that hatchling green turtles imprinted on the beach whence they fledged. On the basis of this belief, Carr and his colleagues delivered thousands of green turtles throughout the Caribbean with the assistance of a Navy Grumman sea plane. In light of the evidence regarding convergence ecology, the problem of head-starting became significantly more complex:

It is now obvious that when young cultured sea turtles are released in so-called 'head-starting' projects, the release sites ought to be chosen with the greatest care. Shore located at a distance from any major current of its eddy ought to be avoided, no matter how great the convenience or public-relations value of other localities may be. The adaptive utility of the neonatal swimming urge and open-sea guidance sense of hatchlings is that it takes them out of shelf waters, where predation is at maximum, and into an offshore convergence habitat.[75]

It is clear that Carr was most concerned about the possibility that head-starting would inadvertently bring hatchling sea turtles into contact with areas of pollution. Yet, his comments had profound implications for head-starting, which he criticized. To have a chance of success, any head-starting program would have to take into account the currents, because an unfavorable current could take a hatchling far from its goal of the convergence zones with their abundant food supply. Alternatively, the absence of a current could leave a hatchling effectively stranded in a near-shore zone of low resources and high predation. Another problem was the sex ratio of the head-started turtles. The eggs were hatched in styrofoam containers and were shaded from the sun. Since sea turtle gender is determined by temperature, it is likely that in the shaded areas (generally cooler than the hot sand of the beach) most of the hatchlings would have been males.

Conclusion

Nearly five decades had passed since Carr first considered the natural history of sea turtles in his dissertation, and he had spent more than thirty years dedicated to the study of their ecology and migration. Many of the stories he had heard from the turtle captains had been confirmed through tag returns from all over the Caribbean. The riddle of the ridley had been solved and documented by an old, grainy film that showed the *arribada* at Rancho Nuevo. Carr's enthusiasm for the discovery was tempered by the virtual disappearance of the *arribada* during the 1960s. In addition to long-term studies of the green turtle, Carr and his students published more restricted studies about other species such as the hawksbill. The scope of Carr's research extended beyond Tortuguero and the Caribbean to include Ascension Island, and he collaborated with other scientists to produce ambitious theories regarding olfaction, vision, and the role of seafloor spreading. In addition, his wife, his brother, and several of his sons made significant contributions to his research program. A major grant from NMFS facilitated an ambitious program to survey the turtle colonies of the Western Atlantic. Carr and his students participated in the Western Atlantic Turtle Symposium, an event that suggested growing interest in the ecology and conservation of sea turtles. Nevertheless, questions still remained. Although conservation efforts had protected many populations from deliberate slaughter, new threats seemed to proliferate, such as the nets of shrimp boats. Sargassum mats had initially seemed promising as a refuge for sea turtles during their lost year, but oceanic zones of convergence (including sargassum) seemed to be a more promising explanation. Though the intricacies of sea turtle biology drew him away, Carr always returned home to Florida.

CHAPTER 10

Home to Florida

Of all the places Carr visited, he chose to live in one of the most beautiful. From the window of his kitchen, he and his family could study a number of animals that lived in or passed through their small lake, known as Wewa Pond. Herons and egrets, frogs and turtles (including Jasper the alligator snapping turtle), ducks and gallinules, alligators, and even the occasional bobcat appeared near Wewa. The Carrs created a sanctuary for wildlife and for themselves. The driveway was long enough so that Archie could spot a car full of well-wishers and have time to escape through the back door and into the woods where he could write uninterrupted. Marjorie and the children would entertain the guests. On special occasions, he would amuse visitors by calling in Jasper with a piece of catfish or by visiting the "buzzard" feeder, where he would dump fresh roadkill to attract vultures or eagles. The name "Wewa," carried special meaning. The Carrs bought the property in the early 1950s after they had returned from Honduras. After a lengthy consideration of names drawing on the lexicons of Native Americans, the Spanish, and Floridians, they settled on Wewa, meaning water in the language of the Seminoles. Wewa's location (south of Gainesville in Micanopy) meant that Carr crossed Payne's Prairie twice on most days, a commute enlivened by the prospect of seeing additional natural spectacles.[1]

By the 1960s, Carr's study of the ecology and conservation of sea turtles had expanded to a worldwide enterprise as a result of his affiliation with the Caribbean Conservation Corporation (CCC) and the Marine Turtle Group of the International Union for the Conservation of Nature. We have seen the passion with which Carr pursued his travels in Honduras, Costa Rica, Africa, and many other places renowned for biodiversity. Nevertheless, even when he was in Africa or Central America, the Florida naturalist was never too far from home. Florida continued to serve as his biological and cultural touchstone, grounding his observations and analyses. If place is fundamental to the naturalist tradition, as I have suggested throughout this book, nowhere was more central to Carr's experience than Florida, his home. Late in his career, Carr returned to his roots and devoted most of his writing for popular audiences to the nature of Florida. He also contributed to several conservation campaigns within the state for wildlands and wildlife, including the American alligator. One of the most notable of these was the fight to save the Ocklawaha River, which was led by Marjorie Carr.

Yet Carr's experiences around the world now informed his view of Florida. In 1973, he wrote *The Everglades* for Time-Life Books. Sometime earlier, he had planned a book devoted to Florida natural history, and he wrote two dozen or so articles that could form the basis of such a book. Still, Carr worried that his book on Florida would degenerate into a bitter jeremiad regarding the ruin of his home. Given that he was unable to complete the book before he died in 1987, it seems plausible that the nature of Florida and its loss was too close to Carr's heart or that there was simply too much to say about the subject. Nevertheless, Marjorie collected the articles and other notes from Archie's five decades of writing and published them posthumously as *A Naturalist in Florida: A Celebration of Eden.* The last book Carr published in his lifetime was *The Everglades*.

The Everglades

Like other books in the Time-Life American Wilderness series, Carr's *The Everglades* is lavishly illustrated. The writing is no less distinguished. For the most part, Carr restricted his descriptions to the state of the Florida Everglades in the 1970s rather than allowing himself to wax nostalgic about what the unique wet prairie had been in times past. Inevitably, environmental degradation became Carr's subject. Rather than focusing on the obvious and direct effects of smog or pollution, Carr emphasized the subtle yet equally pernicious changes brought about by the introduction of exotic species of plants and animals: "There are more ways to pollute a landscape than by loading it with sewage, smog and beer cans. You can load it with exotic plants and animals. Whether brought in intentionally or accidentally, these are likely to change the landscape. The changes they make are rarely good, and often are atrocious."[2] Carr then cited several examples of exotic species: armadillos, walking catfish, cajeput, Brazilian pepper, and Australian pines, each of which initiated significant changes in the landscapes of Florida.

Development posed less of a problem in the Everglades for a simple reason in Carr's opinion: mosquitoes. He wrote: "Thank the Lord for the mosquitoes. The world owes them a lot for their part in preserving Cape Sable. A heroic statue of a mosquito in bronze ought to be set up on a hurricane-proof pedestal, a huge plinth of Key Largo limestone perhaps, at some commanding point on the cape."[3] The mosquito swarms in the Everglades were as bad as anywhere in the world, and according to Carr much worse at certain times of the year in places like Madeira Bay. Mosquitoes posed a serious deterrent to colonization and development.

Nevertheless, mosquitoes failed to halt development of other places such as Marco Island, the site of a recent (in 1973) development project that cost half a billion dollars. Carr reduced the crisis of environmental degradation in Florida to a basic choice: "I went away confident that the choice that southern Florida faces is not between water for birds and water for people, as short-sighted boosters were proclaiming a little while ago. The question is, rather, whether both shall survive on a shared water ration in a magic but pitifully fragile land."[4]

Saving Florida Landscapes

During the 1960s and 1970s, Carr devoted considerable time and energy to international conservation efforts, but these activities did not preclude advocacy on behalf of the landscapes and species of Florida. In 1969, Carr fought against the construction of a major road across the University of Florida campus that would pass near the north shore of Lake Alice (formerly called Jonah's Pond). In a letter to the president of the university, Stephen C. O'Connell, Carr enclosed excerpts from two of his books, *The Reptiles* and *Ulendo*. In *The Reptiles*, he wrote: "Jonah's Pond is one of the solid assets of the University of Florida. It is a sinkhole lake with tree-swamp at one end and open water at the other, and all through it a grand confusion of marsh creatures and of floating and emergent plants. The place is a little relic of a vanishing past, and incredibly, it lies on the campus of a university with 13,000 students, less than half a mile from where I'm writing now."[5] This passage went on to reveal the destruction of thousands of reptiles killed by speeding cars on highways (like the one planned for the border of Lake Alice). From *Ulendo*, Carr offered a comparable description of Lake Alice as a remnant of old Florida:

> There is a little pond I know—Jonah's Pond we used to call it, until that seemed to some less elegant than "Lake Alice"—a few priceless acres of marsh and swamp and pond water at the edge of the University of Florida campus, where any spring morning you can to this day have a glimpse of how Florida used to be. There will be five kinds of herons nesting there in the good seasons, croaking and chuckling over the sharing of the home space; gallinules and boat-tailed grackles running the pads, over and around basking turtles and alligators; white ibis, glossy ibis, anhingas, swamp rabbits, and people coming and going; and fat watersnakes that do not bother to whip off when cars full of children stop and gibber ten feet up the fill from the edge of the pond. The flooding sun, the bullfrog talk, the song of redwing blackbirds—it is all there still, saved somehow almost as it was when I first saw it thirty-two years ago. It is a little island of old times.[6]

Carr acknowledged the practical arguments for the roadway, but he countered such claims with several practical reasons for the preservation of Lake Alice, such as the lake's contribution to the material resources of the university, its value as a teaching resource, its ameliorative effect on the stress of faculty and staff, and its status as a Florida showcase for visitors to the university. But the single most practical justification for the preservation of Lake Alice was as a model for students: "Perhaps the most important practical by-product of this exchange of a priceless suburban wildlife preserve for a high-speed road through the campus is the damage it will do to the sense of values of our students. At a time when the dismal signs of hand-to-mouth urbanization are everywhere so clear, it would appear an inescapable responsibility of any university to show, by every possible example, that there are better ways to live."[7] In each of these passages, Carr advocated for the preservation of two remnants of Florida: natural and historical, both of which held inherent value. As a result of Carr's writings and contributions from other faculty members, Lake Alice continues to provide an oasis of natural tranquility on an increasingly developed campus with a student body of more than 46,000, more than triple the 13,000 cited by Carr.

Carr also participated in a less direct way in several other major campaigns on behalf of Florida wild lands throughout the state. In 1970, E. O. Wilson, the Harvard biologist, solicited Carr's support for a Nature Conservancy initiative to purchase and preserve Lignumvitae Key in South Florida. That same year, Carr advocated for the transfer of Payne's Prairie (the site of "The Bird and the Behemoth" in *Ulendo*) into public ownership by writing letters to the governor's assistant for conservation. Carr's efforts for Florida wildlands were matched by those on behalf of wildlife such as alligators.

Alligator Conservation

Both as a herpetologist and as a long-time Florida resident, Carr had an affinity for alligators, but even in his dissertation he noted that the American alligator (*Alligator mississippiensis*) had undergone a dramatic population decline over the previous century. He cited Bartram's classic account of alligators so abundant in the St. John's River that it would have been possible to cross the river on their heads.[8] As the graduate assistant for an ichthyology class in about 1935, Carr had one of several close encounters with alligators. The identification labs required undergraduates to seine fish using a 30-foot long net. On one occasion, a group led by Carr dragged in what they thought to be a large log along with the fish. In the process of collecting the fish from the net, the group uncovered the log, which of course hissed and opened its mouth wide to reveal its sharp teeth. After some quick reorganization, Carr and his students bound the jaws of the alligator and transported it back to town where it served as the mascot for homecoming. Carr's interest in alligators continued even as his research narrowed to sea turtles.

When David Ehrenfeld first met Carr, dressed in coat and tie with his diploma from Harvard Medical School in his briefcase, the wise professor asked, "Have you ever seen an alligator nest?" Five minutes later, Ehrenfeld found himself fighting through the vegetation on the edge of Lake Alice to reach a clearing where his new mentor showed him his first alligator nest. Years passed before Ehrenfeld fully appreciated the significance of Carr's gesture: "Only years later did it occur to me that this was the proper introduction to a place—visiting a typical part of the landscape and meeting its oldest inhabitants."[9] For Carr, alligators were an essential element of the place he called home. But the professor had more insights to share with his new disciple. Carr began to make a sound that was equal parts grunting, chuckling, and croaking. As the mother alligator turned toward them, he explained,

> That's the noise alligator hatchlings make.... When they hatch out, the mother hears them and comes to release them from the nest and protect them. They stay with her for a year or two. Old E. A. McIlhenny, the Louisiana tabasco sauce king, was the first person to write about the fierce maternal behavior of alligators, back in 1935. He wasn't a trained zoologist. Nobody believed him. Reptiles weren't supposed to be maternal. But everything he said was right.[10]

With this comment, Carr impressed on Ehrenfeld the value of local knowledge.

In 1967, Carr published an article in *National Geographic* titled, "Alligators: Dragons in Distress." Most of the article examined the life history of alligators including their preferred habitats, omnivorous diet, nesting behavior, and ecological significance. Carr also included a warning about the dangers of feeding alligators, but his chief concern was the demise of the species across most of its range. Poachers were systematically slaughtering alligators for their hides, which sold for up to six dollars per foot. Even in the Everglades, one of the greatest alligator strongholds, poachers were driving them toward extinction. Carr hoped that a change in fashion might bring about a change in fortune for the alligator, just as the National Audubon Society had successfully lobbied against plume hats, thereby averting the extirpation of egrets and herons in the early decades of the twentieth century: "If the vogue for alligator bags, belts, and shoes should pass, the profit would go out of poaching, and it would stop."[11] Still, the survival of alligators depended upon more than cessation of poaching. Somehow people had to learn to live near alligators, which would be difficult but would offer rewards: "It may not be all easy, living on into the future with the alligator. But by protecting him, we will show that we have the sense and soul to cherish a wild creature that was here before any warm-blooded animal walked the earth, and that, given only a little room, would live on with us and help keep up the fading color of our land."[12]

In addition to his writings on the subject, Carr also lobbied for alligator conservation on a limited basis. On February 21, 1968, he wrote to Claude Pepper in the U.S. House of Representatives to urge his support of H.R. 11618, which would regulate interstate trade of alligator hides. Carr wrote, "The steadily increasing demand for leather and other products derived from reptiles makes regulation of interstate traffic a vital factor in the world campaign to prevent extinction of some of the exploited species."[13] Carr also argued against feeding alligators (a fairly popular pastime in Florida) for the simple reason that anything that brought people into closer proximity with these large predators increased the odds that an alligator might try to eat anything that fell into the water, including pets and children. He even helped write signs for Lake Alice at the University of Florida, warning people not to feed alligators. Yet, when his former student, Peter

Pritchard, invited him to join a Florida Audubon committee to develop a policy for alligator management, he declined. Instead, Carr said he would retreat to his ivory tower to write about endangered reptiles and leave real-world alligator policy to young bloods like Pritchard.[14] With the protection of the Endangered Species Act (1973), alligators recovered rapidly, while alligator farms produced enough hides to meet the demand for leather products. During this time, Carr also lent a hand to Marjorie's ongoing campaign to save the Ocklawaha River and stop the Army Corps of Engineers from completing the Cross-Florida Barge Canal.

Marjorie Carr and the Fight to Save the Ocklawaha River

Early in their careers, Archie and Marjorie Carr worked on related subjects. For much of the 1940s and the 1950s, Marjorie concentrated her efforts on her family of five children while remaining active in the Gainesville Garden Club and co-founding the Alachua Audubon Society (the Gainesville chapter of the Florida Audubon Society and the National Audubon Society). As Archie contended with the demands of his teaching, research, and burgeoning writing career, Marjorie kept the children out of his hair by organizing a wide range of activities for the children.

Figure 33. Marjorie Carr at the shore with her children, 1953. Courtesy of Mimi Carr.

Still, Marjorie's interest in science and conservation continued to grow and develop, and she participated in several conservation-oriented campaigns. In the late 1950s, she served as the roadside development chairman of the Gainesville Garden Club.[15] The Florida Department of Transportation had initiated a program of setting aside roadsides as preserves. With the help of other garden club members, Marjorie Carr managed to have the roadside along S.R. 441 through Paynes Prairie established as a preserve.[16]

Along with H. K. Wallace (one of Archie Carr's colleagues in the Department of Zoology), and Enid and John Mahon, Marjorie Carr founded the Alachua Audubon Society in 1960. At about the same time, the University of Florida had decided to drain Lake Alice. Along with Zoology faculty members, Marjorie and the Alachua Audubon Society managed to stop the proposal.[17] On November 8, 1962, the Alachua Audubon Society sponsored a program devoted to environmental problems in Florida called: "The Effects of the Cross-Florida Barge Canal on Wildlife and Wilderness."[18] The talk was polished and included slides and charts, but the audience included many professors, and they remained skeptical. Marjorie would not have predicted that the Cross-Florida Barge Canal would ultimately place her and the organization she would establish at the center of a major grassroots environmental campaign.[19]

After the initial Alachua Audubon Society meeting in the fall of 1962, Marjorie and the other board members started gathering facts regarding the proposed canal. When they heard that the proponents of the Cross-Florida Barge Canal were seeking the endorsement of sports and recreation groups, the board members lobbied for support from other conservation organizations. As Alachua Audubon continued its investigations, another group formed with the specific objective to investigate and report information on the waste of the natural and economic resources of Florida. Incorporated in 1964, this new group called itself Citizens for the Conservation of Florida's Natural and Economic Resources, Inc. Some of the Alachua Audubon members who had been researching the canal joined the new organization.[20]

On June 15, 1965, Marjorie wrote to Claude Pepper, who had been elected to the U.S. House of Representatives in 1963 (a position he would hold until 1983). In her handwritten letter on Alachua Audubon Society stationery, she enumerated the unique natural characteristics of the Ocklawaha River, the destruction that the proposed Cross-Florida Barge Canal would cause, the high value placed by conservation organizations on the river, and the reason the river was "threatened with obliteration." She signed the letter, "Mrs. Archie Carr, Co-Chairman for Conservation."[21] Included in the letter was a brief paper Marjorie had written on the natural history of the Ocklawaha River Wilderness Area. In a passage that reflects the literary influence of her husband, Marjorie wrote:

> From earliest times the Oklawaha has served man as a pathway through the jungle fastness of its great tree swamp. The Indians of the time of William Bartram, Florida's first visiting naturalist, called the river Ockli-Waha—Great River—and it was for them an important highway

and hunting ground. A hundred years ago, when most of Florida was wild, naturalists and hunters alike regarded a trip up the Oklawaha as an exciting and rewarding venture into wilderness. Today, when so many of the diverse original Florida landscapes are threatened with obliteration, the Oklawaha, in its lower reaches, remains as it was, a dark beautiful stream, clear and free-flowing, and now as in past times, noted for its fine fishing.[22]

In his response, Pepper expressed his strong commitment to the canal.[23] Through what must have been an inadvertent misreading of Marjorie's hand-writing, Pepper (or a member of his staff) addressed the letter to "Mrs. Ardill Carr." From that point on, Marjorie had her official correspondence typed. Despite environmentalists' efforts to the contrary, in August 1965, the Senate approved the appropriation for $10 million for continued construction on the Cross-Florida Barge Canal, bringing the total apportionment to $15 million. The next two projects for the Army Corps of Engineers would be the construction of the Rodman and Eureka dams in the Ocklawaha River Valley. Given that the Corps would fill these contracts by January 1966, the environmentalists intensified their campaign.

Despite repeated calls for a public hearing, state officials refused to acknowl-edge the environmentalists until the annual water resources meeting in Tallahassee on January 25, 1966. Roughly 350 people made the trip to Tallahassee at their own expense to oppose the canal, while a smaller number of supporters participated in the hearing. Some of the environmentalists presented evidence for the alternate route and argued eloquently for the preservation of the Ocklawaha River and the wilderness that surrounded it.[24] To the considerable chagrin of Marjorie and other environmentalists, while returning to Gainesville they heard a report that the committee had met at eleven in the morning prior to the public hearing and had voted to continue the barge canal in the same route. The committee had dismissed the testimony of conservation groups before they even heard it.[25] Here was a serious blow to the confidence of the environmentalists. Nevertheless, this hearing served as a watershed in the history of the environmental movement in Florida, given that environmentalists from all over the state had descended upon the state capitol to air their opinions on a threat to Florida wilderness.

Given the general lack of response from various elected officials, Marjorie decided to take the case for the preservation of the Ocklawaha to a higher level. In March 1966, she and the Alachua Audubon conservation board wrote to President Lyndon B. Johnson, but they did not receive a response to either letter. On June 1, 1966, Marion S. Hodge sent a letter on behalf of the conservation committee of Alachua Audubon to the first lady, Lady Bird Johnson, and included excerpts of the 1,000 letters they had received supporting preservation of the Ocklawaha. The first lady represented a strong potential contact given her exten-sive work on the Highway Beautification Act of 1965.[26] While Johnson expressed a certain degree of empathy, she noted that she was more concerned with her husband's health. Later that same month, the Army Corps of Engineers exploded red, white, and blue smoke bombs to celebrate resuming construction on the Cross-Florida Barge Canal. From 1966 to 1968, construction continued. As the conservationists had predicted, the completion and filling of the Rodman Dam in November 1968 destroyed 5,000 acres of forest.

Despite the considerable damage wrought on the Ocklawaha, Marjorie Carr continued to collect documents and develop strategy for a renewed effort. In 1968, an election year, she and the Alachua Audubon Conservation Committee launched a new campaign that was explicitly political in nature. After initial inquiries, the Environmental Defense Fund (EDF) agreed to argue the legal case if the Audubon Society would build the scientific case. In July 1969, the Florida Defenders of the Environment was established. The name pleased the founders since its acronym (FDE) was the opposite of EDF. William Partington took a leave of absence from his position as assistant executive director of the Florida Audubon Society to become FDE's first chairman, and Marjorie agreed to serve as the vice-chairperson. A well-known Florida radio personality, Arthur Godfrey, became the honorary chairman for publicity purposes. As its first task, FDE prepared a major report titled, "The Environmental Impact of the Cross-Florida Barge Canal on the Ocklawaha River Regional Ecosystem," that presented a study of the Cross-Florida Barge Canal from the viewpoints of geology, hydrology, ecology, economics, land-use planning, anthropology, and environmental quality. Most of the chapters were written by professors at the University of Florida, including Archie Carr, who co-wrote the chapter on the vegetation of the Ocklawaha Regional Ecosystem. Despite the credentials of the authors, they wrote in a way to make the report accessible to the lay reader. The FDE report would serve as a model for later environmental impact statements.

Meanwhile, EDF filed its suit in the U.S. District Court in Washington, D.C., on September 16, 1969 on behalf of the people of the United States against the Secretary of the Army and the U.S. Army Corps of Engineers to restrain the Corps from further construction on the barge canal until all evidence had been heard. EDF's suit raised numerous issues that were developed in the FDE's report, such as the impact of invasive, exotic aquatic vegetation, geological problems, and value of the wilderness area.[27]

Another facet of the renewed campaign to save the Ocklawaha involved public outreach through the media. The original campaign focused on politicians, and it neglected the critical component of popular support. In 1969, several articles addressed the impact of the canal on the Ocklawaha. The most widely distributed of these appeared in *Reader's Digest* under the title, "Rape on the Oklawaha."[28] Another article quoted Marjorie Carr's devastating critique of the claims of Army Corps Engineers that the river would be left intact "except for the part between Sharps Ferry and Rodman Dam." Majorie replied, "It was like saying that one is just going to cut off the rooster's tail—right behind the head. That 45-mile stretch is the heart of the river."[29]

Coupled with the support from national environmental groups, FDE increased its dedication to developing the Cross-Florida Barge Canal as a political issue with political consequences. To that end, on August 19, 1970, William

Partington sent a letter to every candidate running for office in Florida to determine his or her official position regarding Florida's problems of environmental quality, with particular emphasis on the canal controversy.[30] Of the 390 questionnaires sent out on August 19, 1970, 123 were returned over the next fifteen days. With 80 percent of the candidates for office in Florida in support of a moratorium or complete abandonment, William Partington wrote to President Richard M. Nixon and called upon him to issue a moratorium.[31] The environment provided a cornerstone in Nixon's State of the Union Address, which he gave three weeks later. In the speech, Nixon argued for an America with a cleaner environment. He suggested that economic growth was only desirable if it improved quality of life.[32] The FDE capitalized on Nixon's statements in his State of the Union Address by sending a letter to the president dated January 27, 1970 regarding the canal that had been signed by 162 environmental scientists.

Both the legal and political fights soon reached a conclusion favorable to the environmentalists. On January 15, 1971, Judge Barrington D. Parker of the U.S. District Court for the District of Columbia temporarily enjoined further work on the Cross-Florida Barge Canal in its summit reach and in the undisturbed part of the Ocklawaha Valley. The president accepted the advice of his Council on

Figure 34. Marjorie Carr receiving recognition from Governor Claude Kirk for her efforts with Florida Defenders of the Environment in October 1970. Archie Carr looks on. Courtesy of Florida Photograph Collection.

Environmental Quality and halted construction on the barge canal. President Nixon suggested that his action contained broad implications for the prevention of environmental damage.[33]

In a membership letter on behalf of FDE, Marjorie Carr exulted over both the court decision and the executive order halting construction on the canal, referring to them as "conservation milestones." But she noted that there was "a great deal yet to be done before the problems related to the barge canal project are finally solved." FDE had recommended to the president that the Department of the Interior undertake an extended, eighteen-month study of the western end of the project area, that the water in Rodman Dam should be lowered to the natural river level, and that the Ocklawaha River be returned to conditions that would qualify it for "scenic river" status.[34] The last recommendation proved to be the greatest challenge. Marjorie Carr and FDE fought for the restoration of the Ocklawaha right up to the time she passed away in October 1997. Two monuments stand as symbols of the Cross-Florida Barge Canal. The first is the continuing effort of the FDE to restore the Ocklawaha and the legacy of FDE's successful attempt to stop the canal. The other monument, standing like a mausoleum to a once-beautiful river, is the Rodman Dam and the lake it created. Despite FDE's ongoing efforts, the wheels of the state bureaucracy continue to grind slowly. On a brighter note, on January 22, 1991, the governor of Florida signed a resolution agreeing to the terms of the federal deauthorization bill, thereby officially deauthorizing the Cross-Florida Barge Canal project. This action ultimately led to the creation of the Cross-Florida Greenway State Recreation and Conservation Area, which was officially renamed the Marjorie Harris Carr Cross-Florida Greenway in honor of the "Micanopy housewife," who led the fight to stop the Cross-Florida Barge Canal project. Archie contributed to Marjorie's campaign in fairly minor ways while he continued his research and writing as his department underwent a major reorganization.

More Changes at the University of Florida

In the early 1970s, the biological sciences underwent a significant reorganization as a result of more than a decade of internal and external analysis. In November 1959, an external consultant critiqued the Department of Biology at the University of Florida (UF). It was clear that the department was primarily one of "animal biology," since botany and microbiology had migrated to other units within the university. He noted that the department "enjoyed" the worst physical housing he had seen in years, despite significant demand for courses. Administratively, the department was split between introductory courses designed to expose every UF student to biology (the C-6 program) and upper-level undergraduate and graduate courses. Since sections for courses were limited to fifty or less, the C-6 program placed extraordinary demands on professors while contributing little in the way of resources.

In February 1960, a committee of department faculty supported some of the consultant's recommendations, particularly the reorganization of the divided department into a single administrative unit, changing the name of the department to Department of Zoology, and promptly improving the facilities and buildings, particularly those of the Florida Museum of Natural History, which would house all research collections. The committee also stressed the strengths of the department:

> The physical location of the University makes it natural and most proper that from a geographic standpoint we emphasize the biology of the southeastern U.S. and the Caribbean area. We can do this better than any other institution in the United States. Tropical biology is a tremendously rich and rewarding field of study and research. The rapidly increasing realization of the intimate social and political relationships of the U.S.

Figure 35. Archie Carr with his German shepherd, Ben, investigating a sea turtle (ca. 1975). Courtesy of Mimi Carr.

and our Latin American-West Indian neighbors is cause enough for us to devote effort to arriving at a more thorough knowledge of the biology of the region.[35]

In 1964, UF created the Division of Biological Sciences as an initial attempt to unify the biological sciences. Though successful in terms of collaboration among individuals, the division represented a compromise solution and was not regarded as a success. The new division would unite the biological sciences under the College of Liberal Arts, where the department of zoology was already housed. Other departments such as botany, bacteriology, and microbiology would join zoology in the College of Liberal Arts while maintaining their programs in other colleges such as agriculture or medicine.[36]

Despite more than a decade of discussions surrounding reorganization of the biological sciences at UF, the Department of Zoology was one of the programs least affected by the changes. Partially as a result of his nonconfrontational nature and his many commitments, Carr stayed out of the debates. He even failed to show up at some of the most controversy-filled department meetings, where junior colleagues hoped he would take a stand; instead, Carr could disappear at even the slightest hint of controversy. When cornered, Carr would acknowledge both sides of most debates. At UF, zoology continued to enjoy a central position within the Division of Biological Sciences, whereas at other universities, such as Harvard and particularly Yale, the study of whole organisms was relegated to museums.[37]

A Naturalist in Florida: A Celebration of Eden

Although *The Everglades* was the last book that Carr published during his life-time (save for second editions), he continued to write about Florida. In the late 1960s, he began to write a book about Florida, but he was mindful of the risk that this book could become a polemic on environmental degradation:

> When I set out to write this book I immediately sensed a danger looming. It was that I was almost bound to fall into the trap of nostalgia and indignation, of turning this book into a diatribe against the passing of original Florida. Because to anyone who has known Florida as long as I have, and whose main interest in the place has been its wild landscapes and wild creatures, the losses have been the most spectacular events of the past three decades.[38]

With his extensive knowledge of Floridiana, Carr could cite several prior lamentations: John K. Small's *Eden to Sahara* and Thomas Barbour's *That Vanishing Eden*.[39] Carr knew the latter work particularly well since he discussed the destruc-tion of Florida's wildlands at length with Barbour, his old friend and mentor. In reflecting on these works, Carr had an epiphany:

> So being a naturalist, living in the woods, and having the peculiar back-ground I have, I am especially susceptible to the disease of bitterness over

the ruin of Florida—over the partly aimless, partly avaricious ruin of unequaled natural riches of the most nearly tropical state. But in my case I decided simply, "What the hell, you cry the blues and soon nobody listens." And that made me see that there was really no sense writing another vanishing Eden book at all.[40]

So rather than lamenting what had been, Carr celebrated what was left. His essays on Florida appeared in *Audubon* (the magazine for the National Audubon Society) and *Animal Kingdom* (the magazine for the New York Zoological Society). Several years after Carr passed away, Marjorie collected and published an edited volume of these and other essays about Florida as *A Naturalist in Florida: A Celebration of Eden*.[41] In these essays, most of which he wrote during his final years, Carr returned to the place he knew best. "All the Way Down upon the Suwannee River" skips lightly from past to present to prehistoric and back as Carr cites William Bartram's description of the river and its Indian inhabitants, recounts battles of the War of 1812 and the Civil War, revels in the pleasure of floating down the cool, spring-fed river, exults in the discovery of teeth of mammoths and mastodons, and warns of the perils of overuse for recreation.[42] Carr's narrative of the Suwannee flows smoothly between nature and culture like the best environmental history. Similarly, in telling the tale of Florida jubilees (spectacular congregations of fish and other aquatic animals), Carr weaves together strands of biology, paleogeography, Florida cracker culture, Bartram's *Travels*, and inevitably environmental decline. As promised, Carr kept his sense of regret in check, only occasionally lapsing into despair over the loss of this landscape or that animal. For example, he pondered the demise of the Wacahoota jubilees:

> the Wacahoota jubilees are finished. As they became more widely known in town people started coming out in greater numbers. Their trespassing annoyed the owner and moved him to modify the landscape in ways that changed the character of the flow in the jubilee stream out of Moore's Prairie. I don't know what he did back there, but since then there has been no Wacahoota jubilee, and apparently there will never be another.[43]

In writing about Florida, Carr often stood up for the most disliked and misunderstood of animals. He believed that animosity toward snakes and alligators was largely undeserved. Carr cited Rachel Carson's *Silent Spring* and the beneficial effect it had on bird populations by warning people of the environmental consequences of chemical insecticides. As yet, snakes had no such advocate, even though their populations were also in a state of extreme decline. Carr argued that the allegedly innate human fear of snakes had outlived its usefulness. Following trends of popular natural history, he suggested that people take up snake watching just as so many enjoyed bird watching. For those less apt to follow snakes in the wild, Carr recommended keeping one around the house for a pet, noting which snakes were appropriate for domestication. Ultimately, however, Carr's story about snakes was an attempt to avert a largely unnoticed catastrophe. He hoped that by alerting people to the plight of snakes, he could help save them:

> Although keeping snakes in homes and zoos is fun, it does not discharge our obligation to save them as wild species in the world around us. Most sensible people have now worked up a healthy fright over the future the world holds for their children, and birds are often drawn in under the umbrella of their concern. But I have heard little worrying over the future of snakes, and this to me is depressing. Snakes are not degenerate beings, punished with leglessness for ancient sins, as people once said. A snake is an elegant product of a hundred million years of natural selection. Its loss of legs was an evolutionary advance, a means of living successfully in unexploited ways. But because those ways are secret, the decline of snakes in our changing world has gone on almost unmonitored. Others have spoken for cranes and whales, and I hasten to say these words in praise of snakes, whose silent spring is also far along.[44]

Carr's largesse toward snakes was hardly naïve. In 1968, while at Tortuguero, he had been struck by a fer-de-lance (*Bothrops atrox*), known locally as terciopelo. Despite three decades as a research herpetologist and considerable knowledge of venomous snakes, Carr initially hoped that he had been struck by a branch or a nasty boa constrictor. Finding the fer-de-lance laid to rest that hope. He was some distance from the beach, and he knew he should try to relax rather than running for help. Nevertheless, he started running but soon reined himself in. As he walked toward Marjorie and his friends, he alternately whistled and yelled until his wife heard and went to find antivenin. In the meantime, Carr tied a tourniquet above the wound, wondering if tourniquets were still part of the recommended treatment. By the time Marjorie had determined how to administer the antivenin and the test for sensitivity to horse serum, Carr realized that an hour had passed with no noticeable intensification of symptoms, and his hope that he had suffered a "false strike" began to grow. In the meantime, many friends and passersby had stopped to offer prognoses (from imminent death to possible survival) and advice (from drinking a shot of kerosene to chewing a leaf of tobacco to avoiding the gaze of a pregnant woman). As time passed and Carr became more confident that he would survive the bite of the most dangerous snake in the Americas, he wondered about false strikes. Rattlesnakes typically rattled before striking and delivered a deadly strike through relatively short fangs. The fer-de-lance, on the other hand, had no rattle, so it might deliver a false strike and protect its long fangs from damage. That made sense to Carr, and he concluded his reverie: "Anyway, that is what I have been stirred to think; and I hope it is not just moonshine from excess adrenaline let loose by the recollection of my *Bothrops* bite."[45] Years later, Carr was bitten by a rattlesnake that someone had left in an unmarked bag in the Department of Zoology office. Once again, he survived the bite, in part due to the efforts of his colleagues who rushed him to the hospital. The episode prompted a stern memorandum from the department chair, Thomas Emmel.[46]

In 1969, Carr wrote an article for *Audubon* titled, "Thoughts on Wilderness Preservation and a Central American Ethic," in which he acknowledged his strong connections to both Florida and Central America. Florida was where he lived, but he had spent nearly a third of his time in Central America over the previous twenty years. Development was destroying Florida: "Florida is now so wholly in the grasp of the developers that despite the growth of a powerful conservation conscience in the state, there seems little hope of saving more than a few scraps of the original landscape."[47] Prospects for conservation in Central America, however, were considerably more hopeful: "The bulldozers are ripping up Central America too, but the land is a little bigger and rougher down there, and the ruin began later. There still may be time to search out and save samples of most of the kinds of original environments of the Isthmus, if only the motivation can be spread and maintenance techniques worked out before development brings complete destruction."[48] Carr reviewed the Central American wilderness assets including the Atlantic rainforest and its species. He imagined that rainforest preserves might be extended to the coast, where most of the land was under government control. The *Milla Maritima* law excluded private ownership in these areas (to the great benefit of places like Tortuguero). Despite the stunning ecological diversity of the forests of the Caribbean slope, scientists had not for the most part studied them ecologically or even listed their species.

The status of rainforests in Central America and elsewhere led Carr to ponder the alleged demise of natural history in the biological sciences:

> It is this sort of situation, which can be matched in many parts of the world, that explains why practitioners of natural history get irked when cellular and molecular biologists suggest that study of whole animals should be soft-pedaled while sounder ground for understanding them is being worked out at more fundamental levels of the discipline.[49]

Carr suspected that postponing studies of organisms for studies of molecules would have dire consequences for the rainforest: "If we wait for the great gap to close, the original landscapes of Earth will simply disappear."[50] He called for action on the part of conservationists even given limited data:

> From time to time conservationists are bound to feel their position comparable to that of a man overly imbued with reverence for life, who decides at last that he must stay in this chair to avoid doing damage to the small unseen beings under his feet. But the untenability of that man's extreme position in no way negates the validity of his original reverence. In the same way, the impossibility of preserving either species or biological organizations completely intact in no way justifies our relaxing the effort to keep whatever can be saved, and to go on haggling with the forces remaking the Earth over each separate relic of the old landscapes.[51]

In 1973, again in the pages of *Audubon*, Carr worried about Spanish moss, which appeared to be diseased across much of Florida. Carr discussed the moss (an epiphyte, not a parasite as some believed) and its historical values as a resource.

Native Americans, Spanish settlers, and American homesteaders all recognized a valuable plant in Spanish moss. Its uses ranged from bedding to clothing. Next Carr turned to the oak trees the moss graced and the long history of live oaking (lumbering) in Florida. Live oaks represented the climax forest type in ecological succession. According to Carr, pulp and paper mills were converting forest to pine plantation, which brought to mind a crass advertisement for the chemical and paper company Olin. The ad showed a stand of pines under the caption: "If you think it's beautiful now, wait until we chop it all down." In a letter to the editor, Carr had challenged both the letter and the spirit of this ad. A host of animals depended on the moss forest from spadefoot toads to bats and flying squirrels, signs of which evoked the moss forest:

> The rustle of a spadefoot foraging was like a buck in bushes; the repeated squeak of a flying squirrel, in lingering concern over the barred owl's bellow, seemed a strident noise. The moon blazed and faded, I remember, as separate high clouds went by on a wind too high to stir the moss; and one moment the vaulted rooms of the moss forest were flooded with silver light, and the next the glowworms down at the pond edge were torch bright in the dark.[52]

At the age of 73, Carr embarked on a new endeavor: a regular column for *Animal Kingdom*, the magazine for the members of the New York Zoological Society (the Bronx Zoo). In his first article, "Armadillo Dilemma," Carr explained why he agreed to write the column:

> When I was invited to hold forth in this little corner of *Animal Kingdom*, one inducement was that my themes could be wide-ranging as long as they had some logical connection with animals. I was told that personal views and observations were welcome, which I believe means that occasional departures from rigorous objectivity would not be taken amiss. This, too, attracted me because I have always thought that a reasonable amount of emotion is inevitable in zoology, provided the worker is involved with species other than white mice.[53]

The reference to mice was a subtle jab at the rise of genetics and molecular biology as dominant fields within the biological sciences. As a subject for his inaugural essay, Carr chose his mixed feelings regarding the nine-banded armadillo (*Dasypus novemcinctus*) and his ambivalence toward the species. It was an animal he discovered in his youth when he would beg his father to read stories from Rudyard Kipling's *Just So Stories*. In his mammalogy classes as a student at the University of Florida during the Depression, armadillos were rare in Florida. Carr knew of three separate accidental introductions of the armadillo to Florida during the 1920s and 1930s. Compared with the natural rate of increase for the species, the armadillo population had exploded in Florida. This struck Carr as strange given that three separate lines of armadillos including one in the genus *Dasypus* had abandoned Florida after the Pleistocene. Why did the immigrant species thrive? Mainly because major predators such as panthers, bears, and bobcats had declined in Florida (although Carr's German shepherd strove to fulfill his role in the natural order by killing so many armadillos that his teeth wore down).

The naturalist in Carr was troubled by the proliferation of armadillos, but it was the rigorous ecologist in Carr who could explain why. Before armadillos arrived in Florida, Carr's ecology classes would sample leaf litter with a Berlese funnel. Such samples would typically yield a host of terrestrial invertebrates from insects to spiders. Similar samples taken in the same forests after the arrival of armadillos produced little or nothing. Moreover, the smaller snakes, salamanders, skinks, and toads were all gone as well. They had disappeared down the gullets of armadillos. Moreover, incessant rooting by armadillos was destroying the leaf-mold layer and conversion zone for detritus and nutrients. Already Carr was telling a narrative of environmental degradation, so he attempted a positive conclusion: "But there are useful morals in the melancholy mischief of the Florida armadillos. One is that animals and plants ought not be moved around to places outside their natural range. Another is that trapping bobcats in Florida ought to stop. A third is that the natural world can be spoiled in very subtle ways."[54] In this article, Carr established a template for many of his "Life Line" articles for *Animal Kingdom*, creating a seamless flow from personal to general, from prehistoric to present, from fiction to nonfiction, and from natural to cultural.

Wewa, the Carr's homestead in Micanopy, Florida, was the subject of several articles and was mentioned in several more. Carr told of his buzzard feeder. By collecting road-killed animals (often armadillos), the Carrs attracted two species of vultures, bald eagles, and even the alligator from their pond to their yard. It was the alligator that convinced them to abandon the feeder due to a firm rule against feeding alligators, even inadvertently.[55] In other articles, he addressed the adaptive significance of the peninsula cooter's (*Pseudemys floridana peninsularis*) odd tendency to lay eggs in three holes (two side pockets adjacent to the main chamber), the spectacular aggregations of freshwater fishes he called jubilees, and autophagy in blue-tailed skinks (*Eumeces laticeps*), in which a skink ate its own tail.

Several "Life Line" articles focused on the wildlife of Wewa Pond on the Carrs' farm. The sight of eight ducks (two shovelers and six blue-winged teal) on Wewa sent Carr searching through his memories of ducks. He could recall hunting with his father in the frigid cold of Texas in winter and the exquisite meals of duck his mother created. He remembered arriving at the University of Florida in the 1930s and finding hundreds of ducks as he crossed Paynes Prairie on the newly constructed U.S. Route 441, which neatly bisected the wet prairie. In 1986, Carr found no ducks as he crossed Paynes Prairie twice daily for the entire winter. Ducks were in a state of decline across North America, along with their prairie pothole breeding grounds. Carr remembered that in the 1960s ducks flocked to Wewa Pond. In the aftermath of Christmas, the whole Carr family was lounging around enjoying their own company while idly watching the ducks on the pond. On a brief trip to the barn, Archie found a fresh bobcat track, and upon his return, his family told him they had seen a bobcat stalking ducks on the pond. He immediately followed the trails of the departed bobcat and found the remains of a female wood duck, which elicited mild remorse from Carr: "I felt a pang of regret that the duck had met her end, but I was not in a position to hold the bobcat's violence against him. As I said at the start, I had eaten many wild ducks myself in my younger, undisciplined days."[56] Carr's comment evokes Aldo Leopold's ambivalent musings on hunting in general and duck hunting in particular, most notably in the essay, "Red Legs Kicking": "I cannot remember the shot; I remember only my unspeakable delight when my first duck hit the snowy ice with a thud and lay there, belly up, red legs kicking."[57]

If Carr saw in culture the roots of the destruction of nature, ultimately he considered culture to hold the seeds of hope for natural Florida as well. In an essay for *Born of the Sun* (the official Florida Bicentennial Commemorative Book), he took heart from the partial recovery of several species: manatees, alligators, otters, beavers, and wading birds (herons and egrets). All of these animals had been hunted to the brink of extinction in Florida during Carr's lifetime. Offsetting such optimism was the realization that unique ecosystems such as the large Florida springs were in a state of decline. Carr attributed such deterioration to a variety of factors, but chiefly to the efforts of subsistence fishermen who used dynamite to catch fish, euphemistically called "cut-bait fishermen." Cut-bait fishing had left the majority of springs denuded of life, with little hope of recovery. And yet, Carr gazed into the hearts of Floridians and found good reasons for hope:

> In listing some reasons for optimism over the state of nature and man in Florida, one favorable development outweighs all the rest. It is not another species on the mend or a new park or preserve or sanctuary established. It is rather a change in the hearts of the people. Although original Florida is still undergoing degradation, an assessment of the trends would show the rate of loss being overtaken by growth of a system of ecologic ethics, by a new public consciousness and conscience ... The rise of this new stewardship gives heart to opponents of ecologic ruin everywhere and brings promise of better times for man and nature in Florida.[58]

Such a statement reflects a considerable shift in Carr's thoughts on Florida, nature, and history. In several examples noted above, Carr equated historical Florida with natural Florida. This new configuration suggests that the rise of American environmentalism transformed the relationship between nature and culture, causing Carr to ponder the role of human agency in the past, present, and future of natural Florida.[59]

Conclusion

A profound sense of place resonates throughout Carr's writings. Just as he constructed lasting images of Florida, that sense of Florida shaped Carr. No matter how far he traveled, Carr remained devoted to Florida's landscapes and wildlife. In his writings and conservation efforts, Carr worked to preserve the natural history of Florida. As their children reached the age of independence,

Marjorie was free to pursue her own passion for conservation. In addition to her important contributions to Archie's research at Tortuguero, she led the fight to save the Ocklawaha. Each of the Carr's five children recalls vibrant conversations around the dinner table about the future of Florida's wildlands and wildlife. Just as their time in Honduras had cemented their relationship, Archie and Marjorie Carr's joint and independent efforts for the conservation of Florida defined their unique relationship and collaboration.

CHAPTER 11

Conclusion

Throughout his long career, Carr acknowledged the value of local knowledge. This theme resonates in many of his scientific papers, his popular writings, and conservation campaigns. In preparing a preface to the revised edition of *The Windward Road*, Carr was pleased to recall how many pieces of folklore had evolved into working hypotheses for decades of research. Across the Caribbean, turtle fishermen, the most famous of which were the Cayman turtle captains, knew where the turtles nested and foraged. At Tortuguero, it was common knowledge that individual turtles nested three to six times over the course of the breeding season and generally within the vicinity of the other nests. Carr and his students spent years substantiating such claims.

But one piece of folk wisdom worried Carr: "In one important way the wisdom of the Caribbean people seems to go unaccountably awry. That is in the widespread belief that the green turtle is an inexhaustible resource. My first season at Tortuguero, when I asked Sibella how long the turtles could stand the slaughter then going on at the nesting beach, she said, 'Dey never finish don Archie. The tet-tel [turtle] never finish.' "[1] For years, Carr studied the ecology of the sea turtles and advocated for their conservation, convincing government officials and even presidents, but Sibella remained steadfast in her view. So when the nesting season of 1978 had nearly passed, Carr knew he would hear from Sibella: "Then 1978 came. The beach piled up with more turtles than we had seen in two decades at Tortuguero, and our insight was not deep enough to explain the jubilee. I knew I was in trouble with Sibella; and sure enough, one morning toward the end of the season when I went in to breakfast Junie [Sibella's daughter] said, 'My mother write I should chide you.' When I asked why—knowing all the time—Junie said: 'Because you always say the tet-tel finish. She tell me I must ask you what you can say now.' "[2] Ever gracious, Carr modestly deferred to Sibella and the power of folk wisdom: "All I could say was, I was happy so many turtles were coming in."[3] Of all Carr's many legacies, including the Caribbean Conservation Corporation, successful students, books, and scientific papers, one of the most satisfying must have been the resurgence of sea turtle populations at Tortuguero and elsewhere.

On May 8, 1987, Carr received one of the greatest honors of his long career: Eminent Ecologist of the Ecological Society of America (ESA). The award indicated the high regard Carr's peers had for him. In granting the designation, the ESA recognized Carr's studies of the ecology of sea turtles and his gift for sharing those findings with a wider public. The ESA's highest honor arrived just in time. Carr, the man who saved sea turtles, was nearing the end of his remarkable life. It was only in the previous year that Carr limited his schedule (even his final performance evaluation from the Department of Zoology at the University of Florida acknowledged his continued productivity in terms of publications and grants). Along with his graduate students and other collaborators, he continued to study and publish on the ecology and conservation of sea turtles. Before Carr began his research program, scientists had little more knowledge of sea turtles than that the adult females came from the sea to nest on beaches and the hatchlings returned to the sea upon leaving the nest. In addition to the taxonomy of sea turtles (which was at best confused when Carr took on the subject in 1942), Carr, his students, and other collaborators had filled in many of the gaps between hatching and nesting. Moreover, Carr's indefatigable efforts for the conservation of sea turtles had addressed some of the threats faced by the various species and preserved many of the nesting beaches and thus future generations of sea turtles. Through his students, Carr contributed in a significant way to the coalescence of a new discipline, conservation biology.

Conservation Biology Emerges

In the 1970s, as more and more biologists became interested in conservation, they began to develop a new subfield in biology: conservation biology—biological study explicitly concerned with conservation. In many respects, Carr had practiced conservation biology throughout much of his career. Conservation was certainly on his mind in 1952 when he wrote that the green turtle faced extinction within two decades. His earliest visits to Tortuguero convinced him that without significant protection, the colony would disappear. Carr's concern with conservation had a prominent place in most of his books and in many of his articles, both technical and popular. The Caribbean Conservation Corporation (CCC) and the Marine Turtle Group of the International Union for the Conservation of Nature gave him significant platforms from which to publicize and address the plight of sea turtles. CCC certainly deserves some of the credit for the establishment of Tortuguero National Park. These are some of Carr's important legacies. Yet he also had a profound effect on the continued conservation of sea turtles through his students. No fewer than five of them

(Caldwell, Ehrenfeld, Hirth, Ogren, and Pritchard) continued to study sea turtles and advocate for their conservation. Another cohort would soon follow, including Bjorndal, Mortimer, and Meylan (see below). David Ehrenfeld was particularly active in the development of conservation biology. In 1969, as a professor of biology at Barnard College, he completed a manuscript for a book titled *Biological Conservation*. Before publication, Ehrenfeld sent the manuscript to his former mentor. Carr complimented Ehrenfeld on his writing, but gently questioned his treatment of the field of conservation:

> As to your coverage of the field of conservation as I conceive it, you somewhat surprise me in appearing to fall partly in with the tendency of most conservationists to soft-pedal the humanistic value of sample areas of intact wilderness. Most people, who do this soft-pedaling knowingly, do it simply because some congressmen scorn the wild notion that there are real values involved, and because other considerations are easier to consider. Knowing your feelings along this line, I kept looking for a bit on the subject and was surprised how somewhat timidly you allowed yourself to speak of it. Somewhere I wrote "A reverence for wilderness is one of the humanities. It was the first humanity." I think that got edited out of whatever I had it in, but I still believe it and think you do too. I just don't think you spelled out the idea quite strongly enough. Would you consider doing one more paragraph along these lines, to be put in, say toward the end, maybe under "The Good Life" or "Collective Interest" or somewhere. I realize that the idea is implicit in a large part of your book, but I still think it's your duty as author of what will probably be an important book on conservation to suggest somewhere *the real reason for saving tuataras is so people can sing them out of their holes.*
> Meanwhile, congratulations on a good job.[1]

Carr's comment regarding "reverence for wilderness" had appeared in print in *Ulendo* in 1964. When Ehrenfeld's book *Biological Conservation* was published in 1970, he had added a paragraph in line with Carr's suggestion:

> Even if there were no ecologic crises, conservation would still have its most important mission before it: preserving wilderness for those human beings who are fortunate enough to know now that it is part of them and that they enjoy it, and also for those in the future who may learn to use technology in a way that does not subvert the human heritage. Carr must have had something like this in mind when he said: "The real reason for saving tuataras is so people can continue to sing them out of their holes."[5]

Ehrenfeld also recognized the Carrs in the preface: "Above all, I am grateful to Dr. Archie Carr and Mrs. Marjorie Carr for having started and encouraged my interest in conservation and my delight in the things that conservationists enjoy."[6] In addition, he dedicated the book to the Carr family: "For the Carrs: Archie, Margie, Mimi, Chuck, Steve, Tom, Jasper, and David."[7] Astute readers and friends of the Carrs noticed an addition to the family: Jasper (the Carrs had five children). Jasper, the alligator snapping turtle, had been given to the Carrs by a fisherman from Jasper, Florida. Wewa Pond became his new home. In the 1960s, when Ehrenfeld was in graduate school, Archie enjoyed entertaining guests by calling Jasper and dangling a piece of catfish or calf liver by the edge of the pond and then waiting for the great turtle to snatch the food from his hand.

As the environmental movement grew during the 1970s and conservation of all forms concerned Americans, biologists reexamined their role in the study and protection of rare species. The ESA and the American Society of Naturalists addressed issues of conservation as part of a much larger agenda, but no society existed in America that was dedicated wholly to the study of biological conservation. That changed on May 8, 1985 at the conclusion of the Second Annual Conference on Conservation Biology in Ann Arbor, Michigan. After reports by Jared Diamond and Peter Brussard, chairmen of two ad hoc committees, an informal motion to organize a society was approved by acclamation. After three additional meetings and the establishment of an interim Board of Governors and the election of officers, Michael Soulé was elected pro tem president. A search committee led by Robert May and Daniel Simberloff ultimately appointed Ehrenfeld editor for the society's journal, *Conservation Biology*. In his editorial for the inaugural issue, Ehrenfeld heralded the rise of conservation biology as a discipline. "But why," Ehrenfeld asked, 'conservation biology'? Why not simply 'conservation'?" In answer, he wrote: "Conservation and biology are interdependent and inseparable because biology is at the heart of all phases of conservation and is the ultimate arbiter of its success and failure."[8] Ehrenfeld next divided conservation into three stages:

1. The delineation of diversity (the responsibility of all too few taxonomists and systematists).

2. The tasks of measuring the increase, decline, and change in distribution of populations and predicting the imminence of extinction (the responsibility of population biologists joined by ecologists, physiologists, behaviorists).

3. Management, which integrates the other stages.[9]

In addition to biologists, many others in a range of academic disciplines contributed to conservation biology, as well as nongovernmental environmental organizations, federal and regional agencies, lawyers, and fund raisers.

While Ehrenfeld's list provided a taxonomy of professionals who contributed to conservation biology, it could also serve as a guide to the stages of Carr's long career. In fact, Carr embodied each of Ehrenfeld's stages. As a graduate student and young professor, Carr studied the diversity and taxonomy of reptiles and amphibians. Over the course of seven summers at the Museum of Comparative Zoology at Harvard, Carr wrote many species and subspecies descriptions, devised taxonomic keys, and revised many previous taxonomies. His debate with Leonhard Stejneger revealed his development as an evolutionary biologist and ecologist. Even his long-time mentor Thomas Barbour deferred to his younger colleague in most aspects of classification and systematics. As Carr investigated the ecology and migrations of sea turtles, he regularly returned to the problem of classification to distinguish among various species, subspecies, and populations. As technical director for the CCC and chair of the Marine Turtle Group, he identified sea turtle systematics as one of the most critical initial stages in the development of a conservation agenda. Questions of evolution and speciation continued to interest Carr, particularly in his travels to Central America and Africa. Like many naturalists he often framed his observations in light of evolutionary theory and considered how particular species evolved.

In the 1950s, when Carr concentrated his research efforts on sea turtles, he set out to determine their ecology and migrations. Although evolution and classification remained among Carr's concerns, his main interest was ecology and the behavior of sea turtles in the wild. He sought local nonscientists in finding interesting turtle species, and he interviewed turtle captains throughout the Caribbean in search of information regarding turtle nesting and migration. Carr was able to focus his research agenda based on information from locals. Carr's tagging program in Tortuguero depended on Caribbean fishermen to return tags for a five-dollar reward, and Caribbean culture became a significant aspect of Carr's writing for popular audiences. The Tortuguero tagging program established that turtles did in fact migrate great distances from their nesting beaches before returning to breed. It also showed that sea turtles exhibited great site fidelity by consistently returning to the same beach to nest every two or three years. When turtles returned to lay a second or third clutch of eggs, Carr and his wife demonstrated that often the second or third nest was within a hundred meters of the original. Moreover, no turtle that nested at Tortuguero was ever found nesting at any other beach.

One of the perennial problems of sea turtle biology that continued to challenge Carr was the mystery of the lost year. Though Carr's initial suspicions that hatchling sea turtles found their way to beds of sargassum were confirmed, further investigation suggested that young turtles traveled more or less at the whim of oceanic currents and convergences (where sargassum and other material collected). How sea turtles (both juveniles and adults) found their way around the ocean continued to stymie Carr and his research associates. Vision, probably olfaction, and possibly magnetic compass orientation played a role in successful orientation, but the exact mechanism of orientation remained mysterious. Nevertheless, certain populations of green turtles regularly completed trips of more than 1,000 miles to reach their nesting beaches. Most notable in this regard were the Ascension Island turtles. A tagging program demonstrated that the Ascension turtles migrated to and from their feeding grounds off the coast of Brazil. But this discovery, while significant, left open several basic questions such as how and why the turtles traveled such great distances. Carr's sea floor-spreading hypothesis suggested how such an impressive migration may have slowly developed. Late in his career, as the questions regarding sea turtle biology became more complex, Carr sought collaborators in other fields such as oceanography while he strove to educate himself in these disciplines.

Even as he was initiating his formal study of the ecology and migration of sea turtles, a fortuitous sequence of events stemming from the publication of *The Windward Road* drove the development of the Caribbean Conservation Corporation, which provided an organization dedicated to conservation activities. With the help of the CCC, Carr could advocate for the conservation of sea turtles while continuing his scientific studies. In addition, as the chairman of the Marine Turtle Group of the International Union for the Conservation of Nature, Carr directed international conservation efforts. Thus, for more than thirty years, Carr was simultaneously operating in all three of Ehrenfeld's stages of conservation biology: taxonomy and speciation, ecology and evolution, and management. He passed on the torch to a new generation of graduate students.

Archie's Angels

For about a decade, Carr did not accept new doctoral students who were interested in becoming sea turtle biologists, starting sometime before Peter Pritchard completed his doctorate in 1969. Sea turtles, in Carr's view, raised questions that were simply too great to address within the confines of a doctoral thesis. Three young women convinced Carr otherwise during the 1970s. The first was Karen Bjorndal, who remembers writing a long letter about her interest in sea turtles and sending it directly to Carr. She never received a response, but she enrolled in the graduate program nonetheless. Despite rumors that Carr refused to accept doctoral students, Bjorndal was persistent and convinced Carr of her sincerity and dedication to the study of sea turtles:

> I just basically camped on his doorstep for about 6 months and hung around and hung around and computerized his data. This was back in the good old days of card punching. I went through all of his old Tortuguero field books, coded them all in, punched the cards, and read the cards with his secretary to proof them. It was a very long-term labor of love. That started in '74 and I worked on that project for several years and it has been continued and maintained ever since.[10]

Carr did eventually agree to chair Bjorndal's doctoral committee. Her dissertation was titled, "Nutrition and Grazing Behavior of the Green Turtle, *Chelonia mydas*, a Seagrass Herbivore."[11] With her doctorate completed, Bjorndal stayed in Gainesville and worked with Carr until he passed away. In 1987, Bjorndal was appointed associate director of the newly formed Center for Sea Turtle Research, which continues to serve as a major center for research today. After Carr's death in 1987, Bjorndal accepted the directorship of the renamed Archie Carr Center for Sea Turtle Research (ACCSTR). Now a highly regarded professor and chair of the Department of Zoology, she continues to direct the Carr Center. Along with her husband and research associate, Alan Bolten, Bjorndal advances the state of knowledge regarding sea turtles and secures the ongoing role of the University of Florida Department of Zoology and the ACCSTR in sea turtle research.[12]

Shortly after Bjorndal arrived at University of Florida in 1973, she was joined by Jeanne Mortimer, whose path to working with Carr was less direct. After completing her bachelor's degree at Notre Dame, Mortimer was accepted at the University of Florida, but she, too, had heard Carr was not accepting graduate students. She planned to study with another herpetologist, Walter Auffenberg. During the summer before she started graduate school, Mortimer planned to visit several Costa Rican field stations. Before leaving the states, she visited Carr in Gainesville to ask if she could work at Tortuguero for part of the summer. Carr asked a simple question, "Can you walk 5 miles?"[13]

As a result of this meeting, Mortimer spent the last month of the turtle season at Tortuguero and contributed to the tagging project. As luck would have it, Archie and Marjorie Carr were in residence at Tortuguero the entire time she was there. Mortimer recalled how much they seemed to enjoy their work with the sea turtles:

> The first summer I wouldn't know the difference because they were there the entire time I was there. It was quite fun. Marjorie was taking care of the turtle information on cards, on index cards. She was responsible for the index cards, filling them in each day. One card for each turtle. Archie was really fun, always telling stories. One of the things that I liked about him was just the fact that he made it seem like doing this kind of work was both important and fun.[14]

With Bjorndal's encouragement, Mortimer completed her master's thesis on the feeding ecology of green turtles by visiting the slaughterhouses in Nicaragua and examining the gut contents of turtles.[15] Carr then agreed to supervise Mortimer's doctorate and eventually sent her to Ascension Island to study the sea turtles there. Support from the National Geographic Foundation enabled Mortimer to spend two seasons on Ascension Island. Her dissertation was titled, "Reproductive Ecology of the Green Turtle (*Chelonia mydas*) at Ascension Island."[16]

In 1975, Carr accepted yet another graduate student, Anne Meylan, who matriculated in January and spent her first summer at Tortuguero. For her master's thesis, Meylan tracked the internesting behavior of green turtles. This research filled a significant gap in the literature. From the suite of papers by the Carrs on recruitment and remigration of green turtles at Tortuguero, it was clear that turtles returned to the same beach to lay a second or even a third clutch of eggs, often within a few hundred yards of the initial nest. But where did the turtles spend their time between nestings? Meylan answered this question by attaching optical floats that had a blinking light atop a fishing pole mast to monitor the turtle's activity from a tower on the beach. For the optical float to be effective, it had to be visible from great distances, and this requirement led to unintended consequences. Meylan remembered witnessing poachers taking one of the turtles and its float despite all protests, but it was Carr's response that stood out:

> This morning one of the harpoon boats on the way to the site happened to run into my turtle with the nice big red buoy tied behind it, so I'm watching through the telescope getting more and more worried that

they were going to do something with my turtle and started to yell and scream on the walkie talkie to, I think it was Stephen [Carr] at that time, and what do we do, and they approached my turtle and they grabbed the float and started pulling in the turtle, and I ran up to where the Carrs were sleeping because their little building looked right over the runway and then out onto the beach. So I knock on the door, and Marjorie comes to the door just out of bed and I said, "Marjorie, they've got my turtle," and she said "Oh no," and she ran in and got Dr. Carr. Meanwhile I'm back at my tower waiting to see what's happening, and I hear this feeble little gun shot. Dr. Carr had gotten his gun and had come onto the front porch there in his pajamas looking very feeble and small, and Marjorie was standing next to him just shrieking to support him and he fired the pistol in the air to warn them. This was all in the wee dawn hours, and I won't ever forget it.[17]

Carr's seemingly inconsequential act signaled to Meylan his complete dedication to sea turtle conservation. For her doctoral research, Meylan shifted gears and examined spongivory (sponge eating) and the feeding ecology of hawksbill turtles, thus adding a critical piece of data to the life history of the hawksbill.[18] She found that sponges made up the majority of the hawksbill's diet, a trait shared with only a dozen or so vertebrates. Moreover, hawksbills ate sponges that were toxic to other vertebrates.[19]

Thus, after years of redirecting doctoral students away from projects involving sea turtles, Carr supervised three doctoral students during the late 1970s. In November 1979, Bjorndal, Mortimer, and Meylan were all active in the first World Conference on Sea Turtle Conservation, held in Washington, D.C. Each presented one or more papers at the conference. In fact, so closely were they associated with Carr that other scientists at the conference gave them a nickname, according to Mortimer: "[The World Conference] was a big deal, there were hundreds of people, and it was very international—the first time they brought in people from every continent. At the time, it was quite funny because we got the nickname of 'Archie's Angels' because [the popular television show] 'Charlie's Angels' was on then."[20] Bjorndal, Mortimer, and Meylan played a significant role in the success of the conference. Carr and several of his other former students, including Pritchard and Ehrenfeld, also presented papers. Bjorndal collected and edited the proceedings in *Biology and Conservation of Sea Turtles*.[21]

Bjorndal, Mortimer, and Meylan each felt honored to study under Carr. Whenever they struggled with an issue of turtle biology, Carr would exclaim, "Read the classics!" "The classics" was a subtle way of referring to any number of his earlier papers on the biology of turtles. Even near the end of his career, Carr continued to write. Each day he would make notes or handwrite drafts on yellow legal pads.[22] At the end of the day, Carr left the pad by a typewriter. Whichever of his three doctoral students found the pad first would decipher his handwriting and type his notes or drafts. Each felt honored by the opportunity to transcribe Carr's thoughts, and each believed that they learned about the demanding process of writing by assisting Carr. Ehrenfeld said, "Archie Carr was very fortunate with his last three students: they are brilliant, creative, dedicated, and tough—each unique in her own way."[23]

Despite the success of his last three students, Carr remained pessimistic about the prospects for students who hoped to do the kind of biology on which he had built a career, as he noted in one of his last interviews conducted in March 1987:

> I wouldn't accept a graduate student today who wanted to do a paper like [my doctoral dissertation]. In those days they had to be done because we lived in a different time. We didn't even know what we had here. You can't get a job in zoology being the kind of biologist I was any more. You've got to be highly quantitative, highly statistically oriented—lab work preferably. To get in your basic zoology department, you ought to be interested in mitochondrial DNA.[24]

One cannot help but hear a note of regret in Carr's pragmatic advice. Years later, many scientists identified him as a model naturalist and exactly the kind of scientist that conservation demanded.

Thirty Years with Sea Turtles

Carr addressed his perspective on conservation biology in various writings and speeches. In March 1984, he gave the Fairfield Osborn Address at Rockefeller University. Carr titled his speech, "Thirty Years with Sea Turtles: Perspectives for World Conservation." Fittingly, he opened with a tribute to Osborn and his 1948 book, *Our Plundered Planet*. Carr noted:

> He wrote that in 1948—before *Silent Spring*, before *Sand County Almanac*. The dust bowl had not long subsided in those days; the ideas of conservation had not gone far beyond contour plowing and pleas for family planning. David Ehrenfeld had not yet made plain *The Arrogance of Humanism* or catalogued the values of non-resources; Tom Eisner had not explained the potential that humble creatures hold for industrial chemistry and pharmacology; Ed Wilson was still too young to have traced out origins of the conservation ethic in human evolution. Anyway it was mankind's own ecologic predicament that Fairfield Osborn was mainly stirred up about in *Our Plundered Planet*—the overgrowth of the human race, and the heedless ruination of our natural support system.[25]

Having recognized Osborn's prescient views and acknowledging more recent reflections on conservation, Carr turned to a few of the bright spots in the state of sea turtle conservation—the World Conference on Sea Turtle Biology and Conservation in 1979 and the West Atlantic Sea Turtle Symposium in 1983. Both meetings signaled to Carr the widespread development of interest in the biology and survival of sea turtles. In addition to these meetings, Carr was impressed by the National Marine Fisheries Service's efforts to develop Turtle Excluder Devices (TEDs) to reduce the number of turtles drowning in the nets of shrimp trawlers.

Despite these positive developments in the study, conservation, and management of sea turtles, Carr stressed that many gaps in knowledge remained. Among the most basic of problems in Carr's view was sea turtle classification. He wrote, "we don't even know how many kinds of sea turtles there are. Procedures of taxonomy—the science of classification—have advanced greatly in recent times, but the field has lost vogue, and this has happened before the job of classifying living species is anywhere near complete. In the case of marine turtles, nobody has ever made a careful study of the numerous reproductively isolated populations of the turtles of the world."[26] With only eight species of sea turtles recognized with scientific names, Carr worried that named species could dwindle to a single breeding population and thus forsake considerable diversity while technically preserving the species. As Carr had suspected forty years earlier in his correspondence with Thomas Barbour, it was easy to distinguish between Atlantic and Pacific populations in tropical America. More problematic, however, were the many island-breeding colonies spread out across the Pacific Ocean. Given the marked breeding-site fidelity, Carr thought it ridiculous to assume that any single population in the Pacific and Indian Oceans would embody the whole genetic makeup of the group. There was a greater degree of resolution in the Atlantic–Caribbean system, which had four distinct populations, all reproductively isolated with distinct ecologic regimes. Carr believed that genetic analysis would support further division of Chelonia mydas in the Caribbean. Having suggested the continued significance of classification, Carr concluded:

> I go into this, not because I find rooting out this hidden diversity a particular challenge in itself, but because it is exactly what modern conservationists have vowed they are obliged to do. Just how much of our energy and resources should be devoted to taxonomic splitting is hard to say, but it ought to be kept carefully in mind what is being lost if we neglect to tally the multiplicity of living things.[27]

Carr noted other major gaps in the biology of sea turtles with serious implications for management and conservation, such as the time required to reach sexual maturity. In the 1960s, biologists accepted data from aquarists that turtles reached breeding age in just six years. When Carr and the CCC launched Operation Green Turtle, they expected to see signs of breeding within six to eight years. They abandoned the project after nine years. Given new evidence that turtles might require forty to fifty years to reach breeding age, Carr speculated that the turtles from Operation Green Turtle might still reappear to breed at their release sites even as late as the year 2000.

Another discovery had equally profound implications for management. Working at Tortuguero and other sites, James Spotila, Stephen Morreale, and other researchers, basing their study on earlier research showing temperature dependence in aquatic turtles, determined that relatively cool nests (28° C or lower) produced almost all male offspring, whereas warmer nests (29.5° C or higher) produced all females.[28] This information had unsettling implications for Operation Green Turtle and other head-starting efforts. Carr reflected,

> I had been in contact with the investigators all along, and was aware of the trend of their findings; but reading the awful truth in the stark prose of the journal *Science* was unsettling. What the data meant, without any shadow of doubt, was that in the hundreds of thousands of artificially incubated hatchlings that have been hopefully released during the past two decades, and in the tens of thousands of laboriously and expensively raised yearlings that have been produced and put into the sea in head-starting projects around the world, nobody has the vaguest idea what the sex-ratios were.[29]

It is likely, however, that the majority of turtles were males. Carr also realized that the even bigger problem was that no one had the vaguest inkling what natural sex ratios were.

Having reviewed some of the remaining problems of natural history, Carr turned to the problems that resulted from human activity. After a brief summary of his study of sargassum and driftlines, he noted that marine pollution was concentrated along the same convergences in which the turtles drifted. Oil stuck in the mouths of some of the turtles, and others mistook plastic bags for jellyfish, with deadly consequences. Both had grave implications for the "lost year:" "So after all the long wondering about the lost year, we find the little turtles in a man-made trap, and no clear way to help them out."[30] Carr also worried that marine pollution disturbed turtles' ability to navigate in open water. Nevertheless, one of the greatest threats to the continued survival of sea turtles was direct harvest for international commerce. International laws such as the Convention on International Trade in Endangered Species (CITES) made allowances for subsistence hunting. But Carr noted that turtle hunters, slaughterhouse operators, and even freezer-boat captains all claimed exemption from international endangered species laws by identifying their activities as "subsistence use":

> This semantic distortion of the term "subsistence" is a widespread problem in biological conservation. In the case of the Caribbean green turtle, the mainland shores and islands are settling up fast, and there is growing demand for green turtle meat, taken from the pitiful remnants of once teeming populations. But most of the green turtles caught today are sold, not eaten by the turtlers and their families. One's natural reluctance to deprive local people of a hereditary element of their diet fades away when the last of the Caribbean turtles are being purveyed to city restaurants and tourist hotels, or smuggled frozen up to Miami.[31]

Similarly, hawksbill turtles were harvested for subsistence use in Japan, but it was the international market for its shell ("tortoiseshell") that presented the greatest threat to its survival.

Carr closed with the most extreme example of species decline, the Kemp's ridley, by retelling the story of his twenty-year search for the breeding grounds and the subsequent discovery of the 1947 film that showed an *arribada* of as many as 40,000 sea turtles. After showing slides and a short clip of the film, Carr concluded with a quotation from *So Excellent a Fishe*:

> One point stands out, however: in its very redundancy, an *arribada* is one of the wonders of the natural world. The losses that ridleys have suffered have degraded two related natural assets—the wild species involved, *L. kempii* and *L. olivacea*, and the *arribadas* in which both reproduce. Concern over the threat to the existence of wild species is widespread, but the obligation to preserve biological phenomena and organization is less widely recognized. Like the Serengeti fauna, the hawks of Hawk Mountain, and the monarch butterflies of the Sierra Chinqua, the *arribadas* are phenomenal, mind-gripping examples of biologic order, scientific and esthetic treasures of the living world. There is no civilized way to escape the obligation to save them.[32]

Thus Carr extended the responsibility of conservation ethics beyond the conservation of species to the preservation of aggregations and other biological phenomena. In a sense, animals were most vulnerable to threats at points of congregation, particularly in the case of the phenomena Carr cited. In his review of three decades of research on sea turtles, Carr incorporated all aspects of conservation biology, including taxonomy, ecology and population biology, and management. Moreover, he recommended a broader vision of conservation that protected groups of animals at critical sites.

Legacies

Carr's contributions to the biology and conservation of sea turtles are numerous. His scientific research documented many of the aspects of the lives of most sea turtle species, and his conservation efforts likely saved several species from extinction. Tagging and monitoring sea turtles continues at Tortuguero under the management of the Caribbean Conservation Corporation, still based in Gainesville, Florida. During the twilight of his long career, the University of Florida honored Carr in a variety of ways. In 1979, the Archie F. Carr medal was established to honor influential biologists. Edward O. Wilson, the Harvard biologist, was the first recipient. The Archie F. Carr, Jr. Postdoctoral Fellowship was established in 1983. Carr received the University of Florida Presidential Medal in 1987 (the university's highest honor). As a result of Carr's efforts over his long career, the Department of Zoology at the University of Florida had become associated with sea turtle research. Under the direction of Karen Bjorndal, the Archie Carr Center for Sea Turtle Research continues his work.

The CCC also continues its prominent role in sea turtle research and conservation. In Florida, CCC manages several programs, including the Florida Sea Turtle Grants program, which was developed to support research and education programs with funds generated by the "sea turtle" license plate. In addition, CCC led a campaign in Florida to reduce shore lighting ("Lights out for Sea Turtles") and thereby limit the number of hatchling turtles drawn away from the surf by bright lights. The CCC also continues the tagging program that Carr started at Tortuguero in 1956. After five decades, it is the longest continuously running study of any vertebrate. As ecotourism has developed, CCC has created programs for tourists to contribute to the tagging program. Turtle researchers continue to conduct research at Tortuguero; CCC coordinates their efforts and often provides them with housing while they are in residence. CCC scientists now draw on a database with nearly fifty years of continuous records (see figure 37). In addition, CCC trains research assistants to tag turtles and monitor the nesting population. Research assistants hail from throughout Latin America, North America, and Europe. Many of them return to their home countries and apply their training to local turtle conservation programs. In 2003, CCC inaugurated a monitoring and conservation program at Chiriqui Beach, Panama, which is a major hawksbill and leatherback turtle nesting colony. Each year several Ngobe Bugle Indians complete the training program at Tortuguero and return to Chiriqui to staff the conservation project. Carr would surely appreciate how the research station at Tortuguero, now the John H. Phipps Biological Field Station, has evolved into a program that trains students from throughout Latin America, much like the Escuela Agricola Panamericana in Honduras.

Just as the Department of Zoology and the CCC have continued Carr's work since his death, Carr's students have all continued to work with sea turtles through research, conservation, or both. Anne Meylan, Jeanne Mortimer, Peter Pritchard, David Ehrenfeld, Harold Hirth, and Larry Ogren, and Karen Bjorndal have all continued to work with sea turtles. Meylan is a senior research scientist at the Fish and Wildlife Conservation Commission in the State of Florida. In that capacity, she has been able to run a long-term study of green turtles in the waters off Zapatilla Key in Bastimentos Island National Marine Park in Panama, where she catches turtles migrating to Tortuguero to nest. This site is located in Bocas del Toro Province (which Carr referred to as "Mouths of the Bull" in *The Windward Road*). Similarly, Mortimer has expanded her research and conservation efforts to include the Seychelles in the Indian Ocean. As a freelance and contract scientist, Mortimer has studied Seychelles sea turtles for more than two decades. There she has been able to develop a conservation plan with the national government while educating residents about the biology and conservation of sea turtles.

Like Bjorndal, Meylan, and Mortimer, Pritchard has continued to study sea turtles, most recently as the director of Chelonian Research Institute in Oviedo, Florida. During the 1970s, Pritchard was a staff scientist for the Florida Audubon Society, and he also completed a series of contracts, including several for sea turtle research in New Guinea. James Spotila recently noted that Pritchard has "done more than any other living person to make turtles a symbol of conservation."[33]

Carr's other doctoral students have followed traditional academic routes. Ehrenfeld began his career at Columbia University and then became professor of biology at Rutgers University. Like Carr, Ehrenfeld has a second career writing books and articles for popular journals such as *Orion* and *Harper's*. He has also remained active in the Society for Conservation Biology, even since he relinquished his post as the editor of the society's journal, *Conservation Biology*. Hirth was professor of biology at the University of Utah for more than three decades. Ogren has worked for government agencies, particularly the National Marine Fisheries Service, in which capacity he advocated for and supported sea turtle research and conservation.

Through the continued efforts of his students, Carr's legacy has continued to grow. Spotila has connected most active sea turtle biologists and conservationists to Carr: "Archie provided a model for others to follow and a legacy of followers to carry on his work. Most sea turtle biologists trace their roots, either directly or indirectly, to Archie Carr. The older ones were his students or worked with him during their careers. The younger ones studied or worked with Archie's academic offspring. Now the world is filled with Archie's academic grandchildren and great-grandchildren."[34]

Carr's family has also continued to carry on his legacy. Of his five children, four devoted a significant portion of their professional lives to conservation. Archie Carr III, known as "Chuck," became the regional director of Latin America and the Caribbean for the Wildlife Conservation Society, which has a satellite office in Gainesville. David was the executive director of the CCC for several years after his father's death. Tom worked as a biological consultant in Latin America, Africa, and the United States (including Alaska), collecting data on endangered megafauna such as manatees, whales, and sea turtles. He is now the director of development for Holbrook Travel in Gainesville, where he organizes Elderhostel programs in Africa and Latin America. Stephen worked for many years on a project to restore endangered sturgeon to the Suwannee River in northern Florida. Of the five Carr children, only Mimi chose a completely different path as an actress, but she has been active in several films about her parents. David Carr contemplated why he and his brothers gravitated toward careers in conservation: "I think we sort of fell into it for nothing more noble than it was what we were brought up with—a family full of biological conservationists. We would talk about biological conservation around the dinner table. And that was just their normal conversation."[35]

Ehrenfeld, who spent considerable time with the Carr family in Micanopy during the 1960s, saw in each of the Carr children a facet of their father:

> The kids all contain parts of Archie. Mimi is the great actress; she's really good and she got a lot of that from her dad, who was a brilliant actor. She's also got her mom's frankness and strength of character. Chuck is much more like his Uncle Tom in some ways. Steven is shy, and Archie had a good deal of shyness about him in several ways. Tommy is a great naturalist and loved the outdoors and is really a great observer. And David was a little bit more of the academic side of Archie.[36]

On further reflection, Ehrenfeld elaborated: "Tom is the only person that I've ever known who is probably a better naturalist than his father."[37] The legacies of Archie and Majorie Carr flow through the ongoing work of their children.

Marjorie Carr enjoys a considerable legacy in her own right in Florida. The Marjorie Harris Carr Cross-Florida Greenway commemorates her efforts to save the Ocklawaha River by stopping the Cross-Florida Barge Canal. Since her death in 1997, the Florida Defenders of the Environment has continued to advocate for the complete restoration of the Ocklawaha by eliminating the Rodman Dam. The Florida Defenders of the Environment has broadened its mission to include a range of environmental issues such as growth management.

Even in his final months, Carr continued to receive awards for his work in ecology and conservation. When Carr was awarded the University of Florida Presidential Medal in March of 1987, E.O. Wilson saluted him:

> Archie Carr is one of the archangels of the international conservation movement who has turned our thinking around. I hesitate to analyze him psychologically with the man sitting right here, but I suspect that three factors were important in bringing him to the special role. One was being at the University of Florida rather than at say, Harvard, or the University of Wisconsin, or Berkeley. The State of Florida is a microcosm of the disappearing tropical environment, and he and Margie have worked heroically to achieve sanity and balance at home. Second, studying turtles, especially typical marine turtles, which are among the most endangered of all organisms, Archie Carr confronted the problem of species extinction in vivid and practical terms. Finally, he can write beautifully as everyone knows, and he cares... in a way that makes many of our best natural history writers seem like callow journalists.
>
> Rachel Carson, his fellow archangel, once wrote that "those who dwell, as scientists or laymen, among the beauties and mysteries of nature, are never alone or weary of life." That is the spiritual reward I believe Archie enjoys, and he shares with his family and us; and although he is the kind of person who would consider that sufficient, I'm pleased to add one more voice to pay him tribute and to wish his dream come true, from a world that's finally waking up to its deep heritage—we hope not too late.[38]

By the time Wilson delivered this tribute, Carr's stomach cancer was spreading, and he was having difficulty continuing his work. The progress of the disease meant that he could not travel to accept one of the most prestigious awards of his long career: Eminent Ecologist of the Ecological Society of America. Nevertheless, one of Carr's colleagues, Jack Ewel, and the University of Florida Provost Robert A. Bryan presented the award to Carr at his home in Micanopy on the behalf of the ESA on May 8, 1987. The award citation captured Carr's unique contributions:

> In recognition of
>
> —Your outstanding research contributions to the science of ecology, especially your landmark studies on sea turtle reproductive biology and migrations,

—Your gifts to future generations and to the well-being of our planet through your unselfish and tireless dedication to conservation, and

—Your uncanny ability to communicate the excitement and the music of ecology to non-specialists,

THE ECOLOGICAL SOCIETY OF AMERICA

Is proud to bestow upon you, Archie F. Carr,

The award of EMINENT ECOLOGIST[39]

In response, Carr could manage little more than a feeble "Hip, hip, hooray."[40] Less than two weeks later, on May 21, 1987, Archie Carr died peacefully at home. He had no fear of death and marveled at those who did, according to Ehrenfeld. His funeral, held at the Presbyterian Church on University Avenue in Gainesville, drew the university's finest to pay their respects to the man who saved sea turtles. One of Florida's greatest naturalists, the world's authority on the ecology and conservation of sea turtles, and a beloved nature writer left behind a rich legacy of discoveries, organizations, students, family, and friends.

Figure 36. Archie Carr and green sea turtle, Tortuguero, 1984. Courtesy of Jeanne Mortimer.

Sea Turtles Surviving

Carr's greatest legacy may be the continued survival of sea turtles around the world. Spotila has credited Carr with the continued survival of sea turtles:

> Sea turtles survive today in large part because of Archie Carr. He was one with his organism and also one with people. He was as comfortable with the wealthy and learned as he was with Carib, Tico, Miskito Indian or African fishermen. He publicized the plight of sea turtles but shunned publicity and the public eye. He disappeared around the back of the old Casa Verde at Tortuguero upon the approach of strangers, but took time from his busy day to talk to a ten-year-old boy who was fascinated by the little sea turtle swimming in the Carr laboratory. He was a combination of good heart and powerful intellect. His lasting hope was that the conscience of humanity would eventually save the wild things from obliteration.[11]

Threats to turtles have grown, but so has the number of scientists and organizations concerned with their long-term survival. Although it remains uncertain whether sea turtles will survive in the long run, it does seem that Carr's hopes have been partially met. Scientists estimate the worldwide population of green turtles (*Chelonia mydas*) to be greater than 88,000. Most of the population

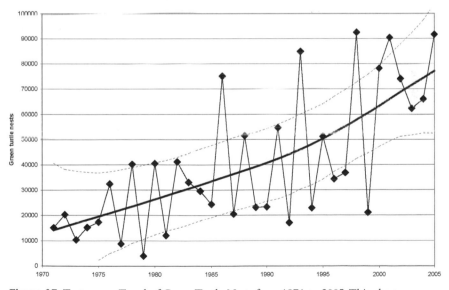

Figure 37. Tortuguero Trend of Green Turtle Nests from 1971 to 2005. This chart illustrates one of Archie Carr's greatest legacies: it clearly indicates an increase in the number of green turtle nests at Tortuguero since 1971. Courtesy of the Caribbean Conservation Corporation.

is concentrated across 11 nesting colonies with 2,000 to 22,500 nesting females on an annual basis. The largest colony in the world (as Carr long suspected) is Tortuguero, with 22,500, followed by Raine Island, Australia, with 18,000. No other colonies support more than 7,000 nesting sea turtles. Nevertheless, poachers hunt green turtles in at least twenty-five countries around the world. Despite this threat and others, green turtles are almost certainly more abundant than they were in 1978, when Carr deferred to Sibella's claim, "The tet-tels never finish."

Protection has also benefited the hawksbill turtle, with an estimated population of between 60,000 and 80,000, but those numbers are only 10 percent of the population a century ago. The most abundant sea turtle species in the world by a considerable margin is the olive ridley, with an estimated 2 million females. Nevertheless, the largest colonies in Costa Rica, Mexico, and India have all dropped precipitously since the 1950s. Meanwhile, the Kemp's ridley sea turtle population has rebounded from its low point of 550 adult females in the entire Gulf of Mexico in 1990 to an estimated 5,000 adult females. Turtle Excluder Devices in shrimp nets have contributed to the resurgence, as have joint Mexican and U.S. efforts to conserve the species at Rancho Nuevo, Mexico and Padre Island, Texas. Though seemingly impressive, 5,000 adult females is still a small fraction of the 40,000 turtles estimated in the Rancho Nuevo *arribada* in Herrera's 1947 film.

Loggerhead and leatherback turtle populations face many threats. In Florida, as many as 60,000 loggerhead nests have been recorded in recent years, with up to half of those concentrated in the Archie Carr National Wildlife Refuge near Merritt Island. Recent estimates, however, indicate a serious decline in Florida and the southeastern United States. In Australia and Japan, loggerhead and leatherback populations have dropped dramatically. The leatherback turtle is one of the most difficult species to estimate because its nesting colonies shift with sand conditions. In 1988 and 1992, approximately 7,000 females laid up to 50,000 nests. The worldwide population of 36,000 nesting females is extremely precarious because virtually all of the eggs are harvested, and many of the adults are killed at many of its former nesting strongholds. Worse, like loggerheads, many leatherbacks become entangled and die in giant drift nets that stretch for hundreds of miles in the open ocean. But the greatest ongoing threat to leatherbacks and loggerheads is longline fishing. Up to a quarter of a million loggerheads and 50,000–60,000 leatherbacks are caught at sea each year. No population can survive that rate of mortality for long.[42] Nonetheless, some significant part of the credit for the continued survival of sea turtles must go to Carr. The same might be said for the survival of the naturalist tradition in America.

"The Naturalists Are Dying Off"

I have suggested that Carr's career demonstrated that the naturalist tradition transformed into related disciplines of ecology and conservation over the course of the twentieth century. Carr's life exhibits many dimensions of the naturalist tradition. Nearly a decade after Carr died, his example inspired one of the most intense responses to any editorial subject in the journal *Conservation Biology*. Inspired by the publication of *A Naturalist in Florida*, Editor Reed Noss lamented the demise of natural history and field biology in an editorial titled, "The Naturalists Are Dying Off." For Noss, some of the fondest memories of college and his research were of field studies and of the mentors who taught him about natural history. Carr had all but stopped teaching, Noss recalled: "But Archie still led the field trips [for the community ecology course] with all the enthusiasm, knowledge, and humor for which he was long known. Whatever natural communities we visited, Archie knew virtually every inhabitant and treated them as friends."[43] Many biologists had voiced their concerns regarding the state of natural history and field studies. Noss exhorted his colleagues to reverse the trend:

> I call on all biologists—ecologists, evolutionary biologists, botanists, zoologists, population geneticists, taxonomists, systematists, and others— to join together in resisting the trend toward indoor biology. Nothing will destroy the science and mission of conservation biology faster than a generation or two of biologists raised on dead facts and technology and lacking direct, personal experience with Nature. In private conversation virtually every biologist I speak with is seriously concerned about the death of natural history.[44]

These comments contradicted the claims of historians of science who cited the demise of natural history at the beginning of the twentieth century. Ironically, Noss and other scientists trained in the naturalist tradition believed the extinction of naturalists was underway by the century's end.

After many impassioned responses to Noss's editorial (only one of which questioned the importance of naturalists), E. O. Wilson called for a return to systematics and natural history in an editorial for *Conservation Biology*. Wilson wrote, "If conservation biology is to mature into an effective science, pure systematics must be accompanied by a massive growth of natural history. For each species, for the higher taxa to which it belongs, and for the populations it comprises, there is value in every scrap of information. Serendipity and pattern recognition are the fruit of encyclopedic knowledge gathered for its own sake."[45] In short, Wilson called for a return to the naturalist tradition. During the course of his career, Archie Carr embodied Wilson's goals for the naturalist tradition and conservation biology in his passion for natural history, his acumen for systematics, his sense for ecology, his dedication to conservation, and his ability to write narratives that captured the hearts and minds of scientists and the public in all of these realms. The story of the man who saved sea turtles should serve as an inspiration to future generations of naturalists and conservationists.

APPENDIX

Archie Carr's Doctoral Students

Belkin, Daniel Arthur. "Anaerobic Mechanisms in the Diving of the Loggerhead Musk Turtle, *Sternothaerus minor*." Ph.D. dissertation, University of Florida, 1961.

Bjorndal, Karen A. "Nutrition and Grazing Behavior of the Green Turtle, *Chelonia mydas*, a Seagrass Herbivore." Ph.D. dissertation, University of Florida, 1979.

Caldwell, David Keller. "The Biology of the Pinfish *Lagodon rhomboides* (L.)." Ph.D. dissertation, University of Florida, 1957.

Christman, Steven P. "Patterns of Geographic Variation in Florida Snakes." Ph.D. dissertation, University of Florida, 1975.

Corn, Michael Jon. "Ecological Separation of *Anolis* Lizards in a Costa Rican Rain Forest." Ph.D. dissertation, University of Florida, 1981.

Crenshaw, John Walden. "The Ecological Geography of the *Pseudemys floridana* Complex in the Southeastern United States." Ph.D. dissertation, University of Florida, 1955.

Deitz, David Charles. "Behavorial Ecology of Young American Alligators." Ph.D. dissertation, University of Florida, 1979.

Ehrenfeld, David William. "The Sea-Finding Orientation of the Green Turtle (*Chelonia mydas*)." Ph.D. dissertation, University of Florida, 1966.

Gourley, Eugene Vincent. "Orientation of the Gopher Tortoise, *Gopherus polyphemus* (Daudin)." Ph.D. dissertation, University of Florida, 1969.

Hirth, Harold Frederick. "The Ecology of Two Lizards on a Tropical Beach." Ph.D. dissertation, University of Florida, 1962.

Howell, John Fincher. "Habitat-Related Variability in the Cave-Dwelling Minnow, *Hybopsis harperi*." Ph.D. dissertation, University of Florida, 1960.

Jackson, Crawford Gardner, Jr. "A Biometrical Study of Form and Growth in *Pseudemys concinna suwanniensis* Carr (Order: Testudinata)." Ph.D. dissertation, University of Florida, 1964.

Johnson, Richard Mann. "A Biogeographic Study of the Herpetofauna of Eastern Tennessee." Ph.D. dissertation, University of Florida, 1958.

Meylan, Anne Barkau. "Feeding Ecology of the Hawksbill Turtle (*Eretmochelys imbricata*): Spongivory as a Feeding Niche in the Coral Reef Community." Ph.D. dissertation, University of Florida, 1984.

Mortimer, Jeanne A. "Reproductive Ecology of the Green Turtle, *Chelonia mydas*, at Ascension Island." Ph.D. dissertation, University of Florida, 1981.

Mount, Robert Hughes. "The Natural History of the Red-Tailed Skink, *Eumeces egregius* Baird." Ph.D. dissertation, University of Florida, 1961.

Pritchard, Peter Charles Howard. "Studies of the Systematics and Reproductive Cycles of the Genus *Lepidochelys*." Ph.D. dissertation, University of Florida, 1969.

NOTES

Chapter 1

1. Anne Meylan, Interview with author via telephone (June 21, 2004), St. Petersburg, Florida.
2. Peter C.H. Pritchard, *Tales from the Thébaïde* (Malabar, Fla.: Krieger, 2006), 10.
3. Brian K. McNab, Eulogy for Archie Carr, in Archie F. Carr, Jr. Papers, Department of Special and Area Collections, George A. Smathers Library, University of Florida (hereafter, "Carr Papers"), 3.
4. David Ehrenfeld, "In Memoriam: Archie Carr," *Conservation Biology* 1, no. 2 (1987): 169–72, p. 169.
5. The other zoology building is called Bartram Hall after William Bartram, thereby honoring two of Florida's greatest naturalists.
6. Ehrenfeld, "In Memoriam," 170.
7. In making this argument, I expand upon the examples presented in Helena M. Pycior, Nancy G. Slack, and Pnina G. Abir-Am, *Creative Couples in the Sciences, Lives of Women in Science* (New Brunswick, N.J.: Rutgers University Press, 1996).
8. Garland E. Allen, *Life Science in the Twentieth Century* (New York: Wiley, 1975), xv.
9. See, for example, Ronald Rainger et al., *The American Development of Biology* (New Brunswick, N.J.: Rutgers University Press, 1991), and Keith Rodney Benson, Jane Maienschein, and Ronald Rainger, *The Expansion of American Biology* (New Brunswick, N.J.: Rutgers University Press, 1991).
10. Keith Rodney Benson, "From Museum Research to Laboratory Research: The Transformation of Natural History into Academic Biology," in *The American Development of Biology*, ed. Ronald Rainger, Keith Rodney Benson, and Jane Maienschein (Philadelphia: University of Pennsylvania Press, 1988), 49–83.
11. Ernst Mayr, *The Growth of Biological Thought: Diversity, Evolution, and Inheritance* (Cambridge, Mass.: Belknap Press, 1982), 143.
12. Edward O. Wilson, *Naturalist* (Washington, D.C.: Island Press, 1994).

13. Edward O. Wilson, *Biophilia* (Cambridge, Mass.: Harvard University Press, 1984).
14. Paul Lawrence Farber, *Finding Order in Nature: The Naturalist Tradition from Linnaeus to E. O. Wilson, Johns Hopkins Introductory Studies in the History of Science* (Baltimore, Md.: Johns Hopkins University Press, 2000), 2.
15. Ibid.
16. See, for example, Nicholas Jardine, James A. Secord, and E. C. Spary, eds., *Cultures of Natural History* (New York: Cambridge University Press, 1996), 8. See also Anne Secord, "Science in the Pub: Artisan Botanists in Early Nineteenth-Century Lancashire," *History of Science* 32 (1994): 269–315.
17. See Philip J. Pauly, *Biologists and the Promise of American Life: From Meriwether Lewis to Alfred Kinsey* (Princeton, N.J.: Princeton University Press, 2000).
18. Mark V. Barrow, "Naturalists as Conservationists: American Scientists, Social Responsibility, and Political Activism after the Bomb," in *Science, History, and Social Activism: A Tribute to Everett Mendelsohn*, ed. Everett Mendelsohn, Garland E. Allen, and Roy M. MacLeod, *Boston Studies in the Philosophy of Science* (Dordrecht: Kluwer Academic Publishers, 2001), 217–33, pp. 227–28.

Chapter 2

1. Archie Fairly Carr, *Handbook of Turtles: The Turtles of the United States, Canada, and Baja California* (Ithaca, N.Y.: Comstock, 1952), 387–88.
2. See Edward O. Wilson, *Naturalist* (Washington, D.C.: Island Press, 1994); Ernst Mayr, *This Is Biology: The Science of the Living World* (Cambridge, Mass.: Belknap Press 1997); and Thomas Eisner, *For Love of Insects* (Cambridge, Mass.: Belknap Press, 2003).
3. Marjorie Harris Carr, Preface, in Archie Carr, *A Naturalist in Florida: A Celebration of Eden*, ed. Marjorie Harris Carr (New Haven, Conn.: Yale University Press, 1994), xiii.
4. Thomas D. Carr, interview with author (January 7, 1998), Gainesville, Florida.
5. In this respect, Carr's experience differed from that of E. O. Wilson, whose religious upbringing was Southern Baptist. For Wilson's reflections, see Wilson, *Naturalist*.
6. I am grateful to Mark Stoll for insights into Presbyterianism and environmentalism in America. See Mark Stoll, *Protestantism, Capitalism, and Nature in America*, 1st ed. (Albuquerque: University of New Mexico Press, 1997).
7. Sidney Lanier, *The Marshes of Glynn* (Darien, Georgia: The Ashantilly Press, 1957).
8. Thomas D. Carr, interview with author.
9. Archie Fairly Carr, "The Ducks of Wewa Pond," *Animal Kingdom* 90, no. 1 (1987): 8. See also Carr, *A Naturalist in Florida*, 1–13.
10. Archie Fairly Carr, *High Jungles and Low* (Gainesville: University Press of Florida, 1953), 140.
11. Thomas D. Carr, Family History, Gainesville, Florida, August 17, 2005, personal papers.
12. Carr, *High Jungles and Low*, 193.
13. Thomas D. Carr, interview with author.
14. For the early history of the University of Florida, see Samuel Proctor, "The University of Florida: Its Early Years, 1853–1906," (Ph.D. dissertation, University of Florida, 1958). Proctor noted that official records of the university indicate a date of 1853 for its foundation, but this date refers to a small college near Ocala, Florida, which catered to adolescents under 15 years old. The true university (as an institution of higher learning for young adults) was founded in Gainesville in 1906.

15. Archie Fairly Carr, "Water Hyacinths—Animal Hideaway," *Animal Kingdom* 87, no. 5 (1984): 55. See also Carr, *A Naturalist in Florida*, 210–19.

16. Theodore H. Hubbell, "Unfinished Business and Beckoning Problems," *Florida Entomologist* 68, no. 1 (1985): 1–10, p. 1.

17. James Speed Rogers and Theodore Huntington Hubbell, *Man and the Biological World* (Gainesville, Fla.: Kallman, 1940).

18. For an analysis of the genesis of academic biology in America, see Philip J. Pauly, *Biologists and the Promise of American Life: From Meriwether Lewis to Alfred Kinsey* (Princeton, N.J.: Princeton University Press, 2000), 126–44. For specific studies of southern science, see Ronald L. Numbers, Todd Lee Savitt, and University of Mississippi Center for the Study of Southern Culture, *Science and Medicine in the Old South* (Baton Rouge: Louisiana State University Press, 1989).

19. Rogers and Hubbell, *Man*, 6.

20. University of Florida, *Annual Catalog 1927–1928* (Gainesville: University of Florida, 1928).

21. Archie F. Carr Animal Ecology—Old Notes, in Carr Papers.

22. For an analysis of the role of life histories in natural history and biology in nineteenth-century Germany, with reference to the United States, see Lynn K. Nyhart, "Natural History and the 'New' Biology," in *Cultures of Natural History*, ed. Nicholas Jardine, James A. Secord, and E. C. Spary (New York: Cambridge University Press, 1996), 426–43.

23. Hubbell, "Unfinished Business," 2.

24. Ibid.

25. Archie Fairly Carr, "All the Way Down Upon the Suwannee River," in *A Naturalist in Florida*, ed. Marjorie Harris Carr, 51–72, pp. 63–64.

26. Hubbell, "Unfinished Business," 2–3.

27. Joshua C. Dickinson, Jr., interview with author (April 19, 1996), Gainesville, Florida.

28. Archie Fairly Carr, "The Plancton and Carbondioxide-Oxygen Cycle in Lake Wauberg, Florida" (M.S. thesis, University of Florida, 1934).

29. O.C. van Hyning, "Reproduction of Some Florida Snakes," *Copeia* 2 (1931): 59–60; van Hyning, "Food of Some Florida Snakes," *Copeia* 1 (1932): 37; van Hyning, "Batrachia and Reptilia of Alachua County, Florida," *Copeia* 1 (1933): 3–7.

30. Thomas D. Carr, interview with author.

31. Archie Fairly Carr, "A Key to the Breeding-Songs of the Florida Frogs," *The Florida Naturalist* 1, no. 2 (1934): 19–23.

32. Ibid., 19.

33. See Paul Lawrence Farber, *Finding Order in Nature: The Naturalist Tradition from Linnaeus to E. O. Wilson*, Johns Hopkins Introductory Studies in the History of Science (Baltimore, Md.: Johns Hopkins University Press, 2000).

34. For institutional and cultural histories of museums and other centers of natural history, see Jardine et al., eds., *Cultures of Natural History*; Sally Gregory Kohlstedt, "Essay Review: Museums: Revisiting Sites in the History of Natural History," *Journal of the History of Biology* 28 (1995): 151–61; Kohlstedt, "Museums on Campus: A Tradition of Inquiry and Teaching," in *The American Development of Biology*, ed. Ronald Rainger, Keith Rodney Benson, and Jane Maienschein (Philadelphia: University of Pennsylvania Press, 1988), 15–47; Mary P. Winsor, *Reading the Shape of Nature: Comparative Zoology at the Agassiz Museum*, Science and Its Conceptual Foundations (Chicago: University of Chicago Press, 1991); Donna

Jeanne Haraway, *Primate Visions: Gender, Race, and Nature in the World of Modern Science* (New York: Routledge, 1989), Paula Findlen, *Possessing Nature: Museums, Collecting, and Scientific Culture in Early Modern Italy*, Studies on the History of Society and Culture 20 (Berkeley: University of California Press, 1994); Ronald Rainger, "Vertebrate Paleontology as Biology: Henry Fairfield Osborn and the American Museum of Natural History," in *The American Development of Biology*, 219–56; and Keith Rodney Benson, "From Museum Research to Laboratory Research: The Transformation of Natural History into Academic Biology," in *The American Development of Biology*, 49–83.

35. Thomas Barbour to Archie Carr, Jr. (January 24, 1934), in Carr Papers.

36. Biographical material on Thomas Barbour can be found in his autobiographical travel narrative, *Naturalist at Large* (Boston: Little Brown and Company, 1943), and in Kraig Adler, "Barbour, Thomas," in *Contributions to the History of Herpetology* (Oxford, Ohio: Society for the Study of Amphibians and Reptiles, 1989), 84–85, Emmet R. Dunn, "Thomas Barbour 1884–1946," *Copeia* (1946), and Arthur Loveridge, "Thomas Barbour Herpetologist," *Herpetologica* 3 (1946), 3.3–39.

37. Frank M. Chapman, *Camps and Cruises of an Ornithologist* (New York: D. Appleton, 1908); Chapman, *My Tropical Air Castle: Nature Studies in Panama* (New York: D. Appleton, 1929); Chapman, *Life in an Air Castle: Nature Studies in the Tropics* (New York: D. Appleton-Century, 1938); and David Fairchild, Elizabeth Kay, and Alfred Kay, *The World Was My Garden: Travels of a Plant Explorer* (New York: C. Scribner's Sons, 1938).

38. Archie F. Carr to Leonhard Stejneger (October 29, 1934), in Carr Papers.

39. Leonhard Stejneger to A.F. Carr, Jr. (November 2, 1934), in Carr Papers. Stejneger's "Checklist" refers to Leonhard Hess Stejneger and Thomas Barbour, *A Check List of North American Amphibians and Reptiles*, 3d ed. (Cambridge, Mass.: Harvard University Press, 1933).

40. For biographical material on Leonhard Stejneger, see Adler, "Stejneger, Leonhard," 62–63; A.K. Fisher, "Leonhard Stejneger," *Copeia* (1931): 74–83; and W.L. Necker, "A Herpetological Bibliography of Leonhard Stejneger (1851–1943)," *Herpetologica* 2 (1943): 137–41.

41. Leonhard Stejneger and Thomas Barbour, *A Check List of North American Amphibians and Reptiles* (Cambridge, Mass.: Harvard University Press, 1917).

42. For a review of correspondence networks emanating from the Smithsonian Institution, see Daniel Goldstein, "'Yours for Science:' the Smithsonian Institution's Correspondents and the Shape of Scientific Community in Nineteenth-Century America," *Isis* 85 (1994): 572–99.

43. Carr spent the summer after completing his master of science degree at the Bass Biological Laboratory in Englewood, Florida. Managed by Stewart Springer, Bass Biological Laboratory served as a clearinghouse for zoological specimens for museums and individuals. It was later called Zoological Research Supply.

44. Helen T. Gaige to Archie Carr (November 23, 1934), in Carr Papers.

45. Ibid.

46. For biographical material on Helen Thompson Gaige, see "Helen Thompson Gaige," in *Contributions the History of Herpetology*, ed. Kraig Adler (Oxford, Ohio: Society for the Study of Amphibians and Reptiles, 1989), 81–82; J.R. Bailey, "Helen Thompson Gaige, 1890–1976," *Copeia* (1977): 609–11; and A.G. Ruthven, "Helen Thompson Gaige," *Herpetologica* 1 (1936): 1–3.

47. Biographical material on the Wrights can be found in W.J. Hamilton, Jr., "Albert Hazen Wright," *Copeia* (1971): 381–82, W.J. Hamilton, "Mrs. Anna A. Wright," *Copeia* (1965): 124; "Wright, Albert H. (1879–1970) and Wright, Anna A. (1882–1964)," in *Contributions to the History of Herpetology*, ed. Kraig Adler (Oxford, Ohio: Society for the Study of Amphibians and Reptiles, 1989), 75–76.

48. Anna Allen Wright and Albert Hazen Wright, *Handbook of Frogs and Toads of the United States and Canada*, 3d ed., vol. 1, in *Handbooks of American Natural History* (Ithaca, N.Y.: Comstock, 1949); Anna Allen Wright and Albert Hazen Wright, *Handbook of Snakes of the United States and Canada*, in *Handbooks of American Natural History* (Ithaca, N.Y.: Comstock, 1957); and Archie Fairly Carr, *Handbook of Turtles*, in *Handbooks of American Natural History* (Ithaca, N.Y.: Comstock, 1957).

49. Edward Harrison Taylor (1889–1978) completed all of his degrees at the University of Kansas (B.A. 1912, M.S. 1916, Ph.D. 1926). After a brief stint with the Civil Service in the Philippines, Taylor returned to Kansas, where he spent the remainder of his career, except during World War I and World War II when he worked in Army intelligence. A prodigious collector, he named approximately 500 species and forms of amphibians and reptiles from Asia, Latin America, and Africa. Biographical material on Taylor can be found in R.G. Webb, "Edward Harrison Taylor 1889–1978," *Herpetologica* 34 (1978): 422–25; W.E. Duellman, "Edward Harrison Taylor 1889–1978," *Copeia* (1978): 737–38; and Edward H. Taylor et al., "Edward H. Taylor: Recollections of an Herpetologist," *Monograph of the Museum of Natural History, The University of Kansas* 4 (1975).

50. For a brief history of herpetology as well as reprints of early studies, see Kraig Adler, *Early Herpetological Studies and Surveys in the Eastern United States, Biologists and Their World* (New York: Arno Press, 1978), and Adler, *Herpetological Explorations of the Great American West*, 2 vols., *Biologists and Their World* (New York: Arno Press, 1978). The historical introductions of these two works were reprinted in a single volume: Kraig Adler, *A Brief History of Herpetology in North America before 1900, Herpetological Circular No. 8* (Oxford, Ohio: Society for the Study of Amphibians and Reptiles, 1979). More recent history can be found in Karl P. Schmidt, "Herpetology," in *A Century of Progress in the Natural Sciences, 1853–1953* (San Francisco, Calif.: California Academy of Sciences, 1955).

51. For biographies of Agassiz, see Edward Lurie, *Louis Agassiz: A Life in Science* (Chicago: University of Chicago Press, 1960) and Lurie, *Nature and the American Mind: Louis Agassiz and the Culture of Science* (New York: Science History Publications, 1974).

52. Louis Agassiz, *Contributions to the Natural History of the United States of America*, 4 vols. (Boston: Little Brown, 1857).

53. For biographies of Baird, see E. F. Rivinus and E. M. Youssef, *Spencer Baird of the Smithsonian* (Washington, D.C.: Smithsonian Institution Press, 1992) and William Healey Dall, *Spencer Fullerton Baird: A Biography, Including Selections from His Correspondence with Audubon, Agassiz, Dana, and Others* (Philadelphia: J.B. Lippincott, 1915).

54. See Goldstein, "Yours for Science," 573–99.

55. Tim M. Berra, "A Chronology of the American Society of Ichthyologists and Herpetologists through 1982," *Special Publication* 2 American Society of Ichthyologists and Herpetologists, Allen Press, Lawrence, Kansas (1984): 1.

56. Note Thomas Barbour's work in the West Indies and Central America and Edward Taylor's studies on five continents. Given the general lack of biological

exploration in Florida before the twentieth century, one could include it with the other sites.

57. Everett Caudle, interview with Marjorie Carr (April 24, 1989), University of Florida Oral History Program, 1.

58. Leslie Kemp Poole, interview with Marjorie Carr (October 18, 1990 and February 4, 1991), Gainesville, Florida.

59. See Mark Derr, *Some Kind of Paradise: A Chronicle of Man and the Land in Florida* (New York: W. Morrow, 1989); Frank M. Chapman and Elizabeth S. Austin, *Frank M. Chapman in Florida: His Journals & Letters* (Gainesville: University of Florida Press, 1967).

60. Marjorie H. Carr, Foreword, in *Ecosystems of Florida*, ed. Ronald L. Myers and John J. Ewel (Orlando: University of Central Florida Press, 1990), ix–xiii, p. xii.

61. Ibid., xiii.

62. Caudle, Marjorie Carr interview, 4.

63. Marjorie H. Carr, "The Breeding Habits, Embryology, and Larval Development of the Large-Mouthed Black Bass in Florida," *Proc. New England Zool. Club* 20 (1942): 43–77.

64. Archie Fairly Carr, "The Geographic and Ecological Distribution of the Reptiles and Amphibians of Florida" (Ph.D. thesis, University of Florida, 1937).

Chapter 3

1. See Howard Gardner, *Creating Minds: An Anatomy of Creativity Seen through the Lives of Freud, Einstein, Picasso, Stravinsky, Eliot, Graham, and Gandhi* (New York: Basic Books, 1993).

2. See Archie Fairly Carr, "The Geographic and Ecological Distribution of the Reptiles and Amphibians of Florida" (Ph.D. thesis, University of Florida, 1937).

3. Thomas Barbour, *Naturalist at Large* (Boston: Little Brown and Company, 1943), 311.

4. For a detailed analysis of Thomas Barbour's science and study of biogeography, see Mary P. Winsor, *Reading the Shape of Nature: Comparative Zoology at the Agassiz Museum, Science and Its Conceptual Foundations* (Chicago: University of Chicago Press, 1991), 245–66.

5. Quoted in Barbour, *Naturalist at Large*, 218.

6. For examples, see Paul Lawrence Farber, *Finding Order in Nature: The Naturalist Tradition from Linnaeus to E. O. Wilson, Johns Hopkins Introductory Studies in the History of Science* (Baltimore, Md.: Johns Hopkins University Press, 2000); Ronald Rainger, *An Agenda for Antiquity: Henry Fairfield Osborn & Vertebrate Paleontology at the American Museum of Natural History, 1890–1935, History of American Science and Technology Series* (Tuscaloosa: University of Alabama Press, 1991); Winsor, *Shape of Nature*.

7. Archie Fairly Carr, "The Identity and Status of Two Turtles of the Genus *Pseudemys*," *Copeia* 3 (1935): 147–48.

8. Archie Fairly Carr, "A New Turtle from Florida, with Notes on *Pseudemys floridana mobiliensis* (Holbrook)," *Occasional Papers for the Museum of Zoology, University of Michigan* 348 (1937): 1–7.

9. Archie Fairly Carr, "The Status of *Pseudemys scripta* and *Pseudemys troostii*," *Herpetologica* 1 (1937): 73–77.

10. Archie Fairly Carr, "Notes on the *Pseudemys scripta* Complex," *Herpetologica* 1 (1937): 131–35.

11. Archie Carr to Thomas Barbour (March 16, 1938), in Carr Papers.

12. Thomas Barbour, Notes on Letter from Archie Carr (March 15, 1938), in Carr Papers.

13. Thomas Barbour to Archie Carr (May 23, 1938), in Carr Papers.

14. Archie Fairly Carr, "*Pseudemys nelsoni*, a New Turtle from Florida," *Occasional Papers of the Boston Society of Natural History* 8 (1938): 305–10.

15. Thomas Barbour and A.F. Carr, "Another Bahamian Fresh-Water Tortoise," *Proceedings of the New England Zoological Club* 17 (1938): 75–76.

16. Archie Fairly Carr, "A New Subspecies of *Pseudemys floridana*, with Notes on the *floridana* Complex," *Copeia* (1938): 105.

17. Archie Carr to Thomas Barbour (May 20, 1939), in Thomas Barbour Papers, Harvard University Archives, Cambridge, Massachusetts (hereafter, "Barbour Papers"), 2.

18. Ibid.

19. Thomas Barbour to Archie Carr (May 24, 1939), in Carr Papers.

20. Archie Fairly Carr, "*Haideotriton wallacei*, a New Subterranean Salamander from Georgia," *Occasional Papers of the Boston Society of Natural History* 8 (1939): 334.

21. Thomas Barbour to Archie Carr (August 5, 1939), in Carr Papers.

22. Archie Carr to Thomas Barbour (August 13, 1939), in Barbour Papers.

23. Thomas Barbour to Archie Carr (August 16, 1939), in Carr Papers.

24. Archie Fairly Carr, "Notes on Escape Behavior in the Florida Marsh Rabbit," *Journal of Mammalogy* 20, no. 3 (1939): 322–25.

25. Archie Fairly Carr, "Notes on the Breeding Habits of the Warmouth Bass," *Proceedings of the Florida Academy of Sciences* 4 (1939): 108–12.

26. Archie Carr to Thomas Barbour (October 21, 1939), in Barbour Papers.

27. Thomas Barbour and Archie Fairly Carr, "Antillean Terrapins," *Memoirs of the Museum of Comparative Zoology* 54, no. 5 (1940): 381–415, p. 385.

28. Thomas Barbour to Archie Carr (November 6, 1939), in Carr Papers. For another perspective on the term "dago dazzler" and the botanist David Fairchild's use of it, see Philip J. Pauly, "The Beauty and Menace of the Japanese Cherry Trees: Conflicting Visions of American Ecological Independence," *Isis* 87, no. 1 (1996): 51–73.

29. Thomas Barbour to Archie Carr (November 16, 1939), in Carr Papers.

30. Archie Carr to Thomas Barbour (January 7, 1940), in Barbour Papers.

31. Ibid., 2.

32. Ibid.

33. Ibid.

34. Thomas Barbour to Archie Carr (January 25, 1940), in Carr Papers.

35. K. Auffenberg and G.C. Gould, *Thomas Farm* (Florida Museum of Natural History, 2004 [cited June 17 2004]); available at http://www.flmnh.ufl.edu/fossilhall/FLPaleo/ThomasFarm/ThomasFarm.htm.

36. Thomas Barbour to Archie Carr (May 10, 1940), in Carr Papers.

37. Archie Fairly Carr, *A Contribution to the Herpetology of Florida, Biological Science Series* (Gainesville: University of Florida, 1940), 1.

38. Ibid.

39. Ibid.

40. Ibid., 96.

41. Ibid., 101.

42. Ibid., 53.

43. George Albert Boulenger and British Museum (Natural History), *Catalogue of the Chelonians, Rhynchocephalians, and Crocodiles in the British Museum (Natural History)*, new ed. (London: Printed by Order of the Trustees, 1889).

44. Thomas Barbour to Archie Carr (May 18, 1940), in Carr Papers.
45. Archie Carr to Thomas Barbour (September 11, 1940), in Barbour Papers.
46. Helene Robinson to Archie Carr (November 30, 1940), in Carr Papers.
47. Archie Carr to Thomas Barbour (December 3, 1940), in Barbour Papers.
48. Ibid.
49. Marjorie H. Carr to Thomas Barbour (December 3, 1940), in Barbour Papers.
50. Thomas Barbour to Archie Carr (December 7, 1940), in Carr Papers.
51. A.F. Carr, "The Fishes of Alachua County, Florida: A Subjective Key," *Dopeia* Ser. B., vol. 3, part Q, no. X (1941): vi. See also Carr, *A Naturalist in Florida*, 125–38.
52. Carr, "The Fishes of Alachua County" 2.
53. Ibid., 1.
54. Thomas Barbour and Archie F. Carr, "Terrapin from Grand Cayman," *Proceedings of the New England Zoological Club* 18 (1941): 57–60.
55. Archie Carr to Thomas Barbour (August 13, 1941), in Barbour Papers.
56. Thomas Barbour to Archie Carr (September 18, 1941), in Carr Papers.
57. Archie Carr to Thomas Barbour (October 18, 1941), in Barbour Papers.
58. See Samuel P. Hays and Barbara D. Hays, *Beauty, Health, and Permanence: Environmental Politics in the United States, 1955–1985* (Cambridge and New York: Cambridge University Press, 1987).
59. Carr to Barbour (October 18, 1941).
60. See G. Baur, "Two New Species of Tortoises from the South," *Science* 16, no. 405 (1890): 263–63.
61. A.F. Carr and Lewis J. Marchand, "A New Turtle from the Chipola River, Florida," *Proceedings of the New England Zoological Club* 20 (1942): 95–100, p. 96.
62. Archie Carr to Thomas Barbour (November 16, 1941), in Barbour Papers.
63. Thomas Barbour to Archie Carr (November 21, 1941), in Carr Papers.
64. Archie Carr to Thomas Barbour (December 5, 1941), in Barbour Papers.
65. Ibid.
66. Marjorie H. Carr to Thomas Barbour (December 23, 1941), in Barbour Papers.
67. Archie Carr to Thomas Barbour (December 23, 1941), in Barbour Papers.
68. Thomas Barbour to Archie Carr (January 24, 1942), in Carr Papers.
69. Helene Robinson to Marjorie Carr (May 6, 1942), in Carr Papers.
70. Archie Carr to Thomas Barbour (May 5, 1942), in Barbour Papers, 2.
71. Thomas Barbour to Archie Carr (May 8, 1942), in Carr Papers.
72. Helene Robinson to Archie Carr (May 27, 1942), in Carr Papers.
73. Carr and Marchand, "A New Turtle."
74. Archie Fairly Carr, "Notes on Sea Turtles," *Proceedings of the New England Zoological Club* XXI (1942): 9.
75. Thomas Barbour, telegram to Archie Carr (July 30, 1942), in Carr Papers.
76. Archie Carr to Thomas Barbour (August 4, 1942), in Barbour Papers.
77. Thomas Barbour to Archie Carr (August 17, 1942), in Carr Papers. See also Thomas Barbour, "Naturalist at Large: The Glory Hole," *The Atlantic* 170 (November 1942): 48–52.
78. Archie Carr to Thomas Barbour (August 19, 1942), in Barbour Papers.
79. Thomas Barbour to Archie Carr (August 22, 1942), in Carr Papers.
80. Thomas Barbour to Archie Carr (August 27, 1942), in Carr Papers.
81. Leonhard Stejneger to Thomas Barbour (August 25, 1942), in Carr Papers.
82. Ibid.
83. Ibid.

84. Ibid.

85. Archie Carr to Thomas Barbour (September 4, 1942), in Barbour Papers, 1–2.

86. Thomas Barbour to Archie Carr (September 9, 1942), in Carr Papers.

87. Marjorie H. Carr to Thomas Barbour (January 15, 1943), in Barbour Papers, 3.

88. Archie Carr to Thomas Barbour (January 18, 1943), in Barbour Papers, 2.

89. Thomas Barbour to Marjorie Carr (January 20, 1943), in Carr Papers.

90. Archie Carr to Thomas Barbour (January 26, 1943), in Barbour Papers.

91. Thomas Barbour to Archie Carr (February 1, 1943), in Carr Papers.

92. Leonhard Stejneger to Thomas Barbour (February 8, 1943), in Carr Papers.

93. Archie Carr to Thomas Barbour (February 14, 1943), in Barbour Papers.

94. Thomas Barbour to Archie Carr (February 18, 1943), in Carr Papers.

95. Archie Carr to Thomas Barbour (March 1, 1943), in Barbour Papers.

96. Thomas Barbour to Archie Carr (March 4, 1943), in Carr Papers.

97. Carr to Barbour (May 5, 1942).

98. Archie Carr to Thomas Barbour (September 6, 1942), in Barbour Papers.

99. Archie Carr to Thomas Barbour (October 9, 1942), in Barbour Papers.

100. Carr to Barbour (March 1, 1943), 1–2.

101. Barbour to Carr (March 4, 1943).

102. Thomas Barbour to Archie Carr (March 6, 1943), in Carr Papers.

103. Archie Carr to Thomas Barbour (March 13, 1943), in Barbour Papers.

104. Thomas Barbour to Archie Carr (April 29, 1943), in Carr Papers. Doris Cochran succeeded Stejneger as curator of the Division of Herpetology at the Smithsonian.

105. Archie Carr to Thomas Barbour (May 3, 1943), in Barbour Papers, 2. It is not clear to whom Carr referred.

106. Thomas Barbour to Archie Carr (April 29, 1943), in Carr Papers.

107. Archie Carr to whom it may concern (May 7, 1943), in Barbour Papers.

108. Archie Carr to Thomas Barbour (July 1943), in Barbour Papers, 2.

109. Thomas Barbour to Archie Carr (July 27, 1943), in Carr Papers.

110. Archie Fairly Carr, Presentation on Behalf of Thomas Barbour, Honorary Doctor of Science, University of Florida (May 29, 1944) in Carr Papers.

111. David E. Duncan, "Capturing Giant Turtles in the Caribbean," *National Geographic* 84, no. 2 (1943): 177–90.

112. Albert Hazen Wright to Archie Carr, (December 1, 1943), in Carr Papers.

113. Archie Carr to Thomas Barbour (December 1943), in Barbour Papers, 1.

114. Ibid., 2.

115. Thomas Barbour to Marjorie Carr (January 1, 1944), in Carr Papers.

116. Archie Carr to Thomas Barbour (November 5, 1944), in Barbour Papers, 1–2.

117. Ibid., 2.

118. Thomas Barbour to Archie Carr (November 8, 1944), in Carr Papers. "Taca plane" refers to a regional airline in Costa Rica.

Chapter 4

1. Archie Carr to Thomas Barbour (December 29, 1944), in Barbour Papers, 1–3.

2. Archie Fairly Carr, *High Jungles and Low*, 2nd ed. (Gainesville: University Press of Florida, 1992), xx–xxi.

3. Carr to Barbour (December 29, 1944), 3.

4. Carr, *High Jungles and Low*, xxi.

5. Archie Carr to Thomas Barbour (January 10, 1945), in Barbour Papers.

6. Ibid.
7. Most of the details in this section are drawn from Simón E. Malo, *El Zamorano: Meeting the Challenge of Tropical America* (Manhattan, Kans.: Simbad Books, 1999).
8. Despite Zemurray's noble objectives in creating the EAP, the legacy of United Fruit in Central America has been critically reviewed by numerous scholars. See, for example, Darío A. Euraque, *Reinterpreting the Banana Republic: Region and State in Honduras, 1870–1972* (Chapel Hill: University of North Carolina Press, 1996).
9. Malo, *El Zamorano*, 20–21.
10. Marjorie H. Carr to Thomas Barbour (March 9, 1945), in Barbour Papers.
11. Ibid.
12. Thomas Barbour to Marjorie Carr (March 12, 1945), in Carr Papers.
13. Ibid. Myriapods refers to any of the many-legged species of centipedes and millipedes.
14. Archie Carr to Thomas Barbour (June 3, 1945), in Barbour Papers.
15. Ibid., 2.
16. Ibid.
17. Marjorie H. Carr to Thomas Barbour (July 17, 1945), in Barbour Papers, 2.
18. Archie Carr to Thomas Barbour (July 24, 1945), in Barbour Papers, 3.
19. Ibid.
20. Marjorie H. Carr to Thomas Barbour (July 17, 1945), in Barbour Papers, 2.
21. Ibid.
22. Ibid.
23. Ibid.
24. Carr, *High Jungles and Low*, p. xxii.
25. Marjorie H. Carr to Thomas Barbour (July 31, 1945), in Barbour Papers.
26. Ibid.
27. Edith D. Oliver to Archie Carr (August 13, 1945), in Carr Papers.
28. Ibid.
29. Archie Carr to Thomas Barbour (September 6, 1945), in Barbour Papers.
30. Archie Carr to Thomas Barbour (October 30, 1945), in Barbour Papers.
31. Carr to Barbour (September 6, 1945).
32. Thomas Barbour to Archie Carr (January 4, 1946), in Carr Papers.
33. See David Ehrenfeld, "In Memoriam: Archie Carr," *Conservation Biology* 1, no. 2 (1987): 169–72.
34. Rosamond P. Barbour to Archie Carr (January 22, 1946), in Carr Papers.
35. Carr, *High Jungles and Low*, 3.
36. Archie Fairly Carr, "Outline for a Classification of Animal Habitats in Honduras," *Bulletin of the American Museum of Natural History* 94, no. 10 (1950): 580.
37. Carr, *High Jungles and Low*, 6.
38. Ibid., 8.
39. Ibid., 9.
40. Marjorie H. Carr, Trip to Uyuca Field Notes (November 23, 1948), in Private Collection of Tom Carr, 35.
41. Carr, *High Jungles and Low*, 10–11.
42. On collecting specimens in ornithology, see Mark V. Barrow, *A Passion for Birds: American Ornithology after Audubon* (Princeton, N.J.: Princeton University Press, 1998).
43. Carr, *High Jungles and Low*, 56.
44. Ibid., 61.

45. Ibid., 63–64.
46. Ibid., 65.
47. Archie F. Carr, Jr., Field Notes—Honduras (Oct. 10, 1947), in Private Collection of Tom Carr, 25–30. Used with permission.
48. Carr, *High Jungles and Low*, 126–27.
49. Ibid., 129.
50. Ibid., 132.
51. Ibid.
52. Ibid., 138.
53. Ibid., 139.
54. Ibid., 141.
55. Ibid., 162.
56. Ibid.
57. Ibid., 184.
58. Ibid., 167.
59. Ibid., 174.
60. Ibid.
61. Ibid., 175.
62. Ibid.
63. Ibid., 220.
64. Ibid., 36.
65. Ibid.
66. Ibid., 37–38.
67. Ibid., 38–43.
68. Ibid., 39.
69. Ibid., 43.
70. Ibid., 43–44.

Chapter 5

1. Joshua C. Dickinson, interview with author (April 19, 1996), Gainesville, Florida.
2. For biographical material on Warder Clyde Allee as well as a study of the ecology group at Chicago, see Gregg Mitman, *The State of Nature: Ecology, Community, and American Social Thought, 1900–1950, Science and Its Conceptual Foundations* (Chicago: University of Chicago Press, 1992) and William C. Kimler, "Allee, Warder Clyde," in *Dictionary of Scientific Biology*, ed. Charles C. Gillispie (New York: Scribner, 1986), 16–18.
3. Warder Clyde Allee, *The Social Life of Animals* (New York: W.W. Norton, 1938).
4. Warder Clyde Allee, Alfred E. Emerson, Orlando Park, Thomas Park, and Karl P. Schmidt, *Principles of Animal Ecology* (Philadelphia: Saunders, 1949).
5. A.F. Carr, Jr. to to W.C. Allee (May 12, 1950), in Carr Papers.
6. Archie Fairly Carr, *Handbook of Turtles: The Turtles of the United States, Canada, and Baja California* in *Handbooks of American Natural History* (Ithaca, N.Y.: Comstock, 1952), 1.
7. Ibid.
8. Ibid., 3–4.
9. Ibid., 5.
10. Ibid., 8.
11. Ibid., 16.
12. Ibid., 17.

13. Ibid., 27.
14. Ibid.
15. Ibid., 28.
16. Ibid.
17. Ibid., 33.
18. Ibid., 35.
19. Ibid., 353.
20. Ibid., 356–57.
21. John Werner to Archie Carr (N.D.), in Carr Papers.
22. Archie Fairly Carr, "The Zoogeography and Migrations of Sea Turtles," *Yearbook of the American Philosophical Society* (1954): 138–40, p. 140.
23. Ibid.
24. Archie Fairly Carr, *The Windward Road: Adventures of a Naturalist on Remote Caribbean Shores* (reissued ed.; Tallahassee: University Presses of Florida, 1979), xxxv.
25. Ibid., 43.
26. Ibid., 47.
27. Ibid., 48.
28. Ibid., 51.
29. Ibid.
30. Ibid.
31. Ibid., 57.
32. Ibid., 58.
33. Ibid., 61.
34. Ibid., 68.
35. Ibid., 76.
36. Ibid., 77.
37. Ibid., 79.
38. Ibid., 86.
39. Ibid., 119.
40. Ibid., 132–33.
41. Ibid., 154.
42. Ibid.
43. Ibid., 215.
44. Ibid., 217.
45. Ibid., 219.
46. Ibid.
47. Ibid.
48. Ibid., 219–20.
49. Ibid., 220.
50. Ibid., 221–22.
51. Ibid.
52. Ibid.
53. Ibid., 238.
54. Ibid., 239.
55. Ibid., 242.
56. Ibid., 244.
57. Ibid., 248.
58. Ibid.

59. Ibid., 252.

60. George Sprugel, Jr. to Archie Carr (October 26, 1954), in Carr Papers.

61. The National Science Foundation was founded on May 10, 1950. From the outset, the support of basic (as opposed to applied) research in the biological sciences was one of NSF's aims. Moreover, Carr's emphasis on conservation suggested the concerns of the Fish and Wildlife Service. For additional details on biological research support at NSF, see Alan T. Waterman, "Federal Support of Fundamental Research in the Biological Sciences," *AIBS Bulletin* October 1951 (1951): 11–17 and Louis Levin, "The Role of the National Science Foundation in Biological Science," *AIBS Bulletin* October 1954 (1954): 19–21. See also Toby A. Appel, *Shaping Biology: The National Science Foundation and American Biological Research, 1945–1975* (Baltimore: The Johns Hopkins University Press, 2000).

62. Archie Fairly Carr, "A Study of the Ecology, Migrations, and Population Levels of Sea Turtles in the Atlantic and Caribbean with Special Reference to the Atlantic Green Turtle, *Chelonia mydas mydas* [Linné]." Research proposal for submission to National Science Foundation, unpublished, in Carr Papers.

63. Ibid., 2.

64. For histories of Audubon societies, see Frank Graham, *The Audubon Ark: A History of the National Audubon Society* (New York: Knopf: Distributed by Random House, 1990). See also Mark V. Barrow, *A Passion for Birds: American Ornithology after Audubon* (Princeton, N.J.: Princeton University Press, 1998).

65. For more on science and sentiment with regard to international conservation of fish, marine mammals, and birds, see Kurkpatrick Dorsey, *The Dawn of Conservation Diplomacy: U.S.-Canadian Wildlife Protection Treaties in the Progressive Era* (Seattle: University of Washington Press, 1998).

66. S.G. Fletcher to Archie Carr (March 11, 1958), in Carr Papers.

67. Archie F. Carr, Jr. to S.G. Fletcher (March 18, 1958), in Carr Papers.

68. For limited biographical material on Joshua B. Powers, see Anon., "J.B. Powers, 96, Publishers' Representative," *The New York Times*, February 23, 1989, and Anon., "Powers, Joshua Bryant," in *Who's Who in America* (Chicago: Marquis, 1977), 2528.

69. Carr's account of Joshua Powers and the founding of the Caribbean Conservation Corporation appeared in Archie Fairly Carr, *The Sea Turtle: So Excellent a Fishe* (Austin: University of Texas Press, 1984).

70. Joshua B. Powers to Archie Carr (January 31, 1958), in Carr Papers.

71. Joshua B. Powers to Archie Carr (March 14, 1958), in Carr Papers.

72. Angel Ramos (*El Mundo*, Puerto Rico), Luis Muñoz Marin (governor of Puerto Rico), John O'Rourke (*The Washington Daily News*), Alfred Stanford, John A. Brogan (Hearst Corporation), James H. Drumm (The Henry Clay Foundation), Cecil Brooks (Incorporated Press Ltd.), Ed Mazzucchi (*Publicidad ARS*, Venezuela), Gale Wallace (United Fruit Company), Franck Magloire (*Le Matin*, Haiti), Ford Baxter (*The Royal Gazette*, Bermuda), G. Martinez Marquez (*El Pais*, Cuba), John R. Herbert (*Quincy Patriot Ledger*), Andrew Heiskell (*Life*), Daniel Morales (*Mañana*, Mexico), James B. Canel (Inter American Press Association), John Klem (Editors Press Service), Richard Dyer (United Fruit Company), John C. McClintock (United Fruit Company), S. G. Fletcher (*The Daily Gleaner*, Jamaica), Herbert L. Matthews (*The New York Times*), and H. Earle Braisted (Joshua B. Powers, Inc.).

73. Joshua B. Powers to Archie Carr (March 14, 1958), 2–3.
74. Archie F. Carr, Jr. to Joshua B. Powers (March 21, 1958), in Carr Papers. Used with permission.
75. Archie F. Carr, Jr. to Joshua B. Powers (January 29, 1959), in Carr Papers, 1.
76. Ibid., 2.
77. Ibid.
78. Joshua B. Powers to John H. Phipps (April 14, 1959), in Carr Papers.
79. Ibid.

Chapter 6

1. Archie Fairly Carr, *The Windward Road: Adventures of a Naturalist on Remote Caribbean Shores* (reissue. ed.; Tallahassee: University Presses of Florida, 1979), xxxiv.
2. Larry Ogren, interview with author (September 12, 2003), Panama City, Florida.
3. Archie Fairly Carr and David K. Caldwell, "The Ecology and Migrations of Sea Turtles, 1: Results of Field Work in Florida, 1955," *American Museum Novitates*, no. 1793 (1956), 1–23.
4. Ibid., 3–4.
5. Archie Fairly Carr, *The Sea Turtle: So Excellent a Fishe* (Austin: University of Texas Press, 1984), 102.
6. Carr and Caldwell, "Ecology and Migrations, 1," 15.
7. Quoted in Archie Fairly Carr, *Handbook of Turtles: The Turtles of the United States, Canada, and Baja California* in Handbooks of American Natural History (Ithaca, N.Y.: Comstock, 1952), 402.
8. Ibid.
9. Carr, *The Windward Road*, 26–27.
10. Carr and Caldwell, "Ecology and Migrations, 1," 19.
11. Tagging crews discontinued the practice of turning turtles during the 1980s, when it became clear that turned turtles suffered considerable distress.
12. Archie Fairly Carr and Leonard Giovannoli, "The Ecology and Migrations of Sea Turtles, 2: Results of Field Work in Costa Rica, 1955," *American Museum Novitates* 1835 (1957): 1–32.
13. Ibid., 10–11.
14. Ibid., 15.
15. Ibid., 16.
16. Carr, *The Windward Road*, 234–35.
17. Tom Harrisson, "The Edible Turtle (*Chelonia mydas*) in Borneo, 1. Breeding Season," *Sarawak Museum Journal* 5, no. 3 (1951): 593–96 and (as cited by Carr) Tom Harrisson, "The Edible Turtle (*Chelonia mydas*) in Borneo, 2. Copulation," *Sarawak Museum Journal* 6, no. 4 (1954): 126–28.
18. Carr and Giovannoli, "Ecology and Migrations, 2," 23.
19. Ibid., 29.
20. Ibid., 24–27.
21. L.D. Gomez and J.M. Savage, "Searchers on That Rich Coast: Costa Rican Field Biology, 1400–1980," in *Costa Rican Natural History*, ed. Daniel H. Janzen (Chicago: University of Chicago Press, 1983), 6.
22. David Ehrenfeld, "In Memoriam: Archie Carr," *Conservation Biology* 1, no. 2 (1987): 169–72, p. 169.
23. Archie Carr, letter to the senior members of the Guiding Counsel of the Faculty of Sciences and Letters (December 11, 1956), in Carr Papers.

24. Archie Carr, La Biologia En Los Estudios Generales (N.D.), in Carr Papers, 1.

25. Ibid., 2.

26. Ibid.

27. Sterling Evans, *The Green Republic: A Conservation History of Costa Rica* (Austin: University of Texas Press, 1999), 22.

28. See P. E. P. Deraniyagala, *The Tetrapod Reptiles of Ceylon*, vol. 1: Testudinates and crocodilians, *Colombo Museum Natural History Series* (England: Colombo, Ceylon, and Dulau and Co., Ltd., 1939).

29. Archie Fairly Carr and Larry Ogren, "The Ecology and Migrations of Sea Turtles, 3: *Dermochelys* in Costa Rica," *American Museum Novitates* 1958 (1959): 1–29. See also T. R. Leary, "A Schooling of Leatherback Turtles, *Dermochelys coriacea coriacea*, on the Texas Coast," *Copeia* 1957, no. 3 (1957): 232.

30. Carr and Ogren, "Ecology and Migrations, 3," 2–3.

31. Ibid., 3.

32. Ibid., 4.

33. Ibid., 22–23.

34. Ibid., 23–24.

35. Archie Carr and Larry Ogren, "The Ecology and Migrations of Sea Turtles, 4: The Green Turtle in the Caribbean Sea," *Bulletin of the American Museum of Natural History* 121, no. 1 (1960): 1–48.

36. Ibid., 20.

37. Ibid., 21. See also K. Schmidt-Nielsen and F. Ragnar, "Salt Glands in Marine Reptiles," *Nature* 182, no. 4638 (1958): 783–85.

38. Archie Fairly Carr and Larry Ogren, "The Ecology and Migrations of Sea Turtles, 4," 47.

39. Archie Fairly Carr and Harold Hirth, "Social Facilitation in Green Turtle Siblings," *Animal Behaviour* 9, no. 1–2 (1961): 68–70.

40. Ibid., 68–69.

41. David K. Caldwell, Archie F. Carr, and Larry H. Ogren, "The Atlantic Loggerhead Sea Turtle, *Caretta caretta* (L.), in America: I. Nesting and Migration of the Atlantic Loggerhead Turtle," *Bulletin of the Florida State Museum* 4, no. 10 (1959): 295–308, p. 296.

42. Ibid., 297.

43. Ibid., 301–3.

44. Ibid., 306–7.

45. David K. Caldwell et al., "*Caretta caretta* (L.), in America: II. Multiple and Group Nesting by the Atlantic Loggerhead Turtle," *Bulletin of the Florida State Museum* 4, no. 10 (1959): 309–318, p. 310.

46. Ibid., 316.

47. Archie Fairly Carr, "Notes on the Zoogeography of the Atlantic Sea Turtles of the Genus *Lepidochelys*," Revista de Biología Tropical 5, no. 1 (1957): 45–61, p. 45.

48. Ibid., 50.

49. Carr included a plot of all recoveries of bottles released in the areas involved furnished by Woods Hole Oceanographic Institution based on U.S. hydrographic records extending from the 1880s to the present.

50. Archie Fairly Carr, "The Ridley Mystery Today," *Animal Kingdom* 64, no. 1 (1961): 7–12, p. 8.

51. Ibid., 9.

52. Ibid., 12.

53. Archie F. Carr, "Pacific Turtle Problem—Mexico's Coast Yields Information on a Matter of Melanism," *Natural History* 70, no. 8 (1961): 64–71, p. 66.

54. Ibid., 67.

55. Ibid., 70. Carr's thoughts on the speciation of the black form of the green sea turtle were soon confirmed by his student David Caldwell, who named the subspecies *Chelonia mydas carrinegra* in Carr's honor. See David K. Caldwell, "Carapace Length—Body Weight Relationship and Size and Sex Ratio of the Northeastern Pacific Green Sea Turtle, *Chelonia mydas carrinegra*," *Los Angeles County Museum, Contributions to Science* 62 (1962): 3–10. Most current taxonomies of the genus *Chelonia* list the status of the black turtle as "uncertain."

56. Carr, "Pacific Turtle Problem, 70."

57. Archie Fairly Carr and Harold Hirth, "The Ecology and Migrations of Sea Turtles, 5: Comparative Features of Isolated Green Turtle Colonies," *American Museum Novitates* 2091 (1962): 1–42, p. 5.

58. Ibid., 37.

59. Ibid.

60. Ibid., 38. By "classic pattern for marine migration," Carr is probably referring to Pacific Salmon, which undertake notoriously difficult upstream migrations to reach spawning waters as adults.

61. Ibid., 39.

62. Ibid., 39–40.

63. Archie F. Carr to Sidney R. Galler (April 4, 1955), in Carr Papers.

64. Archie Carr, "Orientation Cues and Patterns of Mass Travel in Marine Turtles," in Carr Papers.

65. Ibid., 4.

66. Ibid., 6–7.

67. Archie Fairly Carr, "Orientation Problems in the High Seas Travel and Terrestrial Movements of Marine Turtles," *American Scientist* 50, no. 3 (1962): 358–74; Archie Fairly Carr, "Orientation Problems in the High Seas Travel and Terrestrial Movements of Marine Turtles," in *Bio-Telemetry: The Use of Telemetry in Animals Behavior and Physiology in Relation to Ecological Problems*, ed. Lloyd E. Slater (New York: Macmillan, 1963), 179–93.

68. Carr, "Orientation Problems," 363.

69. Archie Carr, "Transoceanic Migrations of the Green Turtle," *BioScience* 14, no. 8 (1964): 50. This paper was reprinted as Archie Fairly Carr, "Transoceanic Migrations of the Green Turtle," *Naval Research Reviews* 17, no. 10 (1964): 12–18.

70. Carr, "Transoceanic Migrations," 51.

71. Ibid., 52.

72. Ehrenfeld, "In Memoriam," 169.

73. Peter C.H. Pritchard, *Tales from the Thébaïde* (Malabar, Fla.: Krieger, 2006), 3.

Chapter 7

1. Lewis Berner, "A Contribution toward a Knowledge of the Mayflies of Florida" (M.S. thesis, University of Florida, 1939); Berner, "The Mayflies of Florida (Ephemeroptera)" (Ph.D. dissertation, University of Florida, 1941).

2. Lewis Berner, *The Mayflies of Florida*, vol. 4, *University of Florida Studies. Biological Science Series* (Gainesville: University of Florida Press, 1950).

3. Lewis Berner, *Entomological Report on Development of the River Volta Basin* (London: Wightman Mountain, 1950).

4. Malaria and schistosomiasis pose significant threats to millions of Africans and Asians even now. Of the 300 million or so cases of malaria that affect people, between 1 and 3 million cases are fatal, and approximately 90 percent of malarial deaths occur in Africa. As of May 1996, the World Health Organization estimated that 200 million people were infected with schisosomiasis or bilharziasis, and 10 percent, or 20 million, of these constituted severe cases; 120 million people were symptomatic, and as many as 500–600 million people were at risk of contracting the disease.

5. Archie Fairly Carr, *Ulendo: Travels of a Naturalist in and out of Africa* (Gainesville: University Press of Florida, 1993), xviii.

6. Marjorie Harris Carr, "1993 Preface to the 1952 Letters," in *Ulendo: Travels of a Naturalist in and out of Africa* (Gainesville: University Press of Florida, 1993), 259.

7. Carr, *Ulendo*, 18–19. "Aerosol" is a rather vague term for an insecticide that was almost undoubtedly DDT, but Carr was probably putting the final touches on *Ulendo* in 1962 when *Silent Spring* alerted Americans to the problem of indiscriminate use of chemical insecticides. Though Carr and Berner conducted their surveys in 1952, Carr may have wished to distance himself from the controversy surrounding the use of DDT, which was full blown by the time *Ulendo* was published in 1964.

8. Ibid., 19.

9. Ibid., 40.

10. Ibid.

11. Ibid., 277.

12. See Archie Fairly Carr, "The Gulf-Island Cottonmouths," *Proceedings of the Florida Academy of Sciences* 1 (1935): 88.

13. Carr, *Ulendo*, 77.

14. Ibid.

15. For analysis of debates regarding wilderness, see William Cronon, "The Trouble with Wilderness; or, Getting Back to the Wrong Nature," in *Uncommon Ground: Toward Reinventing Nature*, ed. William Cronon (New York: W.W. Norton & Company, 1995), 69–90 and Roderick Nash, *Wilderness and the American Mind*, 4th ed. (New Haven, Conn.: Yale University Press, 2001).

16. Carr, *Ulendo*, 79. See also David Livingstone and Charles Livingstone, *Narrative of an Expedition to the Zambesi and Its Tributaries; and of the Discovery of the Lakes Shirwa and Nyassa, 1858–1864* (New York: Harper & Brothers, 1866).

17. Carr, *Ulendo*, 82.

18. Ibid., 87.

19. Ibid., 93.

20. Ibid., 94.

21. Ibid., 95. See also Edward Daniel Young and Horace Waller, *Nyassa: A Journal of Adventures Whilst Exploring Lake Nyassa, Central Africa, and Establishing the Settlement of "Livingstonia,"* 2nd ed. (London: J. Murray, 1877).

22. Carr, *Ulendo*, 101–2.

23. Ibid., 118.

24. Ibid. For recent analyses of cichlid diversity see, Herbert R. Axelrod, *African Cichlids of Lakes Malawi and Tanganyika* (Neptune, N. J.: T. F. H. Publications, 1973); George W. Barlow, *The Cichlid Fishes: Nature's Grand Experiment in Evolution* (Cambridge, Mass.: Perseus, 2000); Geoffrey Fryer and T. D. Iles, *The Cichlid Fishes of the Great Lakes of Africa: Their Biology and Evolution* (Edinburgh: Oliver and

Boyd, 1972); Les Kaufman and Peter Ochumba, "Evolutionary and Conservation Biology of Cichlid Fishes as Revealed by Faunal Remnants in Northern Lake Victoria," *Conservation Biology* 7, no. 3 (1993): 719–30; and Tijs Goldschmidt, *Darwin's Dreampond: Drama in Lake Victoria* (Cambridge, Mass.: MIT Press, 1996).

25. Carr, *Ulendo*, 118–20; Carr, *The Land and Wildlife of Africa* (New York: Time-Life Books, 1964).

26. Carr, *Ulendo*, 119.

27. Ibid., 121.

28. Ibid., 122. See also Ernst Mayr, *Animal Species and Evolution* (Cambridge, Mass.: Belknap Press, 1963). Mayr notes that the term "Jordan's law" was proposed by J. A. Allen in a review of David Starr Jordan (1851–1931), who was an American ichthyologist and became the president of Stanford University. Jordan declined the honor of the appellation on the grounds that claims were based on the work of Moritz Wagner's 1868 work *Migrationsgesetz der Organismen* (rev. 1889). See Moritz Wagner, *Die Darwinische Theorie Und Das Migrationsgesetz Der Organismen* (Leipzig: Duncker & Humblot, 1868), translated as Moritz Wagner and James L. Laird, *The Darwinian Theory and the Law of the Migration of Organisms* (London: E. Stanford, 1873).

29. Carr, *Ulendo*, 130–32. See also Carr, "The Breeding Habits, Embryology, and Larval Development of the Large-Mouthed Black Bass in Florida," 43–77; Marjorie H. Carr, "Notes on the Breeding Habitats of the Eastern Stumpknocker *Lepomis punctatus punctatus* (Cuvier)," *Proceedings of the Florida Academy of Science* (1942): 101–6.

30. Carr, *Ulendo*, 134–35.

31. Ibid., 136–37. See also Geoffrey Fryer, "Some Aspects of Evolution in Lake Nyasa," *Evolution* 13, no. 4 (1959): 440–51; Geoffrey Fryer, "Evolution of Fishes in Lake Nyasa," *Evolution* 14, no. 3 (1960): 396–400.

32. In fact, cichlid diversity is even more complex than Carr suggested, with no fewer than 500 species in Lake Malawi (formerly Nyasa) alone and another 1500 or so species in the Great Lakes of East Africa. For further analysis and development, see Barlow, *The Cichlid Fishes*. Since Carr observed the Nyasa cichlids, scientists have recognized the significant role of sexual selection in speciation, which sometimes proceeds orthogonally to natural selection with dramatic effects.

33. Carr, *Ulendo*, 138–39.

34. See Goldschmidt, *Darwin's Dreampond*.

35. Carr, *Ulendo*, 160.

36. Ibid., 162.

37. Ibid., 205.

38. Ibid., 205–6.

39. Ibid., 225.

40. Ibid., 231–32.

41. Ibid.

42. Ibid., 235.

43. Ibid., 238.

44. Ibid., 240.

45. Ibid., 245.

46. Ibid., 249–52. For further descriptions and diagrammatic pictures of snares in Africa, see Carr, *Wildlife of Africa*, 172–73.

47. Carr, *Ulendo*, 254–55.

48. F. Fraser Darling, *Wildlife in an African Territory: A Study Made for the Game and Tsetse Control Department of Northern Rhodesia* (London: Oxford University Press, 1960), 4. For interdisciplinary analysis of conservation in Africa, see David Anderson and Richard Grove, *Conservation in Africa: People, Policies, and Practice* (Cambridge: Cambridge University Press, 1987).

49. Darling, *Wildlife in an African Territory*, 5.

50. F. Fraser Darling, "An Ecological Reconnaisance of the Mara Plains in Kenya Colony," *Wildlife Monographs* 5 (1960): 41.

51. Julian Huxley and UNESCO, *The Conservation of Wild Life and Natural Habitats in Central and East Africa: Report on a Mission Accomplished for UNESCO, July–September 1960* (Paris: UNESCO, 1961), 13.

52. Ibid., 92.

53. Carr, *Ulendo*, 236.

54. Ibid., 258.

55. Carr, *Wildlife of Africa*, 178.

56. Marston Bates, "A Glimpse into the Pleistocene (Book Review of *Ulendo*)," *New York Times Book Review* (April 19, 1964), 6.

Chapter 8

1. David Ehrenfeld, "In Memoriam: Archie Carr," *Conservation Biology* 1, no. 2 (1987): 169–70, p. 170.

2. Karen A. Bjorndal, interview with author (October 1, 2003), Gainesville, Florida.

3. Archie Fairly Carr, *The Sea Turtle: So Excellent a Fishe* (Austin: University of Texas Press, 1984), 5.

4. Archie Fairly Carr, Report of the Technical Director, CCC, 1961–62, in Carr Papers, 3–4.

5. Archie Fairly Carr, Harold Hirth, and Larry Ogren, "The Ecology and Migrations of Sea Turtles, 6: The Hawksbill Turtle in the Caribbean Sea," *American Museum Novitates* 2248 (1966): 1–29, pp. 7–8.

6. Archie F. Carr, Jr. to Señor Don Adriano Urbina G. (October 13, 1960), in Carr Papers.

7. Ibid.

8. Archie F. Carr, Jr. Report of the Technical Director, CCC, 1964–65, in Carr Papers, 7.

9. Ibid.

10. Archie F. Carr, Jr. Report of the Technical Director, CCC, 1962–63, in Carr Papers, 3–4.

11. Carr, Report of the Technical Director, 1964–65, 7–10.

12. Archie F. Carr, Jr. Report of the Technical Director, CCC, 1965–66, in Carr Papers, 6–7.

13. Archie F. Carr, Jr. Report of the Technical Director, CCC, 1968–69, in Carr Papers, 3.

14. J. Frick, "Orientation and Behavior of Hatchling Green Turtles (*Chelonia mydas*) in the Sea," *Animal Behavior* 24, no. 4 (1976): 849–57.

15. Carr, Report of the Technical Director, 1968–69, 8–9.

16. Ibid., 7.

17. Portions of this account appeared in Frederick R. Davis, "Saving Sea Turtles: The Evolution of IUCN's Marine Turtle Group," *Endeavour* 29, no. 3 (2005): 114–18.

18. For a biography of Peter Scott, see Elspeth Joscelin Grant Huxley, *Peter Scott, Painter and Naturalist* (Golden, Colo.: Fulcrum, 1995).
19. Peter Scott, "Organisation of the Survival Service Commission of the International Union for the Conservation of Nature and Natural Resources," in Carr Papers. Emphasis in original.
20. For a history of IUCN, see Martin W. Holdgate, *The Green Web: A Union for World Conservation* (Cambridge, England: International Union for Conservation of Nature and Natural Resources, 1999).
21. Peter Scott to Archie Carr (December 4, 1963), in Carr Papers.
22. Archie Carr to Peter Scott (June 30, 1964), in Carr Papers.
23. Peter Scott to Archie Carr (February 1, 1966), in Carr Papers.
24. Archie Carr to Peter Scott (February 16, 1966), in Carr Papers.
25. Carr, *So Excellent a Fishe*, 213.
26. Archie Fairly Carr, "Notes on Sea Turtles," *Proceedings of the New England Zoological Club* XXI (1942): 1–16.
27. Carr, *So Excellent a Fishe*, 214.
28. Carr to Scott (June 30, 1964), 2.
29. Ibid.
30. UNESCO was founded on 16 November 1945. Also founded in 1945, the United Nations FAO leads international efforts to end hunger.
31. Archie Fairly Carr, "Sea Turtles: A Vanishing Asset." Paper presented at the Latin American Conference on the Conservation of Renewable Natural Resources, San Carlos de Bariloche, Argentina, 1968, 163.
32. Ibid., 167.
33. Archie Carr to Peter Scott (February 1, 1968), in Carr Papers.
34. Ibid.
35. Ibid.
36. As chairman of WWF and vice president of IUCN, Scott occupied a unique and critical position in international conservation efforts. For additional details on the complex relationship between WWF and IUCN, see Huxley, *Peter Scott*, 227–28.
37. Archie Carr to Colin Holloway (May 3, 1968), in Carr Papers.
38. Ibid., 2.
39. Ibid., 3.
40. Archie Carr, Annual Report—1968 Marine Turtle Group, in Carr Papers.
41. Archie Fairly Carr, "Survival Outlook of the West-Caribbean Green Turtle Colony," *IUCN Publication, New Series, Supplemental Papers* 20 (1969): 13–16.
42. Ibid.
43. On this theme, see Kurkpatrick Dorsey, *The Dawn of Conservation Diplomacy: U.S.-Canadian Wildlife Protection Treaties in the Progressive Era* (Seattle: University of Washington Press, 1998).
44. Carr, "Survival Outlook."
45. Archie Carr to Ing. Guillermo E. Yglesias P. (April 28, 1969), in Carr Papers, 1.
46. Archie Carr to Colin W. Holloway (February 24, 1970), in Carr Papers, 1.
47. Peter Scott to Archie Carr (June 9, 1970), in Carr Papers, 1.
48. Carr to Holloway (February 24, 1970), 1.
49. Archie Carr, "Research and Conservation Problems in Costa Rica." Paper presented at the Second Working Meeting of the Marine Turtle Specialist Group, in Carr Papers.
50. Huxley, *Peter Scott*, 227.

51. William J. Hart to Archie Carr (March 29, 1963), in Carr Papers.
52. Archie F. Carr, Jr. to William J. Hart (April 10, 1963), in Carr Papers, 1.
53. See Roderick Nash, *Wilderness and the American Mind*, 4th ed. (New Haven, Conn.: Yale University Press, 2001).
54. Carr to Hart (April 10, 1963), 1.
55. Ibid.
56. Ibid.
57. President of the Republic (Costa Rica) and The Minister of Agriculture and Livestock Decree Regarding Sea Turtle Fishing, in Carr Papers.
58. Carr, Report of the Technical Director, 1964–65, 2.
59. Ibid., 3.
60. David Ehrenfeld, *Beginning Again: People and Nature in the New Millennium* (New York: Oxford University Press, 1993), 6.
61. Ibid., 7.
62. Carr, *So Excellent a Fishe*, 210.
63. Archie F. Carr, Jr., "Project Plan for a Tagging and Hatchery Station on the Central Tortuguero Beach near Jaloba, to Be Supported by JAPDEVA," in Carr Papers, 1–2.
64. Guillermo Cruz Bolaños, "Official Petition by Guillermo Cruz (on Behalf of the Caribbean Conservation Corporation) for the Lease of Land around Tortuguero," in Carr Papers, 1–2.
65. Ibid., 2.
66. Guillermo Cruz B. to Fernando López-Calleja U. (February 20, 1968), in Carr Papers.
67. Guillermo Cruz B. to Archie Carr (February 21, 1968), in Carr Papers.
68. James A. Oliver to Jose Joaquin Trejos (March 1, 1968), in Carr Papers.
69. Archie F. Carr, Jr., Report of the Technical Director, CCC, 1967–68, in Carr Papers, 7–8.
70. David Rains Wallace, *The Quetzal and the Macaw: The Story of Costa Rica's National Parks* (San Francisco: Sierra Club Books, 1992), 11–15. See also Sterling Evans, *The Green Republic: A Conservation History of Costa Rica* (Austin: University of Texas Press, 1999), 72–93.
71. Mario A. Boza, "The Only Solution to the Problem of the Management of the Green Turtle: Establishment and Development of Tortuguero National Park," *La Nacion* (Costa Rica), December 17, 1968, 41.
72. Mario A. Boza to Archie Carr (March 10, 1969), in Carr Papers.
73. See Evans, *Green Republic*, 74; Wallace, *The Quetzal*, 16–17.
74. Archie Carr, Memorandum to Cruz, Frick, Oliver, Phipps, and Powers (CCC Board) (September 26, 1969), in Carr Papers, 2–3.
75. Ibid., 3.
76. Ibid., 2.
77. Mario A. Boza to Archie Carr (December 11, 1969), in Carr Papers, 1.
78. Archie F. Carr, Jr. to Mario Andres Boza (October 15, 1970), in Carr Papers.
79. Ibid.
80. Archie F. Carr, Jr. to Stephen Harrell (December 3, 1970), in Carr Papers.
81. Walter Auffenberg, Sr. was the curator of reptiles and amphibians at the Florida Museum of Natural History and an acknowledged expert on Komodo dragons.
82. Walter Auffenberg, Jr. to Wayne King (December 28, 1971), in Carr Papers, 1–2.

83. Archie F. Carr, Jr., Report of the Technical Director, CCC, 1974–75, in Carr Papers, 2–3.
84. Archie F. Carr, Jr. to Daniel Oduber, President of Costa Rica (October 3, 1975), in Carr Papers.
85. Archie Carr, Memorandum to Gerardo Budowsky, Richard Fitter, Tom Harrisson, Wayne King, Nathaniel Reed, and Sir Peter Scott (October 6, 1975), in Carr Papers.
86. "Resellada Ley De Parque Nacional De Tortuguero," La Nacion, October 29, 1975, 12.
87. Guillermo Cruz B. to Archie Carr (October 30, 1975), in Carr Papers.
88. Archie Carr, Memorandum to Marine Turtle Group, IUCN, Coral Gables Talk Force, IUCN, and Other Concerned (October 31, 1975), in Carr Papers.
89. Archie F. Carr, Jr., Report of the Technical Director, CCC, 1975–76, in Carr Papers, 1, emphasis added.
90. Archie Fairly Carr, "Caribbean Green Turtle: Imperiled Gift of the Sea," National Geographic Magazine 131, no. 6 (1967): 879.
91. Carr, So Excellent a Fishe, 234–36.
92. Peggy Fosdick and Sam Fosdick, Last Chance Lost? Can and Should Farming Save the Green Sea Turtle? The Story of Mariculture, Ltd., Cayman Turtle Farm (York, Penn.: I.S. Naylor, 1994), 3–13.
93. Ibid., 14–56.
94. Carr, So Excellent a Fishe, 15–16.
95. Fosdick and Fosdick, Last Chance Lost, 57–58.
96. Ibid., 44–45. Eventually, these tanks proved to be insufficient to clear the wastes, and operations had to be moved to land.
97. Archie F. Carr, Jr. to Irvin S. Naylor (December 18, 1973), in Carr Papers.
98. Ibid.
99. D. W. Ehrenfeld, "Conserving the Edible Sea Turtle: Can Mariculture Help?" American Scientist, 1974, 62 (1974): 23–31, p. 30.
100. Peter C.H. Pritchard, Tales from the Thébaïde (Malabar, Fla.: Krieger, 2006), 12.
101. James R. Wood and Fern E. Wood, "Reproductive Biology of Captive Green Sea Turtles Chelonia mydas," American Zoologist 20 (1980): 499–505, p. 499.
102. Anne B. Meylan and David Ehrenfeld, "Conservation of Marine Turtles," in Turtle Conservation (Washington, D.C.: Smithsonian Institution Press, 2000), 96–125, pp. 114–15.
103. Archie F. Carr, Jr. "Alligators: Dragons in Distress," National Geographic 131, no. 1 (1967): 130–48.

Chapter 9

1. Archie Fairly Carr, The Sea Turtle: So Excellent a Fishe (Austin: University of Texas Press, 1984), 115.
2. Ibid., 116.
3. Larry Ogren, interview with author (September 12, 2003), Panama City, Florida.
4. Carr, So Excellent a Fishe, 114.
5. Ibid.
6. Ibid., 114–15.
7. Ibid., 157.

8. Ibid.
9. Archie Fairly Carr, Harold Hirth, and Larry Ogren, "The Ecology and Migrations of Sea Turtles, 6: The Hawksbill Turtle in the Caribbean Sea," *American Museum Novitates* 2248 (1966): 1–29.
10. Ibid., 7–8.
11. Ibid., 9–27.
12. Ibid., 28. See also Carr, *So Excellent a Fishe*, 227–29.
13. David W. Ehrenfeld and Archie Fairly Carr, "The Role of Vision in the Sea-Finding Orientation of the Green Turtle (*Chelonia mydas*)," *Animal Behaviour* 15, no. 1 (1967): 25–36. See also Carr, *So Excellent a Fishe*, 88–91; reprinted as Archie Fairly Carr, "No One Knows Where the Turtles Go. Part 2: 100 Turtle Eggs," *Natural History* 76, no. 8 (1967): 40–43, 52–59.
14. Ehrenfeld and Carr, "Role of Vision," 35.
15. N. Mrosovsky and Archie Fairly Carr, "Preference for Light of Short Wavelengths in Hatchling Green Sea Turtles, *Chelonia mydas*, Tested on Their Natural Nesting Beaches," *Behaviour* 28, no. 3–4 (1967): 217–31.
16. Carr, *So Excellent a Fishe*, 29–30.
17. Ibid., 30.
18. Ibid., 41.
19. Ibid., 42.
20. Ibid., 43.
21. Ibid., 49.
22. Ibid., 229.
23. Ibid., 230.
24. Ibid., 231–32.
25. Archie Fairly Carr and Marjorie H. Carr, "Modulated Reproductive Periodicity in *Chelonia*," *Ecology* 51, no. 2 (1970): 335–37, p. 335.
26. Ibid.
27. Ibid., 336.
28. Archie Fairly Carr and Donald Goodman, "Ecologic Implications of Size and Growth in *Chelonia*," *Copeia* 1970, no. 4 (1970): 783–86, p. 783.
29. Ibid., 783.
30. Ibid., 786.
31. Archie Fairly Carr and Marjorie H. Carr, "Site Fixity in the Caribbean Green Turtle," *Ecology* 53, no. 3 (1972): 425–29, p. 428.
32. Archie Fairly Carr and Marjorie H. Carr, "Recruitment and Remigration in a Green Turtle Nesting Colony," *Biological Conservation* 4, no. 2 (1970): 282–84, p. 282.
33. Arthur L. Koch, Archie F. Carr, and David W. Ehrenfeld, "The Problem of Open-Sea Navigation: The Migration of the Green Turtle to Ascension Island," *Journal of Theoretical Biology* 22, no. 1 (1969): 163–79.
34. David W. Ehrenfeld and Arthur L. Koch, "Visual Accommodation in the Green Turtle," *Science* 155, no. 3764 (1967): 827–28.
35. M. L. Manton, A. Karr, and David W. Ehrenfeld, "An Operant Method for the Study of Chemoreception in the Green Turtle, *Chelonia mydas*," *Brain, Behavior and Evolution* 5, no. 2 (1972): 188–201. (Note: A. Karr was not Archie Carr.)
36. Archie Fairly Carr and Patrick J. Coleman, "Seafloor Spreading Theory and the Odyssey of the Green Turtle," *Nature* 249, no. 5453 (1974): 128–30, p. 129.
37. Ibid., 130.

38. Martin D. Brasier, "Turtle Drift," *Nature* 250, no. 5464 (1974): 351.
39. Stephen Jay Gould, "Senseless Signs of History," in *The Panda's Thumb: More Reflections in Natural History* (New York: W.W. Norton & Company, 1980), 33–34.
40. Brian W. Bowen, Anne B. Meylan, and John C. Avise, "An Odyssey of the Green Sea Turtle: Ascension Island Revisited," *Proceedings of the National Academy of Sciences of the United States of America* 86, no. 2 (1989): 575.
41. Archie Fairly Carr, Perran Ross, and Stephen Carr, "Internesting Behavior of the Green Turtle *Chelonia mydas* at a Mid-Ocean Island Breeding Ground," *Copeia* 1974, no. 3 (1974): 703.
42. Archie Fairly Carr and Stephen Stancyk, "Observations on the Ecology and Survival Outlook of the Hawksbill Turtle," *Biological Conservation* 8, no. 3 (1975): 161–72, p. 165.
43. Ibid., 169.
44. Ibid., 170.
45. Ibid., 171.
46. Archie Fairly Carr, Marjorie Harris Carr, and Anne Barkau Meylan, "The Ecology and Migrations of Sea Turtles, 7; The West Caribbean Green Turtle Colony," *Bulletin of the American Museum of Natural History* 162, no. 1 (1978): 1–46, p. 5.
47. Ibid., 35.
48. Ibid.
49. Ibid., 41.
50. Ibid., 41–42.
51. Ibid., 43.
52. Ibid.
53. Archie Carr to Richard Williams (April 11, 1977), in Carr Papers.
54. Peter C.H. Pritchard, *Tales from the Thébaïde* (Malabar, Fla.: Krieger, 2006), 2.
55. Archie Carr to Paul Hooker, National Marine Fisheries Service (March 27, 1978), in Carr Papers.
56. Archie Carr, Interim Report to the National Marine Fisheries Service on a Survey and Preliminary Census of Marine Turtle Populations and Habitats in the Western Atlantic, in Carr Papers, 1.
57. P. C. H. Pritchard et al., *Sea Turtle Manual of Research and Conservation Techniques* (San Jose, Costa Rica: Western Atlantic Turtle Symposium, IOCARIBE, 1982).
58. Archie Carr and Anne Meylan, Alpha Helix Expedition to Costa Rica and Nicaragua—1978 Participant Report Abstract, in Carr Papers, 1–2.
59. Archie Fairly Carr and A. B. Meylan, "Evidence of Passive Migration of Green Turtle Hatchlings in Sargassum," *Copeia*, no. 2 (1980): 366–68, p. 367.
60. Archie Fairly Carr, "Some Problems of Sea Turtle Ecology," *American Zoologist* 20, no. 3 (1980): 489–98, p. 493.
61. Larry H. Ogren, "Survey and Reconnaissance of Sea Turtles in the Northern Gulf of Mexico," Unpublished report, National Marine Fisheries Service, Panama City, Florida (1978).
62. See Henry H. Hildebrand, "A Historical Review of the Status of Sea Turtle Populations in the Western Gulf of Mexico," in *Biology and Conservation of Sea Turtles: Proceedings of the World Conference on Sea Turtle Conservation*, Washington, D.C., November 26–30, 1979, ed. Karen A. Bjorndal (Washington, D.C.: Smithsonian Institution Press, 1982), 447–53.
63. Carr, "Problems of Sea Turtle Ecology," 495.
64. Ibid., 496.

65. Ibid., 496–97.
66. Archie Fairly Carr, "Thirty Years with Sea Turtles: Perspectives for World Conservation," Fairfield Osborn Address, in Carr Papers, 6.
67. Archie Fairly Carr, "Rips, Fads, and Little Loggerheads," *Bioscience* 36, no. 2 (1986): 92–100, p. 94.
68. Ibid.
69. Ibid.
70. Ibid., 95.
71. Ibid., 96.
72. Ibid., 98.
73. Ibid., 100.
74. Ibid. Carr elaborated slightly on this point in Archie Fairly Carr, "Impact of Nondegradable Marine Debris on the Ecology and Survival Outlook of Sea Turtles," *Marine Pollution Bulletin* 18, no. 6B (1987): 352–56.
75. Archie Fairly Carr, "New Perspectives on the Pelagic Stage of Sea Turtle Development," *NOAA Technical Memorandum NMFS-SEFC-190* (1986): 19. Reprinted as Archie Fairly Carr, "New Perspectives on the Pelagic Stage of Sea Turtle Development," *Conservation Biology* 1, no. 2 (1987): 103–21, p. 118.

Chapter 10

1. Portions of this chapter appeared in Frederick R. Davis, "A Naturalist's Place: Archie Carr and the Nature of Florida," in *Paradise Lost? The Environmental History of Florida*, ed. Jack E. Davis and Raymond Arsenault (Gainesville, Fla.: University Press of Florida, 2005), 72–91.
2. Archie Fairly Carr and Time-Life Books, *The Everglades* (New York: Time-Life Books, 1973), 157.
3. Ibid., 161.
4. Ibid., 169.
5. Archie Fairly Carr, *The Reptiles* (New York: Time-Life Books, 1963), 170.
6. Archie Fairly Carr, *Ulendo: Travels of a Naturalist in and out of Africa* (Gainesville: University Press of Florida, 1993), 235.
7. Archie Carr to Stephen C. O'Connell (October 17, 1969), in Carr Papers, 3.
8. Archie Fairly Carr, *A Contribution to the Herpetology of Florida*, Biological Science Series (Gainesville: University of Florida, 1940), 69.
9. David Ehrenfeld, *Beginning Again: People and Nature in the New Millennium* (New York: Oxford University Press, 1993), 4.
10. Quoted in ibid.
11. Archie Carr, "Alligators: Dragons in Distress," *National Geographic* 131, no. 1 (1967): 130–48, p. 147.
12. Ibid., 148.
13. Archie Carr to Claude Pepper (February 21, 1968), in Carr Papers.
14. Peter C.H. Pritchard, *Tales from the Thébaïde* (Malabar, Fla.: Krieger, 2006), 13.
15. For analysis of garden clubs and environmentalism, see Jack E. Davis, "Up from the Sawgrass: Marjory Stoneman Douglas and the Influence of Female Activism in Florida Conservation," in *Making Waves: Female Activists in Twentieth-Century Florida*, ed. Jack E. Davis and Kari A. Frederickson (Gainesville: University Press of Florida, 2003), 147–76.

16. Everett Caudle, interview with Marjorie Carr (April 24, 1989), University of Florida Oral History Program. In 1970 the Florida Department of the Environment bought the rest of Paynes Prairie and established it as a state park.
17. Ibid., 7.
18. Marjorie H. Carr, "The Fight to Save the Ocklawaha." Paper presented at the 12th Biennial Sierra Club Wilderness Conference, Washington, D.C., September 25, 1971, 2.
19. Unlike Archie Carr, who has attracted little historical analysis, Marjorie Carr's efforts on behalf of the Ocklawaha River have been the focus of several studies. See Frederick R. Davis, "Get the Facts and Then Act: How Marjorie H. Carr and Florida Defenders of the Environment Fought to Save the Ocklawaha River," *Florida Historical Quarterly* 83, no. 1 (2004): 46–69; Lee Irby, "A Passion for Wild Things: Marjorie Harris Carr and the Fight to Free a River," in *Making Waves: Female Activists in Twentieth-Century Florida*, ed. Jack E. Davis and Kari A. Frederickson, *The Florida History and Culture Series* (Gainesville: University Press of Florida, 2003), 177–96; and Sallie R. Middleton, "Cutting through Paradise: A Political History of the Cross-Florida Barge Canal" (Ph.D. thesis Florida International University, 2001).
20. Marjorie H. Carr to Mrs. Forrest, February 12, 1965, in Florida Defenders of the Environment Papers [hereafter cited as "FDE Papers"].
21. Marjorie H. Carr (Mrs. Archie Carr) to Claude D. Pepper, June 15, 1965. Series 301, Box 35, Folder 4, Claude Pepper Library, Florida State University Libraries, Tallahassee [hereafter cited as "CPL"].
22. Marjorie H. Carr Notes on the Natural History of the Oklawaha River Wilderness Area. Series 301, Box 35, Folder 4, CPL.
23. Claude Pepper to Mrs. Ardill [*sic*] Carr, June 29, 1965. Series 301, Box 35, Folder 4, CPL.
24. Leslie Kemp Poole, interview with Marjorie Carr (October 18, 1990), Gainesville, Florida.
25. Ibid.
26. See Lewis L. Gould, *Lady Bird Johnson and the Environment* (Lawrence: University Press of Kansas, 1988); Gould, *Lady Bird Johnson: Our Environmental First Lady*, *Modern First Ladies* (Lawrence: University Press of Kansas, 1999).
27. William M. Partington, "Oklawaha—the Fight Is on Again!" *The Living Wilderness* Autumn (1969): 19–23, p. 22.
28. James Nathan Miller, "Rape on the Oklawaha," *Reader's Digest* January (1970): 2–8, pp. 7–8.
29. Ben Funk and Frank Murray, "Born of Emotion, Canal Still in Furor," *Times-Union & Journal* (Jacksonville, Florida), December 21, 1969, sec. D, p. 1.
30. William M. Partington and FDE Candidate's Questionnaire, in FDE Papers, 1.
31. Florida Defenders of the Environment Press Release Re: Candidates Questionnaire, September 4, 1970, in FDE Papers. President Richard M. Nixon had undertaken several initiatives on behalf of the environment during 1970. On New Year's Day, he signed into law the National Environmental Protection Act. While Nixon played no role in the passage of NEPA, he astutely appreciated the bill's popularity and took credit for it as an expression of his personal concern for environmental quality. For a detailed analysis, see J. Brooks Flippen, *Nixon and the Environment* (Albuquerque: University of New Mexico Press, 2000), 50–51.
32. Flippen, *Nixon and the Environment*, 53–54.

33. Luther J. Carter and Resources for the Future, *The Florida Experience: Land and Water Policy in a Growth State* (Baltimore, Md.: Johns Hopkins University Press, 1974), 312.

34. Marjorie H. Carr to Prospective FDE Members, February 4, 1971, in FDE Papers, 1.

35. J.C. Dickinson, Jr., Draft Interim Report for the Biological Sciences Committee (February 29, 1960), in George Kelso Davis Papers, Department of Special Collections, George A. Smathers Library, University of Florida, Gainesville.

36. University of Florida University Senate Unification of Basic Biological Sciences (Approved May 28, 1970), in George Kelso Davis Papers, Department of Special Collections, George A. Smathers Library, University of Florida, Gainesville, 1.

37. See Edward O. Wilson, *Naturalist* (Washington, D.C.: Island Press, 1994), 218–37; Joseph S. Fruton, *Eighty Years* (New Haven, Conn.: Epikouros Press, 1994). See also Frederick Rowe Davis, "The History of Ornithology at Yale University and the Peabody Museum of Natural History," in *Contributions to the History of North American Ornithology*, ed. William E. Davis and Jerome Jackson (Cambridge, Mass.: Nuttall Ornithological Club, 2000), 83–121.

38. Archie Carr, *A Naturalist in Florida: A Celebration of Eden*, ed. Marjorie Harris Carr, (New Haven, Conn.: Yale University Press, 1994), xv.

39. John Kunkel Small, *From Eden to Sahara: Florida's Tragedy* (Lancaster, Penn.: The Science Press Printing Company, 1929); Thomas Barbour, *That Vanishing Eden: A Naturalist's Florida* (Boston: Little, Brown and Company, 1944).

40. Carr, *A Naturalist in Florida*, xv.

41. Ibid.

42. Archie Fairly Carr, "All the Way Down Upon the Suwannee River," *Audubon* 85, no. 2 (1983): 78–101. Reprinted in Carr, *A Naturalist in Florida*, 51–72.

43. Carr, *A Naturalist in Florida*, 30.

44. Archie Fairly Carr, "In Praise of Snakes," *Audubon* 73, no. 4 (1971): 18, 25–27, p. 27.

45. Archie F. Carr, "A Naturalist at Large," *Natural History* 78, no. 3 (1969): 18–24, 68–70, p. 70.

46. Thomas C. Emmel, Important Memorandum, Re: Poisonous Snakes and Firearms in Bartram Hall (March 23, 1981), in Carr Papers.

47. Archie Fairly Carr, "Thoughts on Wilderness Preservation and a Central American Ethic," *Audubon* September (1969): 51–55, p. 52.

48. Ibid.

49. Ibid., 54.

50. Ibid.

51. Ibid., 55.

52. Archie Fairly Carr, "The Moss Forest," *Audubon* 73, no. 5 (1973): 36–51, p. 51. See also Carr, *A Naturalist in Florida*, 165–186.

53. Archie Fairly Carr, "Armadillo Dilemma," *Animal Kingdom* 85, no. 5 (1982): 40–44, p. 40. See also, Carr, *A Naturalist in Florida*, 204–09.

54. Ibid., 43.

55. Archie Fairly Carr, "Guess Who's Coming to Dinner," *Animal Kingdom* 85, no. 6 (1983): 46–47.

56. Archie Fairly Carr, "The Ducks of Wewa Pond," *Animal Kingdom* 90, no. 1 (1987): 8–10, p. 10. See also Carr, *A Naturalist in Florida*, 1–13.

57. Aldo Leopold, *A Sand County Almanac, and Sketches Here and There* (New York: Oxford, 1949), 121.

58. Carr, *A Naturalist in Florida,* 236–244, p. 244. See also Beth R. Read, Joan E. Gill, and Bicentennial Commission of Florida, *Born of the Sun: The Official Florida Bicentennial Commemorative Book* (Hollywood, Fla.: Florida Bicentennial Commemorative Journal, 1975).

59. For an overview of human agency and nature, see Ted Steinberg, *Down to Earth: Nature's Role in American History* (Oxford: Oxford University Press, 2002).

Chapter 11

1. Archie Fairly Carr, *The Windward Road: Adventures of a Naturalist on Remote Caribbean Shores* (reissued ed.; Tallahassee: University Presses of Florida, 1979), xxxii.

2. Ibid.

3. Ibid.

4. Archie F. Carr, Jr. to David W. Ehrenfeld (August 1, 1969), in Carr Papers, emphasis added.

5. David W. Ehrenfeld, *Biological Conservation, Modern Biology Series* (New York: Holt, Rinehart, and Winston, Inc., 1970), 216.

6. Ibid., ix.

7. Ibid., v.

8. David Ehrenfeld, "Editorial," *Conservation Biology* 1, no. 1 (1987): 6–7, p. 6.

9. Ibid.

10. Karen A. Bjorndal, interview with author (October 1, 2003), Gainesville, Florida.

11. Karen A. Bjorndal, "Nutrition and Grazing Behavior of the Green Turtle, *Chelonia mydas,* a Seagrass Herbivore" (Ph.D. thesis, University of Florida, 1979). It was published the following year as Karen A. Bjorndal, "Nutrition and Grazing Behavior of the Green Turtle *Chelonia mydas*," *Marine Biology* 56, no. 2 (1980): 147–54.

12. See, for example, Karen A. Bjorndal and J.B.C. Jackson, "Roles of Sea Turtles in Marine Ecosystems: Reconstructing the Past," in *The Biology of Sea Turtles, CRC Marine Biology Series* (Boca Raton, Fla.: CRC Press, 2003). See also Alan B. Bolten and Blair E. Witherington, *Loggerhead Sea Turtles* (Washington, D.C.: Smithsonian Books, 2003).

13. Jeanne A. Mortimer, interview with author via telephone (July 16, 2004), Chicago, Illinois.

14. Ibid.

15. J. A. Mortimer, "Observations on the Feeding Ecology of the Green Turtle, *Chelonia mydas,* in the Western Caribbean" (Master's thesis, University of Florida, 1976). Major findings later appeared as Jeanne A. Mortimer, "The Feeding Ecology of the West Caribbean Green Turtle (*Chelonia mydas*) in Nicaragua," *Biotropica* 13, no. 1 (1981): 49–58. Some of the slaughterhouses in Nicaragua continue to operate even to the present day.

16. J. A. Mortimer, "Reproductive Ecology of the Green Turtle, *Chelonia mydas,* at Ascension Island" (Ph.D. dissertation, University of Florida, 1981).

17. Anne Meylan, interview with author via telephone (June 21, 2004), St. Petersburg, Florida. Meylan's master's thesis appeared as Anne B. Meylan, "The Behavioural Ecology of the West Caribbean Green Turtle (*Chelonia mydas*) in the Internesting Habitat" (M.S. thesis, University of Florida, 1978).

18. Anne Meylan, "Feeding Ecology of the Hawksbill Turtle (*Eretmochelys imbricata*): Spongivory as a Feeding Niche in the Coral Reef Community" (Ph.D. dissertation, University of Florida, 1984).

19. Anne Meylan, "Spongivory in Hawksbill Turtles: A Diet of Glass," *Science* 239, no. 4838 (1988): 393–95.
20. Mortimer, interview with author (July 16, 2004).
21. Karen A. Bjorndal, ed., *Biology and Conservation of Sea Turtles: Proceedings of the World Conference on Sea Turtle Conservation, Washington, D.C., 26–30 November 1979* (Washington, D.C.: Smithsonian Institution Press, 1982).
22. See Peter Meylan, "Archie Carr as Taxonomist," in *Crocodilian, Tuatara, and Turtle Species of the World: A Taxonomic and Geographic Reference*, ed. Wayne King and R. Burke (Washington, D.C.: Association of Systematics Collections, 1989), iv–v.
23. David Ehrenfeld, interview with author via telephone (May 24, 2004), New Brunswick, New Jersey.
24. Quoted in B.W. Bowen and A.L. Bass, "Are the Naturalists Dying Off?" *Conservation Biology* 10, no. 4 (1996): 923–24, p. 924.
25. Archie Fairly Carr, "Thirty Years with Sea Turtles: Perspectives for World Conservation," Fairfield Osborn Address, in Carr Papers.
26. Ibid., 3–4.
27. Ibid., 5.
28. Stephen J. Morreale et al., "Temperature-Dependent Sex Determination: Current Practices Threaten Conservation of Sea Turtles," *Science* 216, no. 4551 (1982): 1245–47, p. 1245. This discovery led to a mnemonic among sea turtle researchers: "Hot chicks and cool dudes."
29. Carr, "Thirty Years with Sea Turtles," 8.
30. Ibid., 12.
31. Ibid., 15–16.
32. Ibid., 21–22.
33. James R. Spotila, *Sea Turtles: A Complete Guide to Their Biology, Behavior, and Conservation* (Baltimore, Md.: Johns Hopkins University Press, 2004), 80. In 2000, Pritchard was recognized by *Time* magazine as a "hero of the planet."
34. Ibid., 8.
35. Bill DeYoung, "The Family Carr: Florida's First Family of Conservation," *The Gainesville Sun*, December 6, 1993, sec. D, p. 1.
36. Ehrenfeld, interview with author (May 24, 2004).
37. Ibid.
38. Edward O. Wilson, Address on the Occasion of the Awarding of the University of Florida Presidential Medal to Archie F. Carr, in Carr Papers.
39. Ecological Society of America Citation for the Award of Eminent Ecologist Presented to Archie F. Carr, in Carr Papers.
40. Peter C.H. Pritchard, *Tales from the Thébaïde* (Malabar, Fla.: Krieger, 2006), 2.
41. Spotila, *Sea Turtles*, 8.
42. Ibid., 91–215.
43. Reed F. Noss, "Editorial: The Naturalists Are Dying Off," *Conservation Biology* 10, no. 1 (1996): 1–3, p. 2.
44. Ibid.
45. E. O. Wilson, "On the Future of Conservation Biology," *Conservation Biology* 14, no. 1 (2000): 1–3, p. 2.

SELECTED BIBLIOGRAPHY

Archival Sources

Archie F. Carr, Jr. Papers, Department of Special and Area Collections, George A. Smathers Library, University of Florida, Gainesville, Florida

Cross-Florida Barge Canal Papers, Claude Pepper Library, Florida State University Libraries, Tallahassee, Florida

Florida Defenders of the Environment Papers (unsorted), Department of Special Collections, George A. Smathers Library, University of Florida, Gainesville, Florida

Thomas Barbour Papers, Harvard University Archives, Cambridge, Massachusetts

Interviews Conducted

Family Members

Archie ("Chuck") Carr, III (son) (January 6, 1998)
David Carr (son) (July 13, 2004)
Margaret ("Mimi") Carr (daughter) (January 9, 1998)
Thomas Carr (brother) (January 7, 1998)
Tom Carr (son) (June 8, 2004)

Colleagues

J.C. ("Josh") Dickinson, III (April 19, 1996)
J.C. Dickinson, Jr. (April 19, 1996)
David Godfrey (director, CCC) (January 7, 1998)
Brian McNab (August 29, 2003)
Frank Nordlie (June 30, 2004)
E.O. Wilson (May 1996)

Students

Karen Bjorndal (October 1, 2003)
David Ehrenfeld (May 24–25, 2004)
Anne Meylan (June 21, 2004)
Jeanne Mortimer (July 16, 2004)
Larry Ogren (September 12, 2003)
Peter Pritchard (June 4, 2004)

Books and Articles by Archie Carr

Carr, Archie Fairly. "A Key to the Breeding-Songs of the Florida Frogs." *The Florida Naturalist* 1, no. 2 (1934): 19–23.

———. "The Plancton and Carbondioxide-oxygen Cycle in Lake Wauberg, Florida." M.S. thesis, University of Florida, 1934.

———. "The Gulf-Island Cottonmouths." *Proceedings of the Florida Academy of Sciences* 1 (1935): 88.

———. "The Identity and Status of Two Turtles of the Genus *Pseudemys*." *Copeia* 3 (1935): 147–48.

———. "The Geographic and Ecological Distribution of the Reptiles and Amphibians of Florida." Ph.D. thesis, University of Florida, 1937.

———. "A New Turtle from Florida, with Notes on *Pseudemys floridana mobiliensis* (Holbrook)." *Occasional Papers for the Museum of Zoology, University of Michigan* 348 (1937): 1–7.

———. "Notes on the *Pseudemys scripta* Complex." *Herpetologica* 1 (1937): 131–35.

———. "The Status of *Pseudemys scripta* and *Pseudemys troostii*." *Herpetologica* 1 (1937): 75–77.

———. "A New Subspecies of *Pseudemys floridana*, with Notes on the *floridana* Complex." *Copeia* (1938): 105–9.

———. "*Pseudemys nelsoni*, A New Turtle from Florida." *Occasional Papers of the Boston Society of Natural History* 8 (1938): 305–10.

———. "*Haideotriton wallacei*, a New Subterranean Salamander from Georgia." *Occasional Papers of the Boston Society of Natural History* 8 (1939): 333–36.

———. "Notes on Escape Behavior in the Florida Marsh Rabbit." *Journal of Mammalogy* 20, no. 3 (1939): 322–25.

———. "Notes on the Breeding Habits of the Warmouth Bass." *Proceedings of the Florida Academy of Sciences* 4 (1939): 108–12.

———. *A Contribution to the Herpetology of Florida, Biological Science Series*. Gainesville: University of Florida, 1940.

———. "The Fishes of Alachua County, Florida: A Subjective Key." *Dopeia* Ser. B., Vol. 3, Part Q, No. X (1941).

———. "Notes on Sea Turtles." *Proceedings of the New England Zoological Club* XXI (1942): 1–16.

———. "Outline for a Classification of Animal Habitats in Honduras." *Bulletin of the American Museum of Natural History* 94, no. 10 (1950): 567–94.

———. *Handbook of Turtles: The Turtles of the United States, Canada, and Baja California*. Ithaca, N.Y.: Comstock, 1952.

———. *High Jungles and Low*. Gainesville: University Press of Florida, 1953.

———. "A Study of the Ecology, Migrations, and Population Levels of Sea Turtles in the Atlantic and Caribbean with Special Reference to the Atlantic Green Turtle, *Chelonia mydas mydas* [Linné]. Research Proposal for Submission to National Science Foundation, unpublished." In Archie F. Carr, Jr. Papers, Department of Special Collections, George A. Smathers Library, University of Florida, Gainesville, 1954.

———. "The Zoogeography and Migrations of Sea Turtles." *Yearbook of the American Philosophical Society* (1954): 138–40.

———. "The Riddle of the Ridley." *Animal Kingdom* 58 (1955): 146–56.

. "Notes on the Zoogeography of the Atlantic Sea Turtles of the Genus *Lepidochelys.*" *Rev. Biol. Trop.* 5, no. 1 (1957): 45–61.

———. "Pacific Turtle Problem—Mexico's Coast Yields Information on a Matter of Melanism." *Natural History* 70, no. 8 (1961): 64–71.

———. "Report of the Technical Director, Caribbean Conservation Corporation, 1961–76." In Archie F. Carr, Jr. Papers, Department of Special Collections, George A. Smathers Library, University of Florida, Gainesville, 1961.

———. "The Ridley Mystery Today." *Animal Kingdom* 64, no. 1 (1961): 7–12.

———. "Orientation Problems in the High Seas Travel and Terrestrial Movements of Marine Turtles." *American Scientist* 50, no. 3 (1962): 358–74.

———. "Orientation Problems in the High Seas Travel and Terrestrial Movements of Marine Turtles." In *Bio-Telemetry: The Use of Telemetry in Animals Behavior and Physiology in Relation to Ecological Problems,* ed. Lloyd E. Slater, 179–93. New York: Macmillian, 1963.

———. *The Reptiles.* New York: Time-Life Books, 1963.

———. *The Land and Wildlife of Africa.* New York: Time-Life Books, 1964.

———. "Transoceanic Migrations of the Green Turtle." *Naval Research Reviews* 17, no. 10 (1964): 12–18.

———. "Caribbean Green Turtle: Imperiled Gift of the Sea." *National Geographic Magazine* 131, no. 6 (1967): 876–90.

———. "No One Knows Where the Turtles Go. Part 2: 100 turtle eggs." *Natural History* 76, no. 8 (1967): 40–43, 52–54, 56, 58–59.

———. *So Excellent a Fishe: A Natural History of Sea Turtles.* Garden City, N.Y.: Natural History Press, 1967.

———. "Sea Turtles: a Vanishing Asset." Paper presented at the Latin American Conference on the Conservation of Renewable Natural Resources, San Carlos de Bariloche, Argentina, March 27–April 2, 1968.

———. "A Naturalist at Large." *Natural History* 78, no. 3 (1969): 18–24, 68–70.

———. "In Praise of Snakes." *Audubon* 73, no. 4 (1971): 18, 25–27.

———. "Research and Conservation Problems in Costa Rica. Paper presented at the Second working meeting of the Marine Turtle Specialist Group." In Archie F. Carr, Jr. Papers, Department of Special Collections, George A. Smathers Library, University of Florida, Gainesville, 1971.

———. *The Everglades.* New York: Time-Life Books, 1973.

———. "The Moss Forest." *Audubon* 73, no. 5 (1973): 36–51.

———. *The Windward Road: Adventures of a Naturalist on Remote Caribbean Shores,* 1979 reissue ed. Gainesville: University Press of Florida, 1979.

———. "Some Problems of Sea Turtle Ecology (presented at Symposium on Behavioral and Reproductive Biology of Sea Turtles, Tampa, Florida, Dec. 27, 1979)." *American Zoologist* 20, no. 3 (1980): 489–98.

———. "Armadillo Dilemma." *Animal Kingdom* 85, no. 5 (1982): 40–44.

———. "All the Way Down upon the Suwannee River." *Audubon* 85, no. 2 (1983): 78–101.

———. "Guess Who's Coming to Dinner." *Animal Kingdom* 85, no. 6 (1983): 46–47.

———. *The Sea Turtle: So Excellent a Fishe.* Austin: University of Texas Press, 1984.

———. "Thirty Years with Sea Turtles: Perspectives for World Conservation, Fairfield Osborn Address." In Archie F. Carr, Jr. Papers, Department of Special Collections, George A. Smathers Library, University of Florida, Gainesville, 1984.

———. "Water Hyacinths—Animal Hideaway." *Animal Kingdom* 87, no. 5 (1984): 12, 55.

———. "New Perspectives on the Pelagic Stage of Sea Turtle Development." *NOAA Technical Memorandum NMFS-SEFC-190* (1986): 1–36.

———. "Rips, FADs, and Little Loggerheads." *Bioscience* 36, no. 2 (1986): 92–100.

———. "The Ducks of Wewa Pond." *Animal Kingdom* 90, no. 1 (1987): 8–10.

———. "Impact of Nondegradable Marine Debris on the Ecology and Survival Outlook of Sea Turtles." *Marine Pollution Bulletin* 18, no. 6B (1987): 352–56.

———. "New Perspectives on the Pelagic Stage of Sea Turtle Development." *Conservation Biology* 1, no. 2 (1987): 103–21.

———. *High Jungles and Low,* 2nd ed. Gainesville: University Press of Florida, 1992.

———. *Ulendo: Travels of a Naturalist In and Out of Africa.* Gainesville: University Press of Florida, 1993.

———. "All the Way Down upon the Suwannee River." In *A Naturalist in Florida: A Celebration of Eden,* ed. Marjorie Harris Carr, 51–72. New Haven, Conn.: Yale University Press, 1994.

Carr, Archie Fairly and David K. Caldwell. "The Ecology and Migrations of Sea Turtles, 1: Results of Field Work in Florida, 1955." *American Museum Novitates,* no. 1793 (1956).

Carr, Archie Fairly and Marjorie H. Carr. "Modulated Reproductive Periodicity in *Chelonia.*" *Ecology* 51, no. 2 (1970): 335–37.

———. "Recruitment and Remigration in a Green Turtle Nesting Colony." *Biological Conservation* 4, no. 2 (1970): 282–84.

———. "Site Fixity in the Caribbean Green Turtle." *Ecology* 53, no. 3 (1972): 425–29.

———. *A Naturalist in Florida: A Celebration of Eden.* New Haven, Conn.: Yale University Press, 1994.

Carr, Archie Fairly, Marjorie H. Carr, and Anne B. Meylan. "The Ecology and Migrations of Sea Turtles, 7: The West Caribbean Green Turtle Colony." *Bulletin of the American Museum of Natural History* 162, no. 1 (1978): 1–46.

Carr, Archie Fairly and Patrick J. Coleman. "Seafloor Spreading Theory and the Odyssey of the Green Turtle." *Nature* 249, no. 5453 (1974): 128–30.

Carr, Archie Fairly and Leonard Giovannoli. "The Ecology and Migrations of Sea Turtles, 2: Results of Field Work in Costa Rica, 1955." *American Museum Novitates* 1835 (1957): 1–32.

Carr, Archie Fairly and Donald Goodman. "Ecologic Implications of Size and Growth in *Chelonia.*" *Copeia* 1970, no. 4 (1970): 783–86.

Carr, Archie Fairly and Harold Hirth. "Social Facilitation in Green Turtle Siblings." *Animal Behaviour* 9, no. 1–2 (1961): 68–70.

———. "The Ecology and Migrations of Sea Turtles, 5: Comparative Features of Isolated Green Turtle Colonies." *American Museum Novitates* 2091 (1962): 1–42.

Carr, Archie Fairly, Harold Hirth, and Larry Ogren. "The Ecology and Migrations of Sea Turtles, 6: The Hawksbill Turtle in the Caribbean Sea." *American Museum Novitates* 2248 (1966): 1–29.

———. "Alpha Helix Expedition to Costa Rica and Nicaragua — 1978 Participant Report Abstract." In Archie F. Carr, Jr. Papers, Department of Special Collections, George A. Smathers Library, University of Florida, Gainesville, 1978.

Carr, Archie Fairly and A. B. Meylan. "Evidence of Passive Migration of Green Turtle Hatchlings in Sargassum." *Copeia*, no. 2 (1980): 366–68.

Carr, A. Fairly and Lewis J. Marchand. "A New Turtle from the Chipola River, Florida." *Proceedings of the New England Zoological Club* 20 (1942): 95–100.

Carr, Archie Fairly and Larry Ogren. "The Ecology and Migrations of Sea Turtles, 3: *Dermochelys* in Costa Rica." *American Museum Novitates* 1958 (1959): 1–29.

———. "The Ecology and Migrations of Sea Turtles, 4: The Green Turtle in the Caribbean Sea." *Bulletin of the American Museum of Natural History* (1960).

Carr, Archie Fairly, Perran Ross, and Stephen Carr. "Interesting Behavior of the Green Turtle *Chelonia mydas* at a Mid-Ocean Island Breeding Ground." *Copeia* 1974, no. 3 (1974): 703–6.

Carr, Archie Fairly and Stephen Stancyk. "Observations on the Ecology and Survival Outlook of the Hawksbill Turtle." *Biological Conservation* 8, no. 3 (1975): 161–72.

Books and Articles by Marjorie Carr

Carr, Marjorie H. "The Breeding Habits, Embryology, and Larval Development of the Large-mouthed Black Bass in Florida." *Proc. New England Zool. Club* 20 (1942): 43–77.

———. "Notes on the Breeding Habits of the Eastern Stumpknocker *Lepomis punctatus punctatus* (Cuvier)." *Proc. Fla. Acad. Sci.* (1942): 101–6.

———. "Trip to Uyuca Field Notes (November 23, 1948)." In Private Collection of Tom Carr, Gainesville, Florida, 1948.

———. "The Fight to Save the Ocklawaha." Paper presented at the 12th Biennial Sierra Club Wilderness Conference, Washington, D.C., September 25, 1971.

———. Foreword. In *Ecosystems of Florida*, ed. Ronald L. Myers and John J. Ewel, ix–xiii. Orlando: University of Central Florida Press, 1990.

———. "1993 Preface to the 1952 Letters." In *Ulendo: Travels of a Naturalist in and out of Africa*. Gainesville: University Press of Florida, 1993.

———. Preface. In Archie Carr, *A Naturalist in Florida: A Celebration of Eden*, ed. Marjorie Harris Carr. New Haven, Conn.: Yale University Press, 1994.

INDEX

DATE DUE